Liaison Functions 2013

Core Programming Topic at the 2013 AIChE Spring Meeting & 9th Global Congress on Process Safety

San Antonio, Texas, USA
28 April – 2 May 2013

ISBN: 978-1-62748-465-7

Printed from e-media with permission by:

Curran Associates, Inc.
57 Morehouse Lane
Red Hook, NY 12571

Some format issues inherent in the e-media version may also appear in this print version.

Printed by Curran Associates, Inc. (2013)

For permission requests, please contact AIChE
at the address below.

AIChE
3 Park Avenue
New York, NY 10016-5991

Phone: (203) 702-7660
Fax: (203) 775-5177

www.aiche.org

Additional copies of this publication are available from:

Curran Associates, Inc.
57 Morehouse Lane
Red Hook, NY 12571 USA
Phone: 845-758-0400
Fax: 845-758-2634
Email: curran@proceedings.com
Web: www.proceedings.com

TABLE OF CONTENTS

(7a) Introduction to the MBA

Jon Haughton, Health & Nutrition, Corn Products International, Westchester, IL, Meagan Lewis, UOP, IL and John M. Forgac, BP Naperville Campus, BP Products North America, Naperville, IL

This is a tutorial-like presentation of the MBA-In-A-Day session. An invited speaker will give an overview of the title subject, who could be a faculty member from a regional business school or an expert practitioner.

The session will be of interest to those who want to learn about MBA programs and content: Those who may be considering enrolling in an MBA program, those who want to understand the type of information that is covered, and those who would benefit from having an overview of a typical MBA program. The speakers will direct their presentations to the interests of chemical engineers. The session will be highly of interest to young professionals: The typical MBA student has six years of experience and is 29 years old, and students with engineering or science backgrounds make up 35 to 40% of MBA programs.

(7b) Entrepreneurship and Innovation

John M. Forgac, BP Naperville Campus, BP Products North America, Naperville, IL, Meagan Lewis, UOP, IL and Jon Haughton, Health & Nutrition, Corn Products International, Westchester, IL

This is a tutorial-like presentation of the MBA-In-A-Day session. An invited speaker will give an overview of the title subject, who could be a faculty member from a regional business school or an expert practitioner.

The session will be of interest to those who want to learn about MBA programs and content: Those who may be considering enrolling in an MBA program, those who want to understand the type of information that is covered, and those who would benefit from having an overview of a typical MBA program. The speakers will direct their presentations to the interests of chemical engineers. The session will be highly of interest to young professionals: The typical MBA student has six years of experience and is 29 years old, and students with engineering or science backgrounds make up 35 to 40% of MBA programs.

(7c) Accounting, Finance, and Statistics I

Jon Haughton, Health & Nutrition, Corn Products International, Westchester, IL, Meagan Lewis, UOP, IL and John M. Forgac, BP Naperville Campus, BP Products North America, Naperville, IL

This is a tutorial-like presentation of the MBA-In-A-Day session. An invited speaker will give an overview of the title subject, who could be a faculty member from a regional business school or an expert practitioner.

The session will be of interest to those who want to learn about MBA programs and content: Those who may be considering enrolling in an MBA program, those who want to understand the type of information that is covered, and those who would benefit from having an overview of a typical MBA program. The speakers will direct their presentations to the interests of chemical engineers. The session will be highly of interest to young professionals: The typical MBA student has six years of experience and is 29 years old, and students with engineering or science backgrounds make up 35 to 40% of MBA programs.

(7d) Accounting, Finance, and Statistics II

John M. Forgac, BP Naperville Campus, BP Products North America, Naperville, IL, Meagan Lewis, UOP, IL and Jon Haughton, Health & Nutrition, Corn Products International, Westchester, IL

This is a tutorial-like presentation of the MBA-In-A-Day session. An invited speaker will give an overview of the title subject, who could be a faculty member from a regional business school or an expert practitioner.

The session will be of interest to those who want to learn about MBA programs and content: Those who may be considering enrolling in an MBA program, those who want to understand the type of information that is covered, and those who would benefit from having an overview of a typical MBA program. The speakers will direct their presentations to the interests of chemical engineers. The session will be highly of interest to young professionals: The typical MBA student has six years of experience and is 29 years old, and students with engineering or science backgrounds make up 35 to 40% of MBA programs.

A UNIFIED PID METHODOLOGY TO MEET PLANT OBJECTIVES

Gregory K. McMillan, CDI Process & Industrial, Austin, TX and Héctor H. Torres, Eastman, Tlaxcala, México

Introduction

Plant objectives for basic and advanced regulatory control can be quite diverse. PID control plays an important role in maximizing safety, environmental and equipment protection, process efficiency and process capacity. There can be many different sources of process variability that control loops need to deal with, such as raw material and recycle composition, weather, utility temperature and pressure, operators, interactions, optimization, on-off actions, startups, transitions, shutdowns, measurement and process noise, and limit cycles. Process nonlinearities, process non-self-regulation, dead time dominance, slow measurements and valves, analyzer cycles times, deadband, and threshold sensitivity or resolution limits can make the task much more difficult. The techniques for applying PID control to achieve plant objectives are largely individualistic than systematic. Each application tends to be treated as unique requiring special expertise and tuning. The result has been over 100 tuning rules to meet different perceived and actual needs of the process. Consultants are often required and PID features are not effectively utilized. In some plants PID loops have been pushed into model predictive control (MPC) because PID function application practices are largely heuristic and undocumented.

A unified approach to tuning has been found that enables a common and simplified method for setting PID tuning parameters. Key features can be used to eliminate the need for retuning to deal with different dynamics and objectives. This paper shows a methodology that integrates a unified tuning approach and key features that minimizes implementation and maintenance efforts. A step by step method will be used that will address the myriad of dynamics and objectives in PID applications.

Types of Process Responses

There are 4 major types of process responses. Early tuning rules were developed with one type of process response in mind. Tuning rules developed in the 1990s have expanded the scope of processes addressed but few effectively deal with all types of responses. The tuning requirements are so different for each type of response that users can readily end up confused. The first step to resolving this deficiency is recognition of the distinctive characteristics of each type of process response.

Response Methodology and Terminology

The effect of automation system dynamics will be included and distinguished in the process response. A first order plus dead time (FOPDT) approximation is used that consists of a dead time, a time constant, and a gain. For a near-integrating process the gain is divided by the time constant to given an integrating process gain used for integrating processes. The identification or approximation of a secondary time constant is added to improve the setting of the rate time for the derivative mode.

In the literature these parameters to define the dynamic response are referred to as process dead time, process time constants, and process gain. Since in industrial applications these parameters are determined by the automation system besides process dynamics, more general terms will be used.

The total loop dead time is the sum of all the pure time delays plus the fraction of the small time constants that is equivalent dead time. All time constants smaller than the largest (primary time constant) and next largest if a secondary time constant is identified create an effective dead time. The delays and time constants as shown in Figure 1 in the final control element (e.g. control valve or variable speed drive), process, measurement, and controller must be included.

In the literature primary and secondary time constants are considered to be in the process as shown in Figure 1. In industrial applications these time constant can be in the automation system. In this paper the largest time constant in Figure 1 is termed the open loop time constant.

The open loop gain is the product of the final control element, process, and measurement gains shown in Figure 1. For self-regulating processes, the open loop gain is dimensionless (%/%). For integrating processes, the resulting gain called an integrating process gain has units of 1/sec (%/sec/%). Processes with a large process time constant are termed near-integrating. The near-integrating process gain is attained by dividing a self-regulating process gain by the primary process time constant.

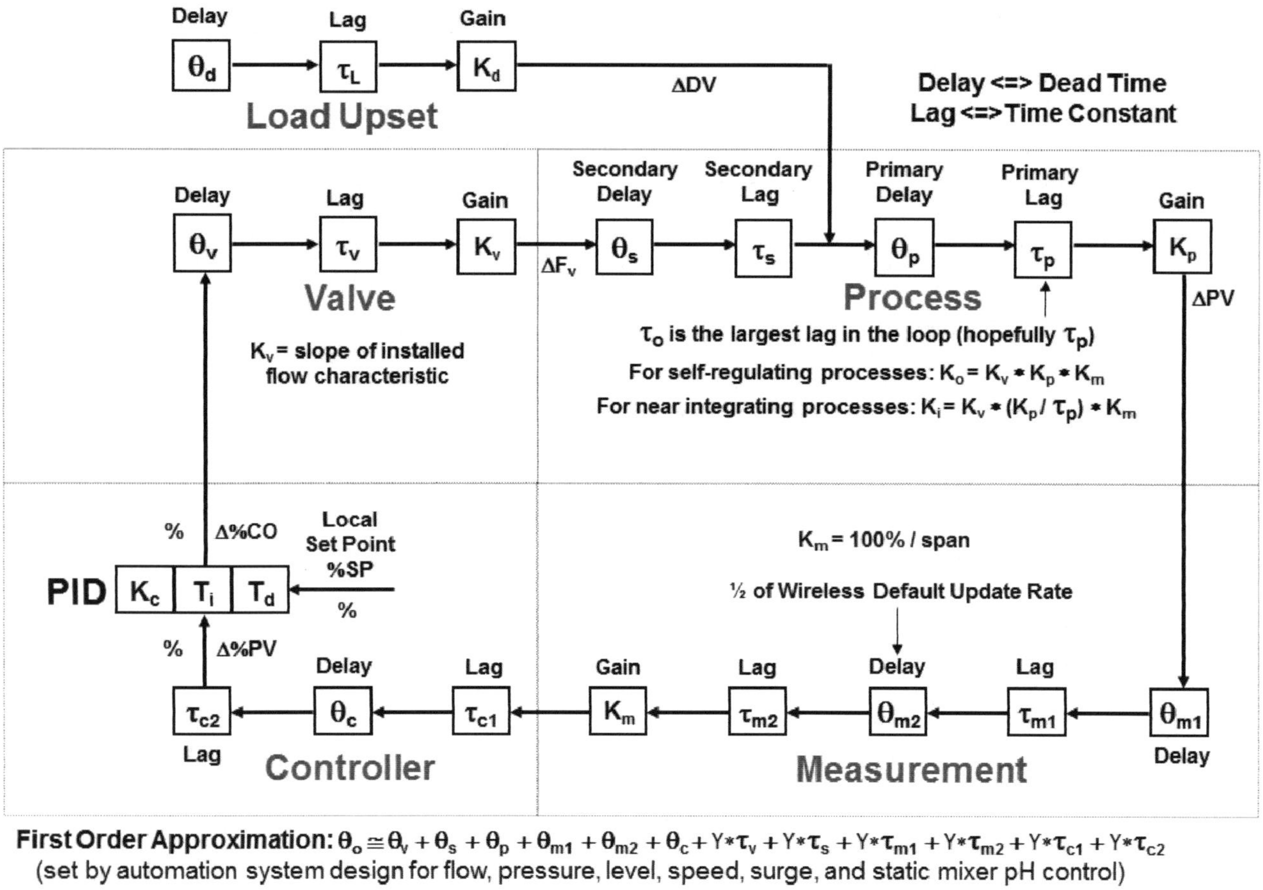

Figure 1 – Block diagram reveals the myriad sources of dynamics in the FOPDT response

Dead time Dominant Self-Regulating Responses

A self-regulating process will reach a new steady state if a manual change in the PID output is made provided there are no disturbances. The steady state is the result of internal self-regulation from a negative feedback process time constant. The actual time to steady state observed will depend upon the open loop time constant (largest time constant in the loop), which may be in the automation system. If the total loop dead time exceeds the open loop time constant, the process is dead time dominant.

The process dynamic response of conveyors, sheets and spin lines is essentially a pure dead time from the transportation delay. However, if die bolt heaters are used instead of hydraulic actuators to control sheet thickness, the thermal lag introduced can be large enough that the process is no longer dead time dominant.

Blenders, desuperheaters, extruders, static mixers, and plug flow reactors have a negligible process time constant due to the lack of back mixing. However, the temperature response is often not dead time dominant due to the thermal lags in the equipment heat transfer surface and thermowell.

The addition of a large wireless default update rate in a flow loop can make the flow loop dead time dominant. A large wireless default update rate in a static mixed pH loop or the use of an at-line or off-line analyzer with a large cycle time can make these composition loops dead time dominant. For these cases where the dead time dominance is due to a discontinuous update of an automation system component, we will see that a PID feature can be enabled to prevent instability and simplify tuning.

Lag Dominant Self-Regulating Response

If the self-regulating process response has an open loop time constant larger than the total loop dead time, the process is termed lag dominant. Continuous operations with mixing such as columns, crystallizers, evaporators, neutralizers, and reactors have a lag dominant response. Liquid flow has a self-regulating response but the process time constant is less than the total loop dead time. A signal filter used to reduce noise can make the process lag dominant. A large automation system lag can make a response lag dominant that would have been dead time dominant based on the process time constant.

If the open loop time constant is much larger than the loop dead time, the process response seen by the PID is a ramp in the control region of 4 dead times. A PID tuned for disturbance rejection will have done nearly all of its corrections in 4 dead times. So far as what the PID sees, the process response is integrating. Such processes are termed "near-integrating" and can use integrating process tuning rules. Temperature and composition control loops on continuous processes are near-integrating.

Integrating Response

Processes that have no negative feedback process time constant and hence no steady state are termed integrating processes. The process response integrates any unbalance in the process. Since the unbalance is nearly never exactly zero, the process will always ramp when the PID is in manual. For most integrating processes the ramp rate is extremely slow (e.g. 0.000001 %/sec). The extreme exception is gas pressure where the ramp rate can be a million to a billion times faster. An electrical furnace had an integrating process gain of 1 %/sec/%. A waste heat incinerator gas pressure had an integrating process gain of 1000 %/sec/% (too fast for any digital controller).

The most common example of an integrating process is level. Gas pressure has an integrating response when changes in operating pressure are a small fraction of the pressure drop across inlet or outlet valves and consequently do not appreciably affect the flow into or out of the system. Batch operations cause an integrating response because there is no discharge flow until the operation is complete. Batch temperature is integrating unless the heat loss changes appreciably with temperature. Batch composition and batch pH has an integrating response unless the reactant or reagent is rapidly consumed in a reaction.

Runaway Response

Processes with a positive feedback time constant can accelerate. These processes are termed open loop unstable. If the PID is put in manual, the process can accelerate and run off-scale. Exothermic reactions where the reaction rate greatly increases with temperature can have runaway response if the heat release exceeds the cooling capability. A large secondary time constant (e.g. heat transfer surface or thermowell lags) or dead time approaching the positive feedback time constant can cause the process to be unstable regardless of PID tuning.

Process Objectives

The main task of the PID is to determine the transfer of variability from a controlled variable such as temperature to the manipulated variable such as coolant flow. The process input most often manipulated is a process or utility flow. For maximum disturbance rejection, the transfer of variability is maximized. This is generally the case for vessel and column temperature, composition, gas pressure, and pH control. An insufficiently aggressive transfer of variability can cause product degradation or an emission, uncontrollable reactions, a relief valve or rupture disc to blow, and a safety instrumented system (SIS) to activate. Safety, equipment and environmental protection, on-stream time, and yield depend upon disturbance rejection.

To meet these objectives the integrated and peak error must be minimized. The ultimate limit to how small these errors are depend upon the total loop dead time (θ_o) and the open loop time constant (τ_o). Figure 2 shows that the ultimate limit to the peak error is proportional to the ratio of dead time to the 63% response time which is the dead time plus the open loop time constant. The peak error is the altitude of the right triangle whose base is the dead time since no correction can be enacted and seen till one dead time. A mirror image of this right triangle placed back to back with the first triangle shows the total recovery from the disturbance. The ultimate limit to the integrated error is the area of the two triangles which ends up being the dead time to 63% response time multiplied by the dead time.

In some applications the transfer must be minimized because changes in the manipulated flow affect other loops and operations. The most common application is surge tank level control where the level PID manipulates the discharge flow to the degree only needed to keep the tank level within low and high limits. The objective is to minimize the disruption of feed changes to downstream operations. This objective is also effective for a distillate receiver level PID manipulating distillate flow. This goal is often termed “maximization of the absorption of variability”.

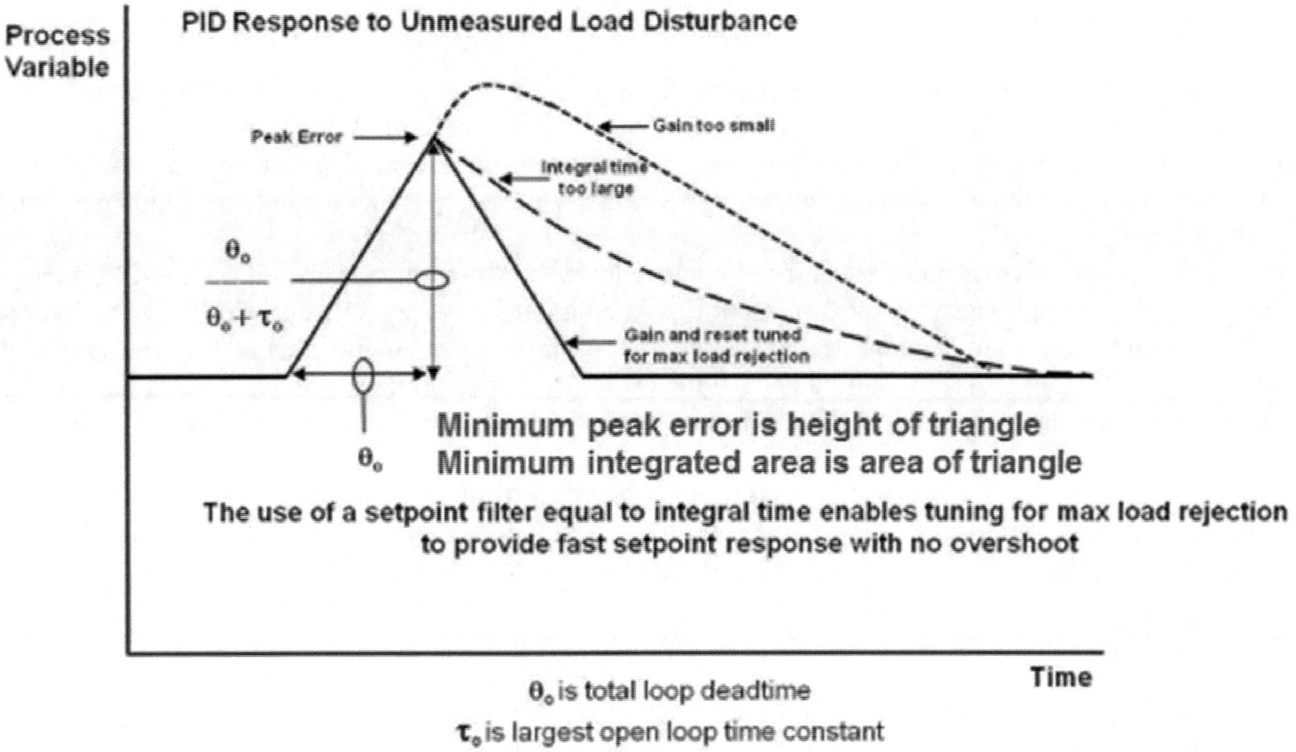

Figure 2 – The ultimate limit to the peak and integrated errors depends upon the total loop dead time and open loop time constant

Aggressive control action can also upset other loops. For interactions that cannot be reduced sufficiently by decoupling techniques, the transfer of variability for the least important loops is minimized to avoid upsetting the more important loops.

When loops are manipulating flows affecting a unit operation, the transfer of variability to these flows is coordinated so the stoichiometric ratio is not momentarily upset from one flow being faster or slower than another flow. In this case the transfer of variability in all loops is regulated to match the slowest loop. This coordination of manipulated flows is particularly important for operations without back mixing to smooth out transients. Coordination is consequently critical for blending and plug flow liquid and gas reactions (e.g. catalyst bed gas reactors).

PID loops must not increase variability. A PID loop can oscillate from overly aggressive action and interactions and in integrating and runaway processes from inadequate immediate action due to a gain setting that is too low. A PID can create variability when there are automation system difficulties.

Automation System Difficulties

The response of the automation system can introduce excessive discontinuities, lag, noise, and dead time. Here we look at the components of the system for the sources of these difficulties. There is only room in this paper to note some of the more significant deficiencies.

Final Control Element

A control valve can have significant stiction from excessive friction often described as a resolution limit. Actuators and positioners can also have a threshold sensitivity limit. One or more integrators in the process or PID in a loop with a resolution or threshold sensitivity limit will cause a limit cycle that cannot be eliminated by tuning. A limit cycle is a sustained equal amplitude oscillation.

A control valve or damper can have significant backlash from play in linkages and connections seen as a dead band upon reversal of direction of the signal. Two or more integrators in the loop will cause a limit cycle than cannot be eliminated by tuning. Thus, an integrating process and a PID controller with integral action can cause a limit cycle. A self-regulating process with a cascade loop where the primary and secondary PID each have integral action will exhibit a limit cycle.

A variable speed drive (VSD) with a resolution limit due to insufficient number of bits in the analog to digital signal input card will result in a limit cycle for one or more integrators. If a dead band is introduced in the VSD setup, limit cycle will develop for two or more integrators.

A loop with a slow final control element that cannot keep up with the rates of change of the controller output will burst into oscillations. The oscillations may not show up during loop tests where typically small changes are made. The burst of oscillations may only occur for large fast disturbances and large setpoint changes causing the rate of change of the PID output to be faster than the valve slewing rate or VSD rate limited response. Large valves and dampers and VSD with marginally sized motors can be too slow.

Measurement

The sensor design, location, and operating conditions can result in excessive measurement noise. Differential head meters and vortex meters are especially sensitive to velocity profile and hence the upstream and downstream piping configuration. Most flow meters get noisy as you approach the low flow limit. pH electrodes are particularly vulnerable to noise from electromagnetic interference and concentration fluctuations.

The measurement can also introduce an excessive lag into the control loop. Fouled electrodes and thermowells can have an excessive lag from a decrease in the mass transfer and heat transfer coefficients, respectively. Pressure and differential pressure transmitters on compressor control can become too slow by simply increasing the transmitter damping setting.

If the measurement lag becomes by far the largest time constant in the loop an insidious situation develops where the trend charts may look smoother because the measurement is showing an extremely attenuated (filtered) view of the actual process oscillations. The controller gain may even be able to be increased further misleading the user into thinking the measurement lag is beneficial. The key is the loop period should show an increase due to the increase in loop dead time.

Wireless measurements can introduce excessive dead time in fast loops by a slow default update rate. At-line analyzers and especially off-line analyzers can cause excessive dead time in almost any loop from the sample transport and processing, analysis cycle time, and multiplex time.

Controller

The controller can introduce an excessive dead time from too large of a module execution time or signal filter time. Even without considering the effect of the increase in dead time on the ultimate limit to errors and tuning, a signal filter and module execution time larger than 10% of the reset time will increase the integrated errors by more than 10% for an unmeasured disturbance.

Too slow of a secondary controller tuning can cause a cascade loop to burst into oscillations when the primary controller output is changing faster the secondary loop can respond. Insufficient gain action can cause slow rolling oscillations in an integrating process and instability in a runaway process.

Unified Tuning

The key breakthrough in thinking is to use the lambda tuning rules for integrating processes for lag dominant self-regulating and runaway processes. The lambda tuning rules automatically prevent the product of the PID gain and reset time from being less than the twice the inverse of the integrating process gain that would result in slow rolling oscillations. The main decision is whether to maximize the transfer of variability or maximize the absorption of the variability. Key features are used rather than tuning to address the automation system difficulties and to meet different process objectives.

Maximization of the Transfer of Variability

The maximum dead time for all possible operating conditions is used as the arrest time, which is the time to start to turn back an excursion from an unmeasured disturbance. The arrest time is lambda. This minimum lambda is the dead time as shown in Figure 2. The minimum lambda plus the maximum integrating process gain (fastest ramp rate) is used in the tuning equations. The lambda tuning rule then gives the same gain and reset time as the Ziegler Nichols Reaction Curve method. For temperature, composition, and pH loops, the rate time should be set equal to the secondary time constant. If this time constant cannot be identified, ½ of the total loop dead time can be used as an approximation. This gives a rate time that is about 1/6 of the reset time. Note that for the ISA Standard Form PID, the rate time must always be less than the reset time to avoid instability (see Appendix A).

If a lower controller gain is used. The equivalent lambda needs to back calculated and used in the calculation of the reset time to prevent slow rolling oscillations. For runaway processes, the PID gain must always be greater than the inverse of the process gain to prevent a runaway.

Maximization of the Absorption of Variability

When the absorption of variability must be maximized, lambda is chosen to be as large as possible. The maximum arrest time lambda depends upon the integrating process gain, the allowable change in the process variable and the available change in the manipulated flow. The following equations show how to calculate lambda for a generic application and then a surge tank level loop.

The maximum lambda (λ) is the maximum allowable % excursion ($\Delta\%PV_{\max}$) divided by the maximum possible PV ramp rate. The maximum possible ramp rate is the PV rate of change per percent output change multiplied by the maximum available percent output change ($\Delta\%CO_{\max}$).

$$\lambda = \frac{\left|\Delta\%PV_{\max}\right|}{\left|\frac{\frac{\Delta\%PV}{\Delta t}}{\Delta\%CO}\right| * \left|\Delta\%CO_{\max}\right|} \tag{1}$$

Realizing that the integrating process gain is the PV ramp rate per percent output change:

$$\lambda = \frac{1}{K_i} * \frac{\left|\Delta\%PV_{\max}\right|}{\left|\Delta\%CO_{\max}\right|} \tag{2}$$

An equivalent setpoint rate limit on the controller output (e.g. flow controller setpoint):

$$\left|\Delta\%CO/\Delta t\right|_{\max} = \frac{\left|\Delta\%CO_{\max}\right|}{\lambda} = K_i * \frac{\left|\Delta\%CO_{\max}\right|^2}{\left|\Delta\%PV_{\max}\right|} \tag{3}$$

For a PV limit ($\%PV_{Limit}$) and corresponding CO limit ($\%CO_{Limit}$) we have:

$$\lambda = \frac{1}{K_i} * \left|\frac{\%PV_{Limit} - \%SP}{\%CO_{Limit} - \%CO}\right| \tag{4}$$

The above calculation would be done for high and low operating limits and various setpoints. The smallest of the arrest times would be used in tuning.

We can obtain the more detailed requirements for surge level tank level control by computing the integrating process gain for level. The integrating process gain is the product of the valve, process, and measurement gains:

$$K_i = K_v * K_p * K_m \tag{5}$$

The valve gain or variable speed drive gain for a linear installed characteristic or flow loop is:

$$K_v = \frac{\Delta F_v}{\Delta\%CO} = \frac{\Delta F_{\max}}{100\%} \tag{6}$$

The level process gain for mass flow is (omit density term for volumetric flow):

$$K_p = \frac{1}{A * \rho} \tag{7}$$

The level measurement gain is:

$$K_m = \frac{\Delta\%PV}{\Delta L} = \frac{100\%}{\Delta L_{\max}} \qquad (8)$$

Substituting in the valve, process, and measurement gains, the integrating process gain is:

$$K_i = K_v * K_p * K_m = \frac{\Delta F_{\max}}{100\%} * \frac{1}{A*\rho} * \frac{100\%}{\Delta L_{\max}} = \frac{\Delta F_{\max}}{\Delta L_{\max}} * \frac{1}{A*\rho} \qquad (9)$$

The consequential arrest time for a level loop is:

$$\lambda = \frac{A*\rho*\Delta L_{\max}}{\Delta F_{\max}} * \frac{\Delta\%PV_{\max}}{\Delta\%CO_{\max}} \qquad (10)$$

An equivalent setpoint rate limit on the controller output (e.g. flow controller setpoint):

$$\left|\Delta\%CO/\Delta t\right|_{\max} = \frac{\left|\Delta\%CO_{\max}\right|}{\lambda} = \frac{\Delta F_{\max}}{A*\rho*\Delta L_{\max}} * \frac{\left|\Delta\%CO_{\max}\right|^2}{\left|\Delta\%PV_{\max}\right|} \qquad (11)$$

Where:

$A =$ cross sectional area (m^2)
$\Delta F_{\max} =$ maximum change in valve or variable speed drive flow (e.g. flow span) (kg/sec)
$\Delta L_{\max} =$ maximum change in level (e.g. level span) (m)
$K_i =$ integrating process gain (1/sec)
$K_m =$ measurement gain (%/m for level)
$K_p =$ process gain (m/kg for level)
$K_v =$ valve or variable speed drive gain (kg/sec/%)
$\lambda =$ arrest time (sec)
$\rho =$ fluid density (kg/m^3)
$\%CO =$ operating point controller output (%)
$\%CO_{Limit} =$ controller output limit (%)
$\Delta\%CO_{\max} =$ maximum available change in controller output (%)
$\%PV_{Limit} =$ process variable limit (%)
$\Delta\%PV_{\max} =$ maximum allowable change in process variable (%)
$\%SP =$ operating point setpoint (%)

Key PID Features

There are seven key related PID control features; structure, external-reset feedback, feedforward, setpoint filter and lead-lag, setpoint rate limits, output tracking, and an enhancement for wireless. These key features synergistically provide capabilities that have not been fully realized. See Appendix A for conversion of tuning settings for different PID Forms and tuning setting units.

PID Structure

Most modern DCS offer a choice of structures shown in Table 1.

Table 1 - Major PID Structure Choices

(1) PID action on error ($\beta = 1$ and $\gamma = 1$)
(2) PI action on error, D action on PV ($\beta = 1$ and $\gamma = 0$)
(3) I action on error, PD action on PV ($\beta = 0$ and $\gamma = 0$)
(4) PD action on error, no I action ($\beta = 1$ and $\gamma = 1$)
(5) P action on error, D action on PV, no I action ($\beta = 1$ and $\gamma = 0$)
(6) ID action on error, no P action ($\gamma = 1$)
(7) I action on error, D action on PV, no P action ($\gamma = 0$)
(8) Two degrees of freedom controller (β and γ adjustable 0 to 1)

The structure choices use β and γ set point weighting factors. A controller that has both factors adjustable is called a "two degree freedom controller." Other structures have the β and γ factors set equal to 0 or 1. The user can also omit a mode entirely to get P-only, I-only, ID, and PD control with various assigned factors. PI control is achieved by simply setting the derivative (rate) time to zero. In general, the user must not set the controller gain equal to zero in an attempt to get I-only or ID control or set the integral (reset) time to zero in an attempt to get P-only or PD control. Note that the use of P-only or PD control requires additional choices of how to set the bias and its ramp time.

In choosing the correct structure it is important to know the relative contribution of the proportional, integral, and derivative modes. Figure 3 shows the contribution of each PID mode for a setpoint change. There is no feedback from the process in this figure (e.g. block valve closed), so we can more clearly see the contribution of each mode.

Structure 1 (PID on error) provides the fastest approach to a new setpoint by virtue of a step and a kick in the controller output from the setpoint change as seen in the second response in Figure 3. For small setpoint changes and low controller gains, the step and kick can help get through significant valve backlash and stick-slip to get the valve moving. The kick appears to be a spike on trend charts with a large time spans. The abrupt change in output is often seen as disruptive by operators when they make setpoint changes. If the burst of flow through the control valve does not affect other users of the process or utility fluid, the step and kick is more of a psychological than a process concern.

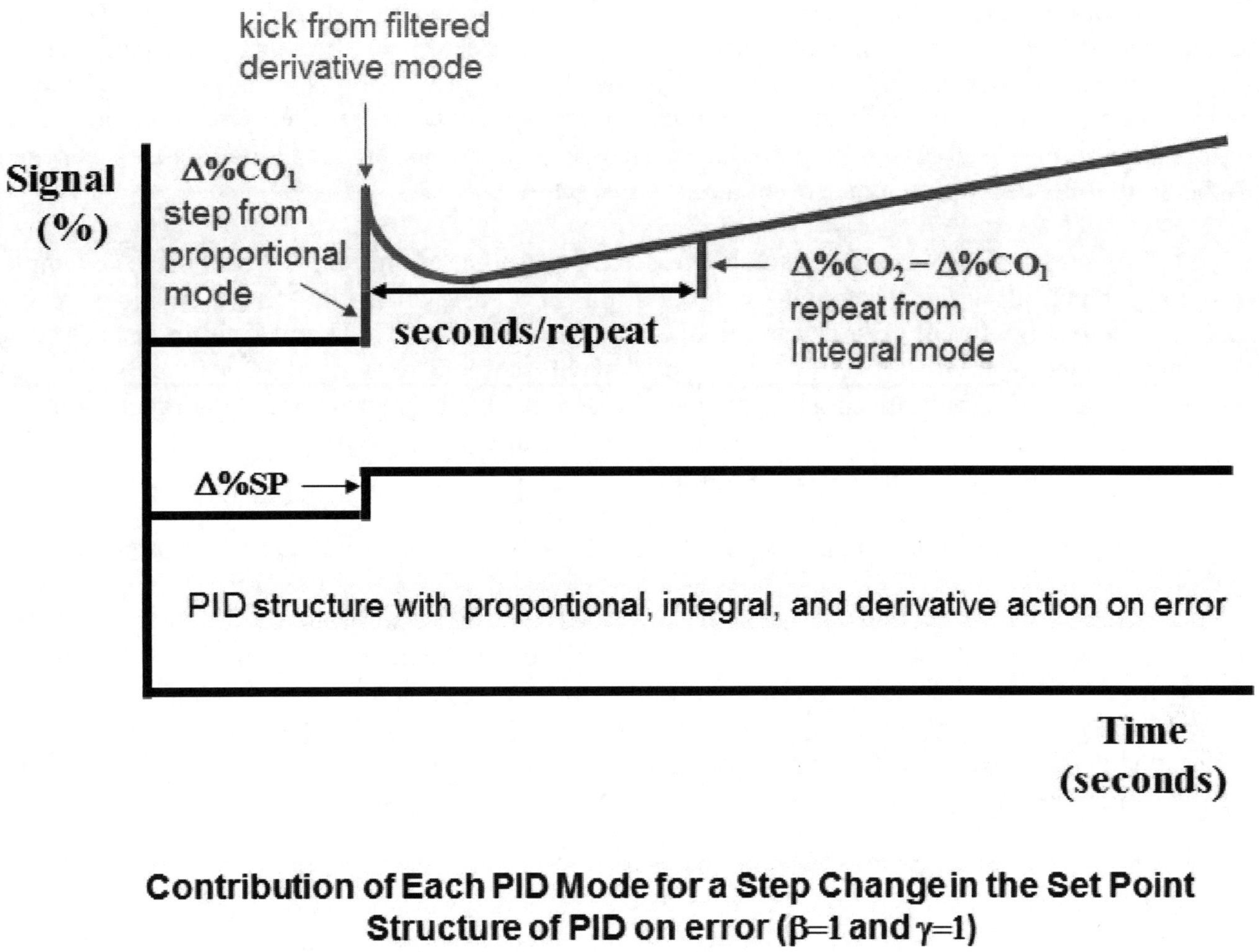

Figure 3 – Contribution of each PID mode for a step change in setpoint for a PID on error structure

Structure 2 (PI on error, D on PV) is the structure most often used. Structure 2 eliminates the kick from derivative action for a setpoint change by setting gamma to zero (γ=0). The increase in rise time going from structure 1 to 2 is negligible for the more important loops, such as column and vessel temperature where derivative action is used. The step in the output from the proportional mode on a setpoint change is large for a high controller gain. Increases in process gain or dead time will increase the overshoot unless the controller gain is decreased accordingly. If the elimination of setpoint overshoot is much more important than rise time (e.g. bioreactor temperature and pH shifts), then structure 3 may be best.

Structure 3 (I on error, PD on PV) eliminates overshoot but with quite a sacrifice in speed of approach to the setpoint. For near and true integrating and runaway processes, structure 3 may have a rise time an order of magnitude larger than the rise time for structure 1 for the same tuning settings and process dynamics. A setpoint filter set equal to the reset time provides the same response.

Structure 4 (PD on error, no I) is used on processes adversely affected by integral action. The temperature control of severely exothermic polymerization reactors use structure 4 because integrating action in the controller increases the risk of a runaway. If integral action is used, the reset time should be increased by a factor of 10 for these positive feedback processes to be safe. Users may not be aware of this requirement leading to overshoot that can trigger a runaway. The bias for structure 4 is set equal to the normal PD controller output when the PV is at setpoint.

Structure 4 is also used on batch processes that respond in only one direction. For example, in bringing a batch pH up to a setpoint by the addition of a base where the base is not consumed in a reaction, the batch will only respond in the direction of increasing pH. The pH will overshoot setpoint if integral action is used. If split ranging is added with an acid reagent, there will be some wasted reagent due to cross neutralization of reagents and limit cycling across the split range point from stiction that is greatest near the closed position. For structure 4 and a single reagent, the bias is set for zero reagent addition when the PV is at setpoint.

Structure 5 (P on error, D on PV, no I) is used for many of the same reasons as structure 4.

Structure 6 (ID on error, no P) is used for valve position controllers (VPC) to eliminate the interaction with process controller whose valve position is being optimized. The VPC could be optimizing the coarse adjustment from a large control valve in parallel with a small control valve manipulated as fine adjustment by the process controller. In Figure 4 the process controller FC215 adjusts the small valve input to the process. The output to the small valve is also used as a PV of the VPC ZC215. The small valve provides precision and the large valve gives rangeability.

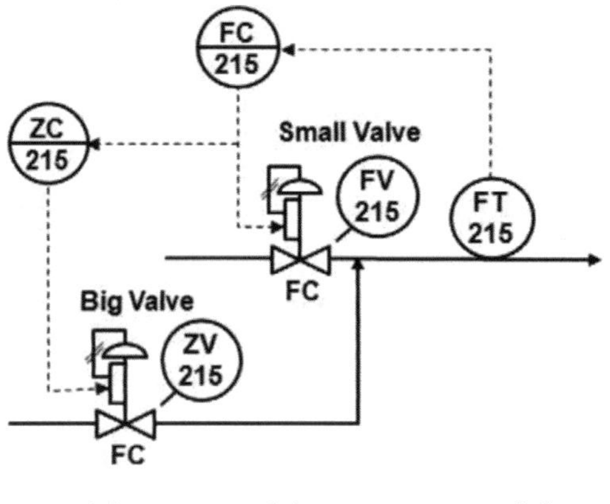

Figure 4 - Valve position control increases precision and rangeability

The VPC could be optimizing utility supply pressure or temperature to minimize energy use or optimizing feed rate to maximize production rate. Normally, the rate time is zero. The tuning of this integral-only controller is problematic. Tuning rules often cited suggest the integral time should be larger than 10 times the product of the gain and reset time of the process controller and 10 times the residence time of the process to eliminate interaction. However, the VPC response for this tuning is too slow to prevent the process controller from getting into trouble for large disturbances. Feedforward action can be added to help. In addition external-reset feedback, setpoint directional rate limits, and an enhanced PID developed for wireless can be used as a flexible and easy to tune solution to deal with unknowns. This solution is described in the Setpoint Rate Limits section.

Structure 7 (I on error, D on PV, no P) is used for the same reasons as structure 6. As with structure 2, the spike from rate action for setpoint changes is eliminated.

Structure 8 (two degrees of freedom via adjustable β and γ) is used to provide a balance between a fast rise time and overshoot. The same effect can be accomplished by the use of a setpoint lead-lag as discussed in the setpoint filter section. This structure can be used to get the structures with proportional or derivative action on PV by zeroing the respective β and γ set point factors.

External-Reset Feedback

PID controllers with the positive feedback implementation of integral action as shown in Figure 5 enable the use of an external reset feedback. When the switch in Figure 5 is in the position to get an external reset feedback to the integral mode from the process variable (PV) of a setpoint being manipulated whether an Analog Output block or a secondary PID, the PID output is prevented from changing faster than the PV of the control valve, VSD, or secondary loop can respond. There is an increasingly larger scope of potential benefits discovered as described in subsequent features and the summary of the unified methodology. The only requirement is that the external reset feedback be properly connected to a fast update of the PV for the valve, VSD, or loop being manipulated. Updates of actual valve position, VSD speed, or secondary loop PV should be at least twice as fast as the the execution time of the controller with external reset feedback.

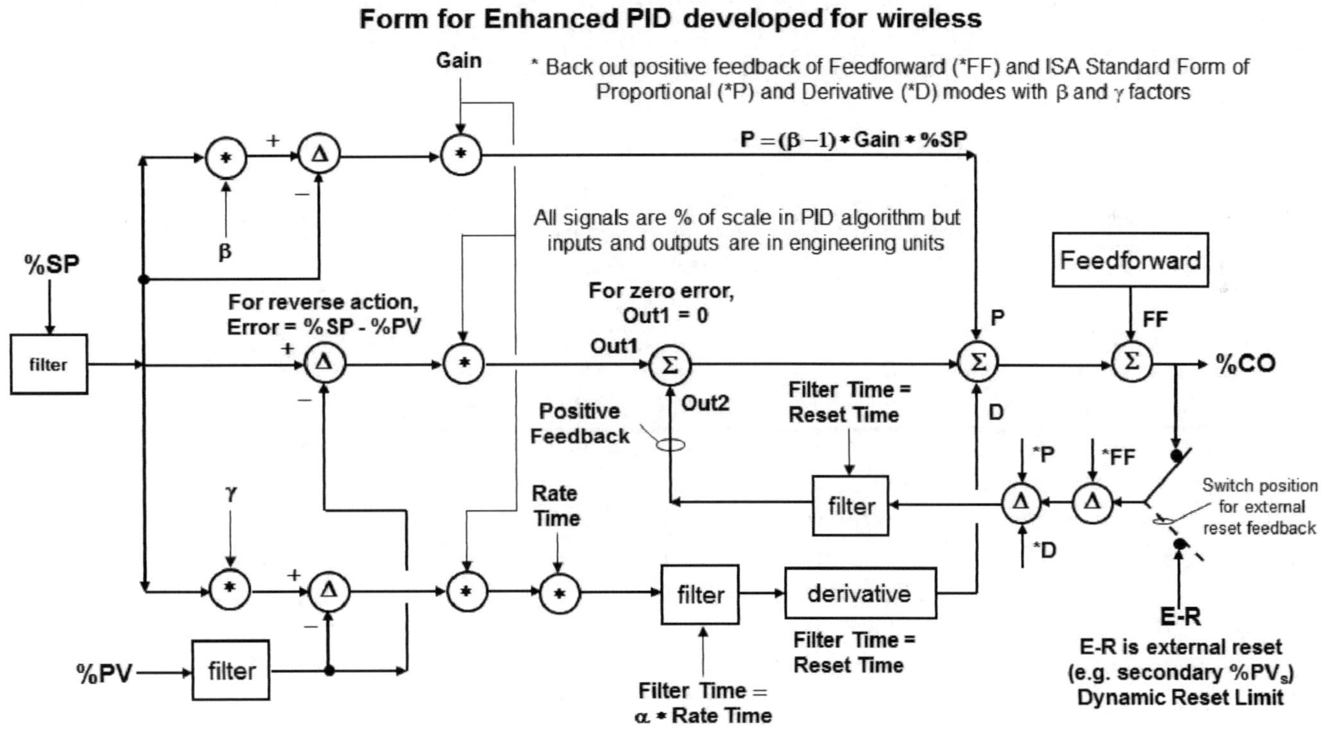

Figure 5 – ISA Standard Form PID with External Reset Feedback

Feedforward

Feedforward control is particularly effective when a disturbance is large, fast, and precisely measured. While theoretically a feedforward multiplier is more effective for changes in the slope of a plot of the manipulated flow versus the measured disturbance, for many practical reasons such as scaling and nonlinearity problems, a feedforward summer is used for feedforward control especially on vessels and columns. The most common feedforward signal is flow because the changes in feed flows are faster than changes in composition and temperature and the flow is the most common process input. A flow feedforward summer is implemented via an external ratio and bias station so the operator can set the desired ratio and see the actual ratio. On startup, the operator may run on ratio control with the process controller in manual until the equipment reaches operating conditions. The initial flow ratios can be set based on the process flow diagram (PFD). The PFD can be put live online on the operator graphics to show the actual ratios used. The technique also enables a plant wide feedforward system to rapidly increase and decrease production rates.

Setpoint Filter and Lead-Lag

A setpoint filter can provide the desired closed loop time constant important for the coordination of loops. The use of secondary PID flow setpoint enables stoichiometric ratios to be enforced without a transient unbalance for production rate changes and setpoint changes and disturbances to the primary PID. External reset feedback must be turned on in the primary PID to prevent it from trying to change the secondary PID set point faster than filter will allow the secondary loop to respond. Turning external reset on eliminates the need to retune the primary PID when the secondary PID filter is added or tuned. If coordination of loops is not necessary, a setpoint filter should not be added to a secondary loop because it will slow down the rejection of primary loop disturbances.

The rise time (time to reach setpoint) and overshoot for a setpoint change can be optimized by the use of a setpoint lead-lag. In this case the setpoint lag time is set equal to the reset time and the setpoint lead time is set equal to about 20% of the lag time. If preventing overshoot is more important than reducing rise time for a setpoint change, the lead time can be decreased.

Setpoint Rate Limit

The setpoint rate limit option in analog output (AO) and PID function blocks in a modern DCS can provide move suppression, a tuning parameter found to be very useful in model predictive control. Since there is often an up and down rate limit, directional move suppression can be used provided the ability needed to enable a slow approach to an optimum and a fast getaway to prevent an abnormal condition. This technique is used to provide a fast opening and slow closing anti-surge control valve. There are many other applications as we will see in the unified methodology. If external reset feedback is turned on in the PID manipulating the setpoint, the PID does not have to be retuned when the rate limits are added or tuned in the manipulated AO or PID. These rate limits can be written to online.

Output Tracking

Disturbances can be too fast and the consequences too severe to rely just on PID control. Here output tracking is used to get the loop out of trouble for equipment and environmental protection. The most common example is compressor surge prevention, where the PID output will track a surge valve position large enough to prevent surge when a crossing of the surge curve is detected or a precipitous drop in compressor suction flow occurs indicating the start of surge. The PID stays in output tracking holding the protective output for several seconds to ensure the system has stabilized.

The fastest possible setpoint response is obtained from a full throttle approach (bang-bang control method) where the PID is put in output tracking at the output limit to drive the PV to the setpoint. The PID output stays at the limit until the PV one dead time into the future is predicted to reach setpoint. A dead time (DT) block with its parameter set equal to the loop dead time is used to create an old PV that is subtracted from the new PV to create a delta PV. The delta PV added to the new PV that is the DT block input gives the PV one dead time into the future. When the future PV is projected to reach setpoint, the PID tracks a final resting value (FRV) captured form previous batches or startups and holds this FRV for one dead time before releasing the PID back to feedback control.

Enhanced PID for Wireless

The enhanced PID option developed for wireless enables the following features:

1. Positive feedback implementation of reset with external-reset feedback
2. Immediate response to a setpoint change or feedforward signal or mode change
3. Suspension of integral action until there is a change in PV
4. Integral action is the exponential response of the positive feedback filter to the change in controller output for the time interval since last update
5. Derivative action is the PV or error change depending upon PID structure divided by the time interval since the last update (instead of PID execution time) multiplied by the gain and rate time

The enhanced PID will go into a holding mode for a loss of communication, stuck control valve, or failure of an electrode, analyzer, or sample system providing a smooth continuation upon restoration of communication and analyzer results. For offline analyzers, late lab results do not cause a problem. The enhanced PID eliminates the need for retuning as discontinuous delays from wireless devices and analyzers are increased. For interacting loops and valve position control and final control elements with resolution limits and deadband, a threshold sensitivity setting enables the enhanced PID to eliminate reactions to insignificant changes that could cause fighting between loops or perpetuate a limit cycle. The threshold sensitivity is set just larger than the PV noise amplitude after filtering.

A Unified Methodology

The following is a generalized step by step procedure to address a wide spectrum of automation system difficulties and application objectives. Note that PID features vary considerably from one DCS manufacturer to another. Any setting of options or tuning must be based on information from the supplier and should be thoroughly tested before and carefully observed after commissioning.

1. Set the output limits to keep the manipulated setpoints in the desired operating range. For variable speed drives set the process PID low output limit so the speed cannot cause the discharge head to approach the static head in order to prevent excessive sensitivity to pressure and to prevent reverse flow. In general, set the anti-reset windup limit to match the output limit. If the output scale is engineering units, the output limits and anti-reset windup must be based on the

output scale range and units.

2. Choose the best structure for your application. Generally the best choice is structure 2 with PI on error and D on PV. For a single direction response (e.g. batch heating or neutralization), use structure 4 or structure 5 so that there is no integral action. For a highly exothermic reaction, you might want structure 5 to help prevent a runaway from integral action.

3. Set the signal filter noise just large enough to keep the controller output fluctuations from exceeding the resolution limit or deadband of the final control element.

4. For near-integrating, true integrating, and runway processes use the lambda integrating process tuning rules. To maximize the transfer of variability from the process variable to the manipulated variable, set the lambda (arrest time) equal to the maximum dead time and use the largest integrating process gain for all possible operating conditions in the tuning. To maximize the absorption of variability (e.g. surge tank level) use the minimum arrest time computed from Equations 1 through 10 for all possible operating conditions. If you decrease the PID gain, proportionally increase the PID reset time to prevent slow rolling oscillations. For runaway processes, the PID gain must not be less than the inverse of the open loop gain. When changing from the Series Form to the ISA Standard Form in newer DCS, convert the tuning settings based on the differences in Form. In the ISA Standard Form the rate time should not be greater than ¼ the reset time. Be sure to realize and convert any difference in Form and tuning setting units between different PID vintages and suppliers (see Appendix A). You must take into account the units and inverse relationship between gain and proportional band and reset time in seconds and minutes per repeat. Not addressing these differences can result in settings off by orders of magnitude.

5. For self-regulating processes with the open loop time constant less than 4 times the dead time, use the lambda self-regulating tuning rules. To maximize the transfer of variability from the process variable to the manipulated variable set the lambda (closed loop time constant) equal to the maximum dead time and use the largest process gain and smallest time constant for all possible operating conditions in the tuning. Schedule tuning settings preferably with an adaptive tuner with the specified lambda to deal with changes in the dynamics.

6. Turn on external reset feedback. Make sure the external reset feedback signal is correctly propagated back to the PID (e.g. BKCAL signal) especially if there are split range, signal characterizer, or signal selector blocks on the PID output.

7. For final control elements that are slow or that have deadband or resolution limit, use a fast readback of the valve position or variable frequency drive speed as the external reset feedback to prevent a burst of oscillations from the PID output changing faster than the final control element can respond.

8. For final control elements that create limit cycles from resolution limits and deadband, use a fast readback of the valve position or variable frequency drive speed to stop the limit cycles.

9. For cascade control, use the PV of the secondary loop as the external reset feedback to prevent a burst of oscillations from violation of the cascade rule where the secondary loop must be significantly faster than the primary loop.

10. For setpoint filters of secondary loops for coordination of flow loops, use the PV of the secondary loop as the external reset feedback to prevent the need to retune the PID.

11. For setpoint rate limits use the PV of the analog output block or secondary loop as the external reset feedback to prevent the need to retune the PID. Add setpoint rate limits to minimize the interaction between loops and in valve position control and to provide directional move suppression to enable a fast getaway for abnormal conditions and a slow approach to optimum. For valve position control, use an enhanced PID developed for wireless with a threshold sensitivity limit to ignore insignificant changes in the valve position to be optimized.

12. Add output tracking for equipment protection and a full throttle (bang-bang control) strategy for the fastest possible time to reach setpoint on startup and for batch operations.

13. Add output tracking logic to momentarily track an output that insures equipment and environmental protection. For compressor surge protection track a sufficiently large opening of the surge valves. To prevent a RCRA pH violation, track a rapidly incrementing reagent valve position to prevent an effluent excursion below 2 pH or above 12 pH.

14. Add feedforward control for large and fast measured disturbances. For flow feedforward, use a ratio and bias station so the operator can enter a desired flow ratio and see the actual flow ratio. Setup the PID to provide a bias correction to the manipulated flow. Add dynamic compensation (dead time and lead-lag blocks) to the feedforward so the manipulated flow arrives at the same point in the process at the same time as the measured disturbance.

15. For wireless devices or analyzers that introduce a large discontinuous PV update delay use an enhanced PID to eliminate the need to retune the controller to prevent oscillations. If the delay is much larger than the 63% process response time, the PID gain can be set as large as the inverse of the open loop gain for self-regulating processes.

16. For valve position control (VPC), make sure the key PID features in Table 3 are used. For examples of VPC for process optimization, see Table 4. The use of an enhanced PID enables the use of integral action in the VPC and simplifies the tuning of the VPC. While model predictive control can provide a more sophisticated optimization, the VPC can be implemented by a simple configuration change that involves setting up a PID as the VPC. The minimal implementation cost and time is attractive for small optimizations.

Table 3 – Key Feature Function and Advantages for Valve Position Control

Feature	Function	Advantage 1	Advantage 2
Direction Velocity Limits	Limit VPC Action Speed Based on Direction	Prevent Running Out of Valve	Minimize Disruption to Process
Dynamic Reset Limit	Limit VPC Action Speed to Process Response	Direction Velocity Limits	Prevent Burst of Oscillations
Adaptive Tuning	Automatically Identify and Schedule Tuning	Eliminate Manual Tuning	Compensation of Nonlinearity
Feedforward	Preemptively Set VPC Out for Upset	Prevent Running Out of Valve	Minimize Disruption
Enhanced PID (PIDPlus)	Suspend Integral Action until PV Update	Eliminate Limit Cycles from Stiction & Backlash	Minimize Oscillations from Interaction & PV Update Delay

Table 4 – Examples of Valve Position Control for Process Optimization

Optimization	VPC PID PV	VPC PID SP	VPC PID Out
Minimize Prime Mover Energy	Reactor Feed Flow PID Out	Max Throttle Position	Compressor or Pump Pressure SP
Minimize Boiler Fuel Cost	Steam Flow PID Out	Max Throttle Position	Boiler Pressure SP
Minimize Boiler Fuel Cost	Equipment Temperature PID Out	Max Throttle Position	Boiler Pressure SP
Minimize Chiller or CTW Energy	Equipment Temperature PID Out	Max Throttle Position	Chiller or CTW Temperature SP
Minimize Purchased Reagent or Fuel Cost	Purchased Reagent or Fuel Flow PID Out	Min Throttle Position	Waste Reagent Or Fuel Flow SP
Minimize Total Reagent Use	Final Neutralization Stage pH PID Out	Min Throttle Position	First Neutralization Stage pH PID SP
Maximize Reactor Production Rate	Reactor or Condenser Temperature PID Out	Max Throttle Position	Feed Flow or Reaction Temperature SP
Maximize Reactor Production Rate	Reactor Vent Pressure PID Out	Max Throttle Position	Feed Flow or Reaction Temperature SP
Maximize Column Production Rate	Reboiler or Condenser Flow PID Out	Max Throttle Position	Feed Flow or Column Pressure SP
Maximize Ratio or Feedforward Accuracy	Process Feedback Correction PID Out	50% (Zero Correction)	Flow Ratio or Feedforward Gain

Not mentioned is the benefit of using cascade control. Fast secondary loops can correct for many disturbances before they affect the primary loop. Secondary loops also reduce nonlinearities seen by the primary loop. A secondary flow loop isolates the nonlinearity of the installed characteristic of the control valve and enables flow feedforward. A secondary jacket or coil temperature loop isolates a crystallizer, evaporator, or reactor temperature loop from cooling or heating system nonlinearities.

Conclusion

The use of key PID features, a unified PID tuning method, and an adaptive tuner enables an effective application of PID control for wide spectrum of process responses. Drastically different sources of variability, automation system difficulties and changing process objectives do not require PID retuning. PID features can be used to tailor the PID response to different application requirements on a much more understandable basis by the user. The role of the PID can be expanded to include process optimization besides regulation.

Resources

1. McMillan, G. "PID Tuning Rules", *InTech*, Jan.-Feb. 2012.
2. McMillan, G. *Good Tuning: A Pocket Guide, 3rd Edition* ISA, 2012.
3. McMillan, G. "Full Throttle Batch and Startup Response" *Control*, May 2006.
4. Shinskey, G. "The Power of External-Reset Feedback" *Control*, May 2006.
5. McMillan, G. "What Is the Key PID Feature for Basic and Advanced Control?" Control Talk blog, *Control*, Apr. 26, 2012.
6. McMillan, G. and Torres, H. "Effective Use of Key PID Features" Paper presented at ISA Automation Week, Orlando, 2012.
7. McMillan, G. "Don't Overlook PID in APC" *Control*, Nov. 2011.
8. Yu, C. and Luyben, W. "Analysis of Valve-Position Control for a Dual Input Process" *Ind. Eng. Chem. Fund.*, pp. 344-350, Vol. 25, No. 3, 1986.
9. Harriott, P. *Process Control* McGraw-Hill, 1964.
10. Bohl, A. and McAvoy, T. "Linear Feedback vs. Time Optimal Control. II. The Regulator Problems", *Ind. Eng. Chem.*, Process Des. Dev., Vol. 15, No. 1, 1976.
11. McMillan, G. *Essentials of Modern Measurements and Final Elements in the Process Industries* ISA, 2010.
12. McMillan, G. "Advances in Flow and Level Measurements Enhance Process Knowledge, Control" *InTech*, Jan.-Feb. 2010.
13. McMillan, G. "Feedforward Control Enables Flexible, Sustainable Manufacturing" *InTech*, Mar.-Apr. 2011.
14. McMillan, G. "Key Design Components of Final Control Elements" *InTech*, Mar.-Apr. 2010.
15. McMillan, G. "What's Your Flow Control Valve Telling You?" *Control Design*, May 2004.
16. McMillan, G. "Improve Control Loop Performance" *Chemical Processing*, Oct. 2007.
17. Shinskey, G. "Flow and Pressure Control Using Variable-Speed Motors" *Proceedings of* Control *Conference*, pp. 161-167, Chicago, 1980.
18. Al-Khalifah, M. and McMillan, G. "Control Valves versus Variable Speed Drive for Flow Control" Paper presented at ISA Automation Week, Orlando, 2012.
19. Smith, C. "Watch Out with Variable Speed Pumping" *Chemical Processing*, May 2008.
20. McMillan, G. "Unlocking the Secret Profiles of Batch Reactors" *Control*, Jul. 2008.
21. McMillan, G. "Is Wireless Process Control Ready for Prime Time" *Control*, May 2009.
22. McMillan, G. "Wireless - Overcoming Challenges of PID Control, Analyzer Applications" *InTech*, Jul.-Aug. 2010.

Appendix A – PID Controller Forms

Conversion of Settings to Correct Units and Form

The methodology in this paper is based on the ISA Standard Form with specific tuning setting units where the controller gain is dimensionless, the reset time is seconds (seconds per repeat), and the rate time is seconds. PID tuning settings must first be checked for units and if necessary undergo a units conversion before being used.

(1) Convert proportional band (%) to controller gain (%/%) (dimensionless)

Gain = 100 % / Proportional Band

(2) Convert reset setting in repeats per minute to reset time in seconds per repeat

Seconds per repeat = 60 / repeats per minute

(3) Convert rate time in minutes to rate time in seconds

Seconds = 60 * minutes

(4) After the tuning setting units are verified, convert the Series or Parallel Form tuning settings to the ISA Standard Form tuning settings per equations below. The primed tuning settings are for the Series Form, the double primed tuning settings are for the Parallel Form, and the unprimed tuning settings are for the ISA Standard Form.

To convert from Series to ISA Standard Form controller gain:

$$K_c = \frac{T_i' + T_d'}{T_i'} * K_c' \tag{A.1}$$

To convert from Series to ISA Standard Form reset (integral) time:

$$T_i = \frac{T_i' + T_d'}{T_i'} * T_i' = T_i' + T_d' \tag{A.2}$$

To convert from Series to ISA Standard Form rate time:

$$T_d = \frac{T_i'}{T_i' + T_d'} * T_d' \tag{A.3}$$

Note that if the rate time is zero, the ISA Standard and Series Form settings are identical. When using the ISA Standard Form, if the rate time is greater than ¼ the reset time the response can become oscillatory. If the rate time exceeds the reset time, the response can become unstable from a reversal of action form these modes. The Series Form inherently prevents this instability by increasing the effective reset time as the rate time is increased.

We can convert from ISA Standard Form to the Series Form using the following equations if the reset time is equal to or greater than 4 times the rate time ($T_i \geq 4 * T_d$).

$$K_c' = \frac{K_c}{2} * \left[1 + \left(1 - 4 * \frac{T_d}{T_i} \right)^{0.5} \right] \quad \text{(A.4)}$$

$$T_i' = \frac{T_i}{2} * \left[1 + \left(1 - 4 * \frac{T_d}{T_i} \right)^{0.5} \right] \quad \text{(A.5)}$$

$$T_d' = \frac{T_i}{2} * \left[1 - \left(1 - 4 * \frac{T_d}{T_i} \right)^{0.5} \right] \quad \text{(A.6)}$$

The Parallel form is used in some older DCS and PLC to isolate the proportional mode tuning setting from the other the modes. The gain setting does not affect the contribution from the integral and derivative modes and is sometimes called non-interacting. We can convert from the Parallel to the ISA Standard Form using the following equations.

$$K_c = K_c'' \quad \text{(A.7)}$$

$$T_i = K_c'' * T_i'' \quad \text{(A.8)}$$

$$T_d = \frac{T_d''}{K_c''} \quad \text{(A.9)}$$

We can convert from the ISA Standard Form to the Parallel Form by the following equations:

$$K_c'' = K_c \quad \text{(A.10)}$$

$$T_i'' = \frac{T_i}{K_c} \quad \text{(A.11)}$$

$$T_d'' = K_c * T_d \quad \text{(A.12)}$$

Where:

K_c = controller gain for ISA Standard Form (%/%) (dimensionless)
$K_c^{'}$ = controller gain for Series Form (%/%) (dimensionless)
$K_c^{''}$ = controller gain for Parallel Form (%/%) (dimensionless)
T_i = integral time (reset time) for ISA Standard Form (seconds)
$T_i^{'}$ = integral time (reset time) for Series Form (seconds)
$T_i^{''}$ = integral time (reset time) for Parallel Form (seconds)
T_d = derivative time (rate time) for ISA Standard Form (seconds)
$T_d^{'}$ = derivative time (rate time) for Series Form (seconds)
$T_d^{''}$ = derivative time (rate time) for Parallel Form (seconds)

Positive Feedback Implementation of Integral Mode

The positive feedback implementation of the integral mode effectively yields the equations above as seen in literature for a PI controller using Laplace transforms. Instead of an integrator, a filter whose input is the controller output and whose output is added to the contribution from the proportional mode in the positive feedback implementation. The filter time is the integral time setting. When external reset feedback is enabled, the input to the filter is switched from the controller output to the external reset feedback signal.

$$O(s) = K_c * E(s) + \frac{1}{1 + T_i * s} * O(s) \qquad \text{(A.13)}$$

$$O(s) * \left(1 - \frac{1}{1 + T_i * s}\right) = K_c * E(s) \qquad \text{(A.14)}$$

$$O(s) * \left(\frac{T_i * s}{1 + T_i * s}\right) = K_c * E(s) \qquad \text{(A.15)}$$

$$\frac{O(s)}{E(s)} = K_c * \left(\frac{1 + T_i * s}{T_i * s}\right) = K_c + \frac{K_c}{T_i * s} \qquad \text{(A.16)}$$

Where:

$E(s)$ = Laplace transform of controller error (%)
$O(s)$ = Laplace transform of controller output (%)

Fundamentals of Model Based Loop Tuning

James Beall

Principal Control Consultant

Emerson Process Management

13 AIChE Spring Meeting, San Antonio, TX

EMERSON. Process Management

James Beall

- Principal Process Control Consultant, Emerson Process Management
- 31 Years experience in process control
  - 12 Years Emerson Process Management
  - 19 Years at Eastman Chemical Company
- Chairman ISA 75.25 Committee - Control Valve Performance Testing
- James.Beall@emerson.com 903-235-7935

13 AIChE Spring Meeting, San Antonio, TX

EMERSON. Process Management

"PID = Proportional, Integral, Derivative"

- **P = proportional** - output contribution is "proportional" to difference between Set Point and PV (error). Therefore, output contribution is "gain" times error. Will have "offset" between Set Point and PV.
- **Integral -** output contribution is "integration" of the difference between Set Point and PV (error). Therefore, always tries to make PV = Set Point (eliminates offset).
- **Derivative -** output contribution is based on the derivative (rate of change) of the difference between Set Point and PV. Actually usually based on PV, not error.

13 AIChE Spring Meeting, San Antonio, TX

EMERSON. Process Management

Caution!!!

- Controllers have different units for their tuning constants. For this presentation:
  - Proportional action will be expressed in **gain**
  - Integral action will be expressed as integral time, **Tr,** in **seconds/repeat**
  - Derivative in units of **seconds.**
- Controllers have different forms (Parallel, Series, Standard) of the PID algorithm. This can significantly affect the actual tuning values to achieve the desired response. The tuning rules in this presentation are for Standard and Series.

13 AIChE Spring Meeting, San Antonio, TX

© 2010, Emerson Process Management

EMERSON. Process Management

Tuning Methods

- First tuning method due to Ziegler & Nichols (1942)
  - Ultimate Gain/Period
  - Reaction Curve
- "little black books"
- Default tuning (gain=1.0, Reset=1 min)
- Many people still do not use any method preferring to "tune-by-feel"
  - Classical control skills now rare
- Most older tuning methods try for "as fast as possible"
- Net result is each loop tuned independently
  - process dynamics not coordinated

13 AIChE Spring Meeting, San Antonio, TX

EMERSON Process Management

Tuning Procedure

- Determine basic process type
  - Self-regulating
  - Integrating
- Determine process dynamics
- Choose desired closed loop response time, "Lambda"
- Calculate tuning constants

13 AIChE Spring Meeting, San Antonio, TX

EMERSON Process Management

Lambda Tuning - Self Regulating Process Open Loop Response

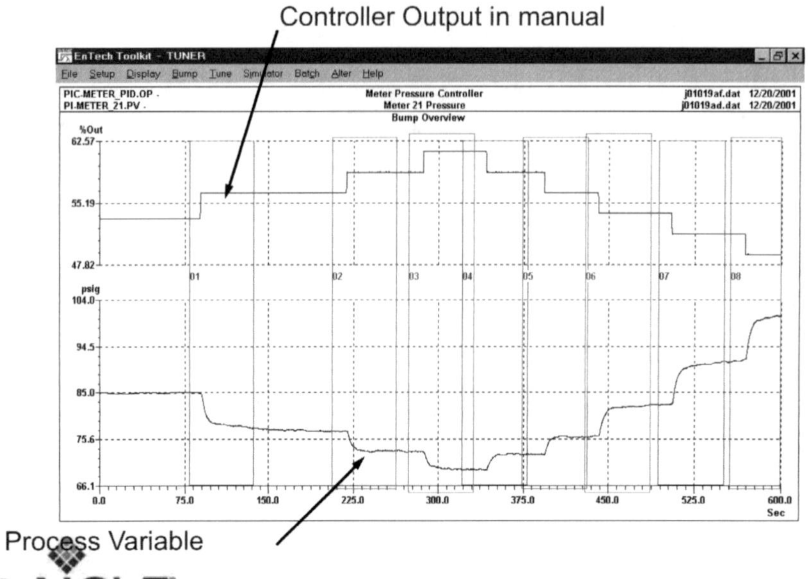

13 AIChE Spring Meeting, San Antonio, TX

EMERSON. Process Management

Characterize the Process Response

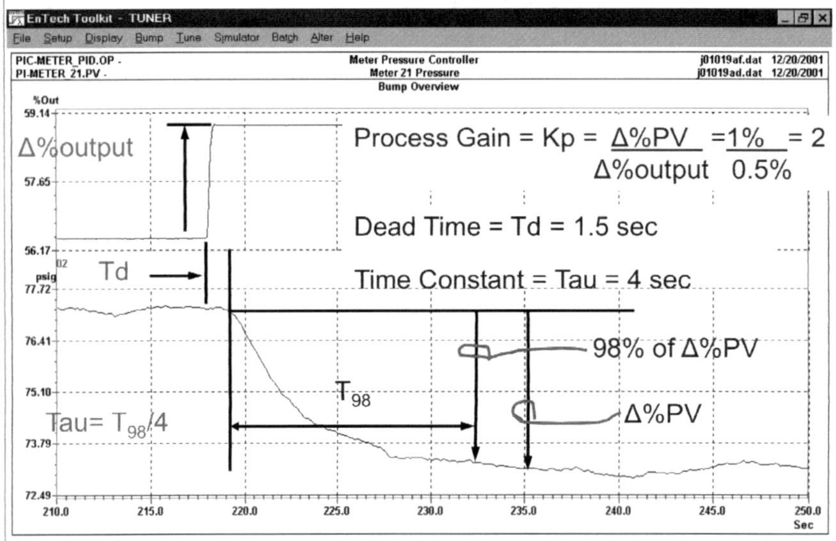

13 AIChE Spring Meeting, San Antonio, TX

EMERSON. Process Management

Lambda Tuning Rules - Self Regulating Process

- Choose Lambda (λ), the "closed loop time constant"
  - A recommended starting point to ensure robustness is 3 * (larger of Td or Tau). This results in stable tuning if Deadtime and Process Gain double.
  - The closed loop "Time to Steady State" (T_{98}) (for a set point change) is approximately 4*Lambda (assuming "P" is on error)
- Tr = Reset Time = Tau (units are time/repeat)
- Kc = Controller Gain = $\frac{Tr}{Kp(\lambda + Td)}$
- NOTE – Tr stays the same, only Kc changes with Lambda

13 AIChE Spring Meeting, San Antonio, TX

EMERSON. Process Management

Lambda Tuning Example

- Choose Lambda = 3 (4 seconds) = 12 seconds
- Integral Time Tr = Tau = 4 seconds/repeat
- Kc = Tr / Kp (λ + Td)

  = 4 seconds /(2*(12 seconds + 1.5 seconds)) = 0.148

Lambda	Gain	Integral Time
12 sec	0.15	4 sec/rep
8 sec	0.21	4 sec/rep
4 sec	0.36	4 sec/rep
1.5 sec	0.67	4 sec/rep

Faster ↓ Higher Gain ↓

13 AIChE Spring Meeting, San Antonio, TX

EMERSON. Process Management

Response – Lambda = 12 sec , 8 sec

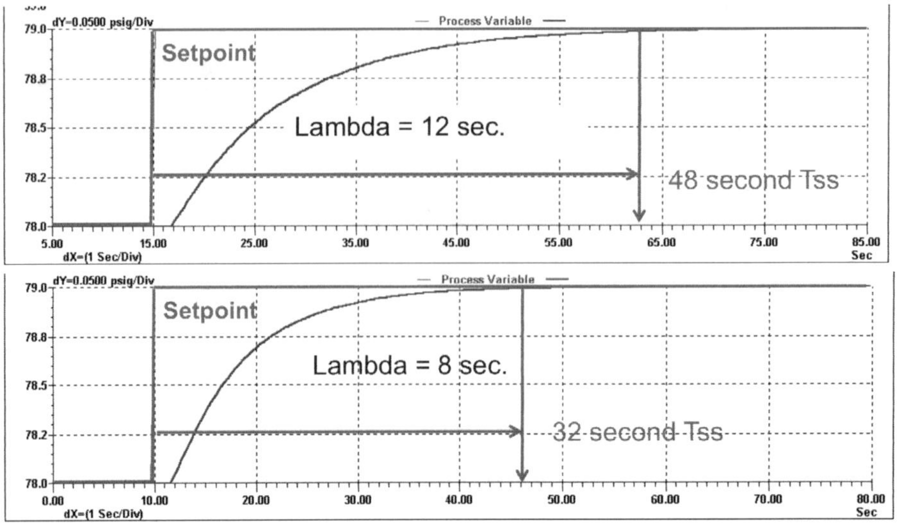

13 AIChE
Spring Meeting, San Antonio, TX

EMERSON.
Process Management

Response – Lambda = 4 sec , 1.5 sec

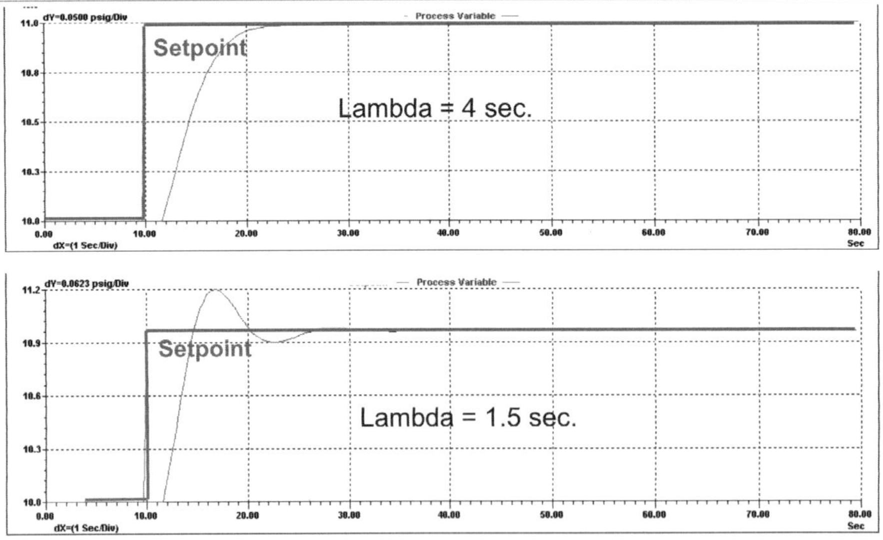

13 AIChE
Spring Meeting, San Antonio, TX

EMERSON.
Process Management

Lambda Tuning - Is it slow? Compared to what?

Typical "Tune by Feel"

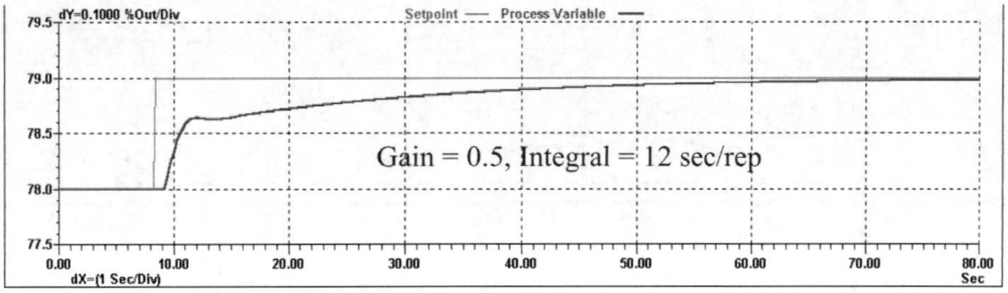

Lambda = 6.6 seconds (normal robustness)

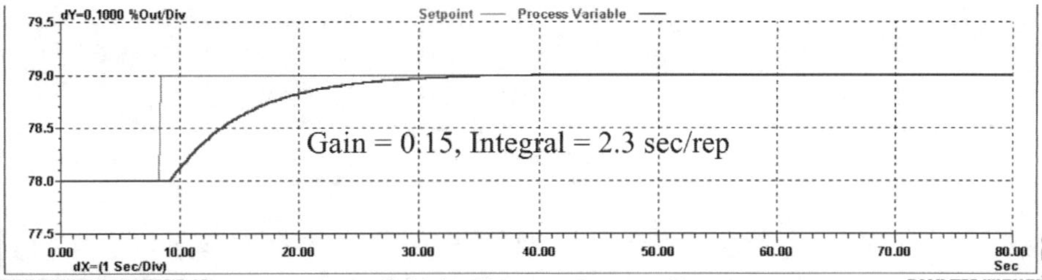

Integrating Process – Identify Open Loop Response

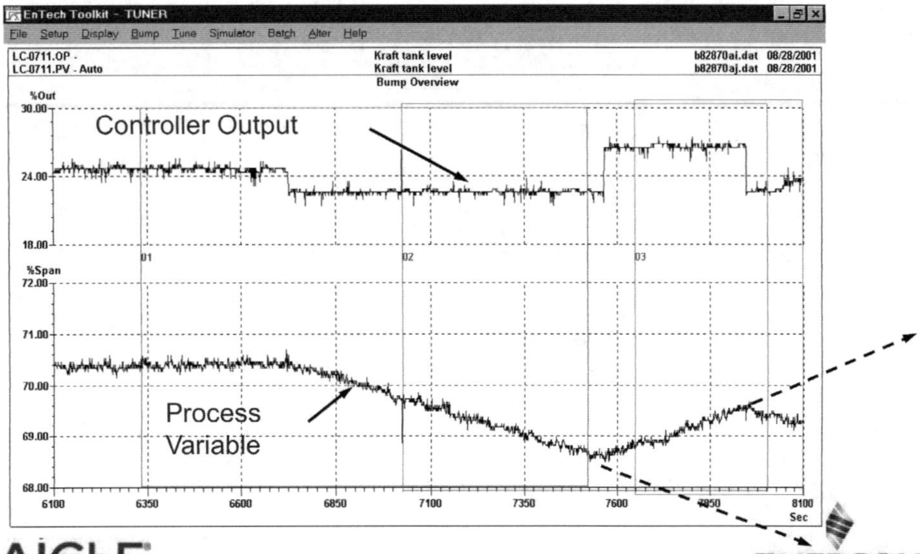

Lambda Tuning for Integrating Processes – Setpoint (Closed Loop) Response

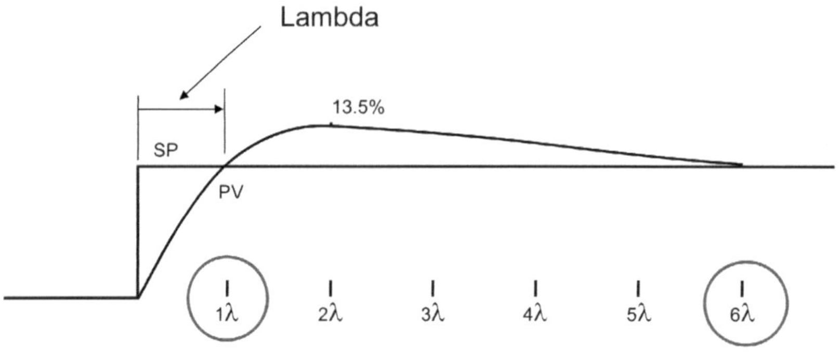

13AIChE Spring Meeting, San Antonio, TX

EMERSON. Process Management

Lambda Tuning for Integrating Processes - Load Disturbance Response

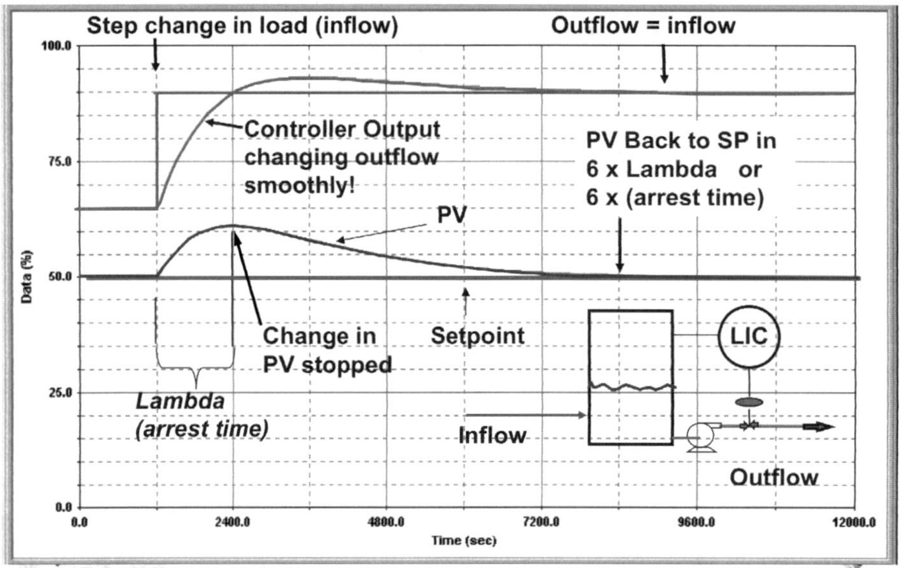

13AIChE Spring Meeting, San Antonio, TX

EMERSON. Process Management

Characterize the Process Response

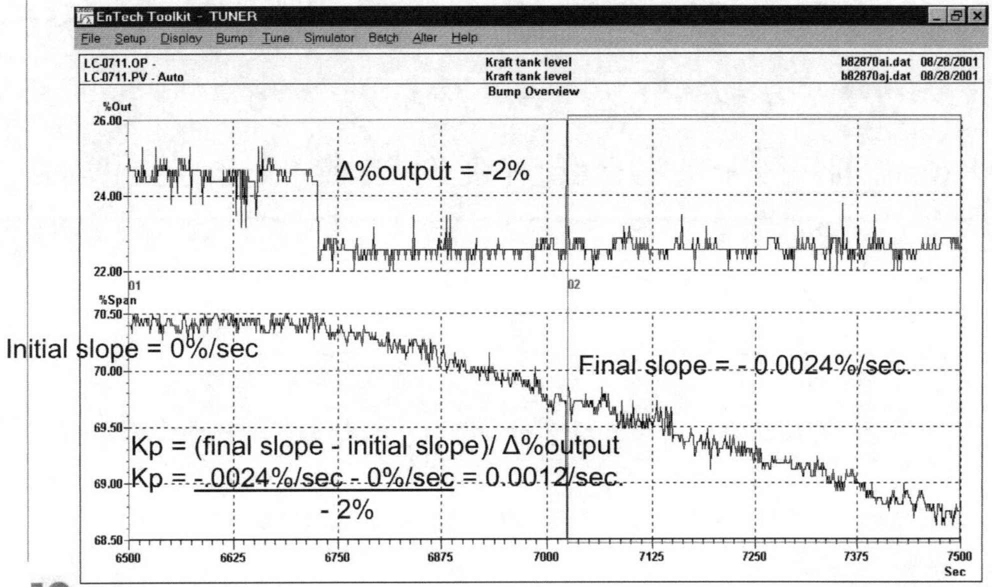

13 AIChE Spring Meeting, San Antonio, TX

EMERSON. Process Management

Lambda Tuning - Integrating Process

- THIS RULE DOES NOT INCLUDE DEAD TIME OR A LAG. Contact the author for applications with dead time or a significant lag.
- Choose Lambda (λ) (more on this later!)
- Tr = (2* λ)
- $Kc = \dfrac{Tr}{Kp(\lambda)^2}$
- NOTE – Gain AND Reset Time change with Lambda

13 AIChE Spring Meeting, San Antonio, TX

EMERSON. Process Management

Lambda Tuning Rules - Integrating Process

- Example – Choose Lambda = 10 minutes = 600 sec
- Tr = (2* λ) = (2*600 sec) = 1200 sec/rep
- $Kc = \frac{Tr}{Kp(\lambda)^2}$ = 1200 sec/((.0012/sec)*(600sec)^2)
- Kc = 2.78

Lambda	Gain	Integral Time
600 sec	2.78	1200 sec/rep
300 sec	5.56	600 sec/rep

Faster   Higher Gain   Faster (shorter) Reset Time

13 AIChE Spring Meeting, San Antonio, TX

EMERSON Process Management

Lambda = 10 min. for Integrating Process

Kc = 2.78, Tr = 1200 sec./rep

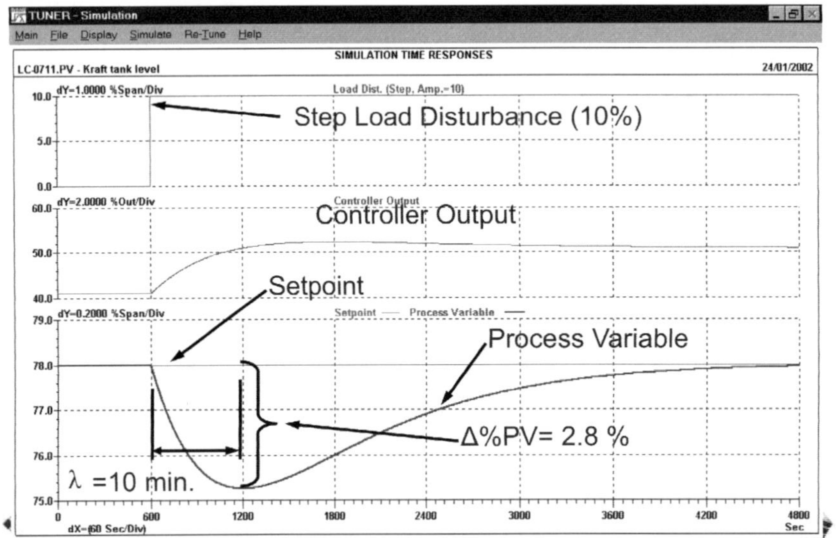

13 AIChE Spring Meeting, San Antonio, TX

EMERSON Process Management

Lambda = 5 min. for Integrating Process

Kc = 5.56, Tr = 600 sec./rep

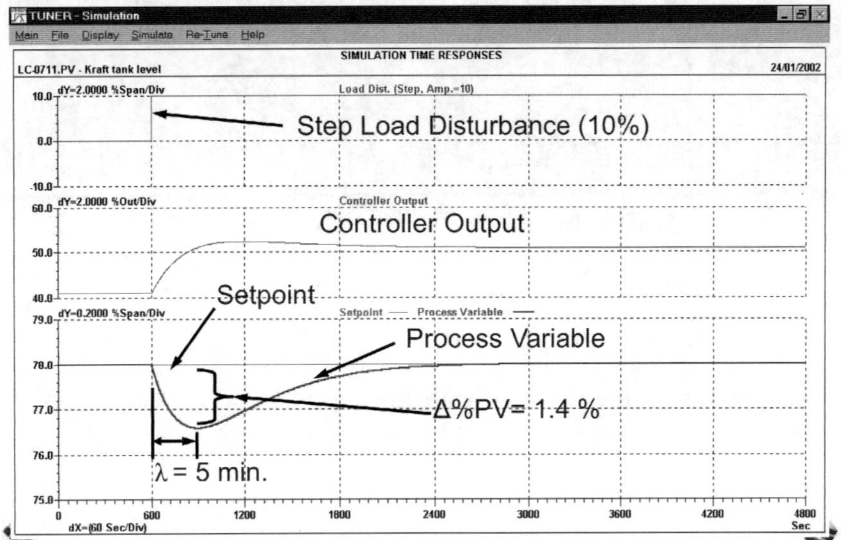

13 AIChE Spring Meeting, San Antonio, TX

EMERSON. Process Management

Attenuate Variability with Control/Equipment

- "Capacity" in the process can be used to attenuate or absorb variability
- Primary source of process capacity is level control
- To utilize level control as a capacity tune the controller only "fast" enough to hold the PV within the allowable level deviation (ALD) for a maximum load change

13 AIChE Spring Meeting, San Antonio, TX

EMERSON. Process Management

Utilize Level Control as Variability Sinks

- Choose the arrest time (Lambda) "slow" enough to provide a variability sink yet maintain level within the allowable variation
- Lambda α $\dfrac{\textbf{ALV}}{\textbf{Kp * MLD}}$
  - ALV = Allowable Level Variation
  - Kp = Integrating process gain
  - MLD = Maximum Load Disturbance (in % of controller output)

13 AIChE Spring Meeting, San Antonio, TX

EMERSON. Process Management

Example of Lambda Calculation

- Previous Example Kp= 0.0012/sec
- Let ALV = +- 15%
- Let MLD = 25%
- Lambda = 1333 seconds = 22.2 minutes
- **Tr = 2666 seconds; Kc =1.25**

13 AIChE Spring Meeting, San Antonio, TX

© 2010, Emerson Process Management

EMERSON. Process Management

Example of Lambda Calculation

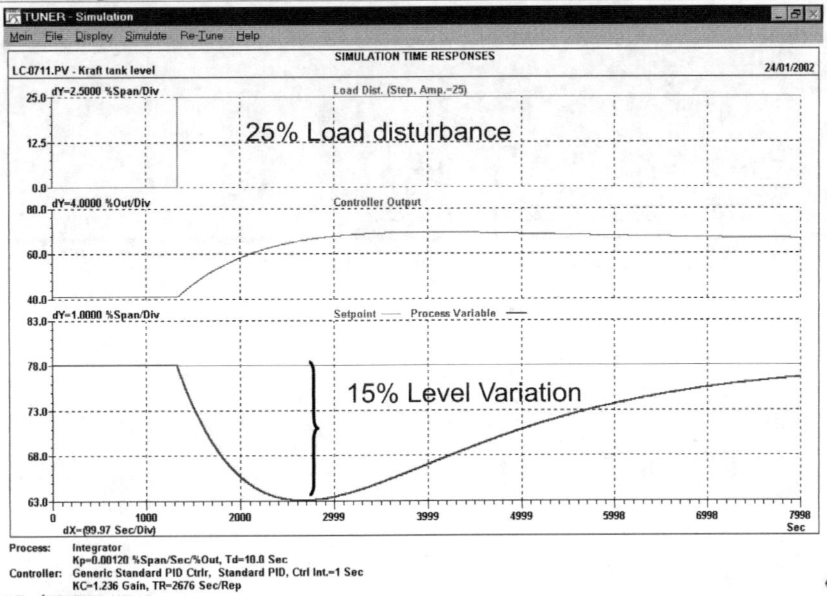

13 AIChE
Spring Meeting, San Antonio, TX

EMERSON.
Process Management

Results – Integrating Process Tuning

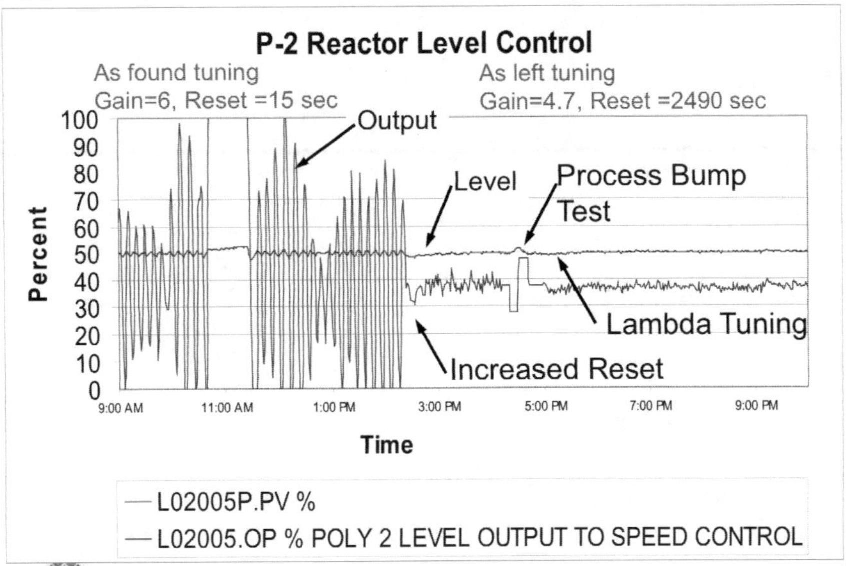

13 AIChE
Spring Meeting, San Antonio, TX

EMERSON.
Process Management

*Process Dynamics – *Model Based Tuning May Include Derivative*

- Pure Gain
- 1st Order
- *2nd Order, Over Damped
- *2nd Order, Under Damped
- 2nd Order, Lead
- 2nd Order, Lead with Overshoot
- 2nd Order, Non-Minimum Phase
- Integrator
- *Integrator, 1st Order Lag
- Integrator, 1st Order Lead
- Integrator, Non-Minimum Phase

13 AIChE Spring Meeting, San Antonio, TX

EMERSON. Process Management

Summary Tuning Procedure

→ Determine basic process type

- Self-regulating
- Integrating

→ Determine process dynamics

→ Choose desired closed loop response time, "Lambda"

→ Calculate tuning constants

→ NOTE – For better load regulation, self-regulation loops with long lags can be tuned as in integrator

13 AIChE Spring Meeting, San Antonio, TX

EMERSON. Process Management

Questions?

James.Beall@Emerson.com

903-235-7935

13 AIChE Spring Meeting, San Antonio, TX

EMERSON Process Management

DETECTING LOSS OF FLAME IN OIL REFINERY FIRED HEATERS USING ADVANCED PRESSURE DIAGNOSTICS

Presented at AIChE Spring Meeting, San Antonio, TX, April 28-May 2, 2013

John P. Miller – Emerson Process Management – Chanhassen, MN, USA
Randy S. Stier – Valero Energy Corporation – San Antonio, TX, USA

Biography

John Miller is a principal engineer of pressure technology for the Rosemount business unit of Emerson Process Management. John works in developing and promoting applications for Rosemount Advanced Diagnostics pressure transmitters and in the research and development of new products. John has a BS in Mechanical Engineering and an MS in Electrical Engineering both from the University of Minnesota. He is inventor on 24 patents and other pending applications.

Randy Stier is a fired heater engineer with a strong track record of improving oil refinery profits through reliability, availability and energy related projects. A chemical engineer by training, Randy earned a BSChE degree from The University of Michigan and an MSChE degree from the Illinois Institute of Technology. Randy is currently Valero's Director of Refinery Heater Best Practices and Chairman of the API Subcommittee on Heat Transfer Equipment.

Abstract

Maintaining stable combustion is critical to the safe operation of fired heaters. When the ratio of air flow to fuel flow to a burner transitions from a point inside the burner's operating envelope to a point outside the burner's operating envelope, the combustion process becomes unstable and then stops. If this loss of flame is not recognized and acted upon properly, an explosion may occur.

One possible way to detect loss of flame may be to monitor pulsations in the difference between the flue gas pressure inside a fired heater and atmospheric pressure at a given elevation (commonly referred to as "draft"). During stable combustion, the draft will pulsate slightly (see Figure 1). This is commonly referred to as combustion rumble. As a burner's air/fuel ratio transitions from inside to outside the burner's operating envelope, the draft pulsations may exhibit a sharp increase in amplitude and frequency, caused by instability in the burner flames. Flame instability is often a precursor to flame out.

Draft transmitters with Advanced Diagnostics Statistical Process Monitoring (SPM) capability monitor the pulsations in the draft measurement and detect the increase in amplitude or frequency associated with flame instability. SPM data may be integrated with heater operations. An alarm limit on the SPM standard deviation may be configured to alert a heater operator to potential combustion problems or flame out.

Valero has worked with Rosemount to evaluate the use of SPM technology for detection of flame instability or flame out. Recent testing was performed on five different burners at three different burner vendors. The technology was also evaluated on a heater at the Valero Texas City Refinery.

Experience gained from these tests may be used to determine where to set the alarm limit for detecting flame instability. There is a correlation between the volumetric firing rate of a burner/heater system and the draft standard deviation. This correlation can be exploited for selecting an appropriate alarm limit in an operating fired heater. Burner vendor tests indicate that a draft standard deviation alarm limit of 30 thousandths inches of water column (m-inWC) may be appropriate.

Additional work is required to validate these observations but initial results indicate the use of advanced pressure diagnostics may be an effective way to detect loss of flame.

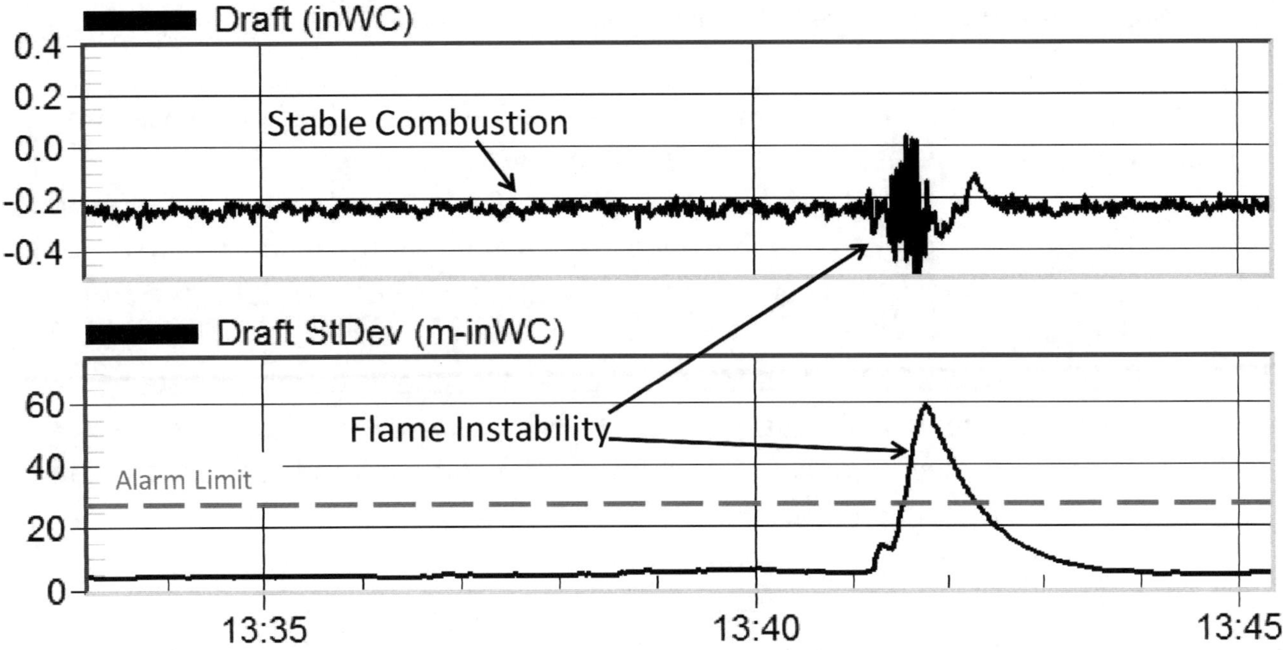

Figure 1 – Detection of Flame Instability with Advanced Pressure Diagnostics

Fired Heaters and Proper Combustion

Fired heaters are a class of indirect-fired equipment used to heat process fluids in oil refineries and other process industries. Many refineries burn fuel gas in their fired heaters in order to get the energy needed to heat the process fluids. These process fluids flow through tubes mounted throughout the fired heaters. Energy is transferred to the process fluids flowing through the heater tubes by conduction, convection, and radiant heat transfer. Process fluids can enter and exit the tubes as a gas, a liquid, or both. Fired heaters account for 2/3 to 3/4 of the energy consumed in a refinery.

Fired heater burners are designed to operate with more combustion air than needed in order to react all of the fuel gas. They are designed this way because burners are not perfect mixing devices. If a burner is only fed the amount of air theoretically required, not all of the fuel gas will find the available oxygen. The fuel gas without oxygen will not burn and combustion will be incomplete.

Maintaining stable combustion is critical to the safe operation of fired heaters. When the ratio of air flow to fuel flow to a burner transitions from a point inside the burner's operating envelope to a point outside the burner's operating envelope, the combustion process will become unstable and then stop. The transition of a burner's air/fuel ratio from inside to outside the burner's operating envelope may be caused by a number of different initiating events including a significant change in fuel gas composition, a sudden decrease in air flow, a sudden increase in fuel flow or a shift in the boundaries of the burner operating envelope.

Incomplete combustion (also known as substoichiometric combustion), caused by too little combustion air, is undesirable in a fired heater for the following reasons:

- A low oxygen-to-fuel gas ratio causes long flames which may touch the tubes. The term for this is flame impingement. If there is flame impingement, it may cause the tubes and other heater parts, such as the tube supports and refractory, to overheat.
- Energy is wasted when unburned fuel gas leaves the heater via the stack.
- Unburned fuel gas has the potential to ignite elsewhere in the fired heater if a source of oxygen is found. This is called "afterburning" and can result in damage to components in the fired heater.
- A low oxygen-to-fuel gas ratio may cause flame instability and potentially flame out.

Refinery fired heaters often use waste gas (hydrocarbon gas produced as a bi-products of other refining processes) as the fuel gas source. Burning waste gas is advantageous because it costs only a fraction per BTU as other fuel sources. The disadvantage of waste gas is that there can be significant variability in the composition, and therefore, a large variability in BTU per standard cubic foot. If the fuel gas BTU content suddenly increases, the heater may go into a state where there is not enough available oxygen for all the fuel. Figure 2 shows an example sequence of events that may result.

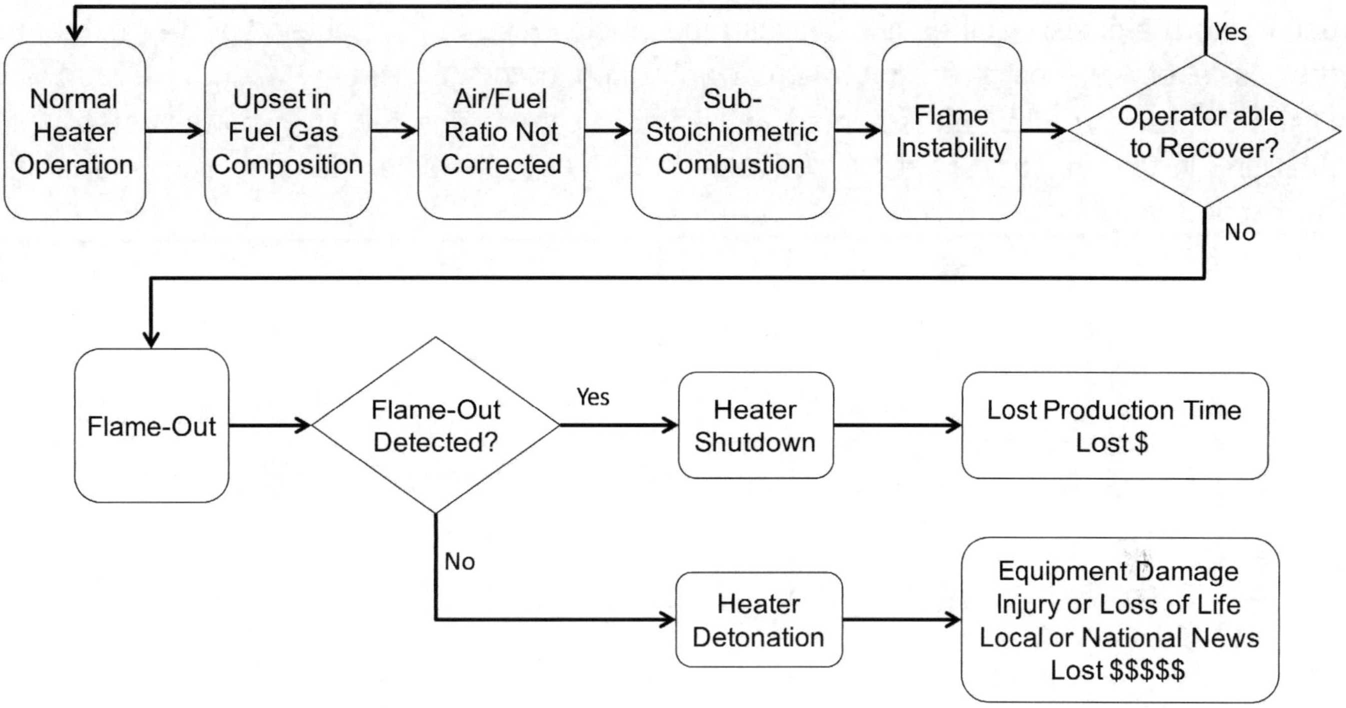

Figure 2 – Example Sequence of Events before and after a Flame out

If a heater experiences an upset in the fuel gas composition (e.g. a surge in the BTU content of the fuel gas or waste gas) and if the air/fuel ratio controller is not able to keep up, the burners may begin operating in substoichiometric, or incomplete, combustion. Depending upon the type of burner, the flames may start going unstable (Flame Instability). If the air/fuel ratio is not corrected (either manually by an operator, by via automatic O_2 control), the burner's flame may be extinguished shortly thereafter. After a flame out, a heater must be shut down to purge out the unburned fuel gas, resulting in lost production time for that operating unit. If a flame out is not detected, it is an especially dangerous situation because unburned fuel and air continues to be fed into the firebox. The slightest spark or ignition source could light this air/fuel mixture, causing a detonation of the heater, resulting in major equipment damage, and possibly, injury or loss of life.

Some refinery fired heaters with are equipped with flame scanners that can detect a loss of flame. However, flame scanners become problematic in multi-burner heaters because a flame scanner cannot detect if just a single burner is extinguished. Thus, most multi-burner heaters do not use flame scanners. For these heaters, current industry practice is to train operators to manually recognize loss of flame via operational variables, such as a sudden drop in firebox temperature and a sudden increase in O_2 or combustibles in the flue gas.

Draft Measurement and Flame Instability

The industry best practice calls for fired heaters to have a draft measurement. Draft is defined as the pressure difference between the ambient atmosphere outside of the heater and the inside of the heater firebox where combustion occurs ($Draft = P_{atmosphere} - P_{firebox}$). Figure 3 illustrates draft measurement on a fired heater. The red line shows the absolute pressure inside the heater at various elevations ($P_{firebox}$), while the blue line shows the corresponding ambient pressure ($P_{atmosphere}$).

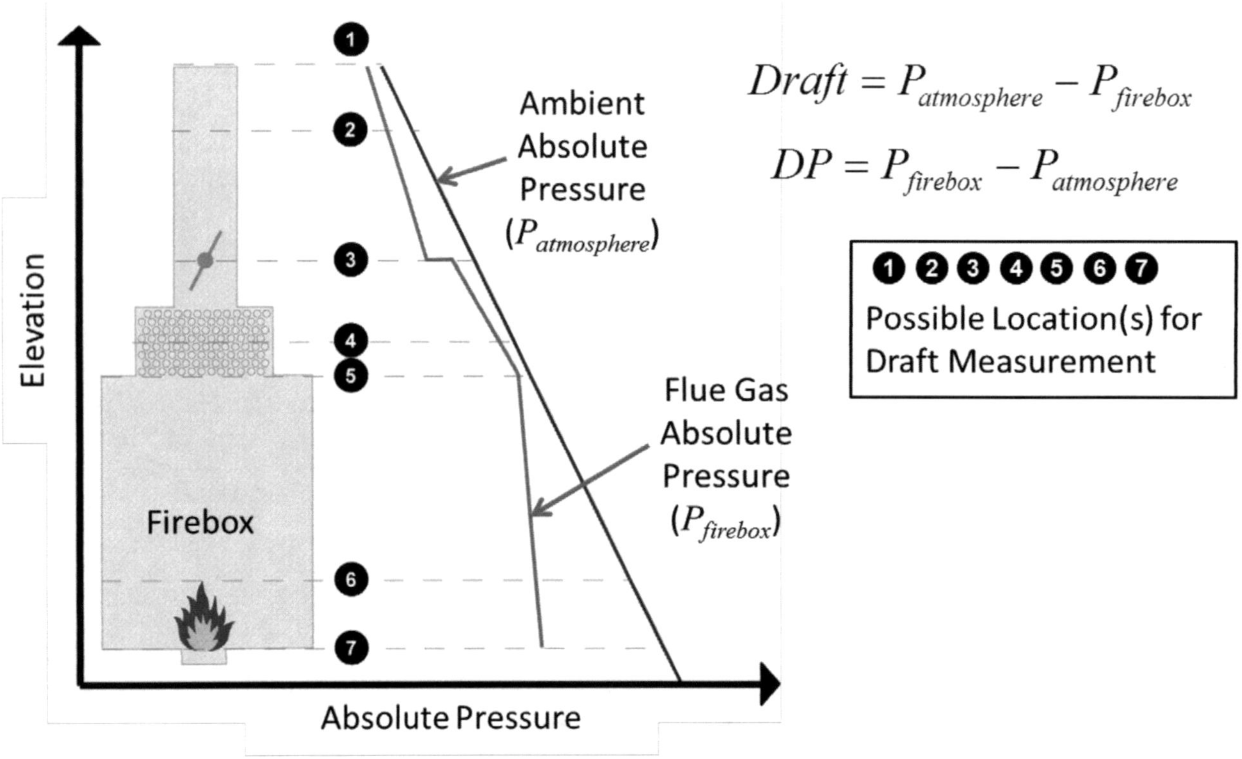

Figure 3 – Draft Measurement on a Fired Heater

Draft can be measured at any one of multiple elevations on the heater including at the stack (1 and 2), near the stack damper (3), at the convection section (4), at the bridgewall (5), near the burner (6), and at the floor of the heater (7). Typically, draft is measured at only one of these elevations. During normal operation, the firebox pressure is less than atmospheric pressure. If the firebox pressure goes higher, it is an indicator that there may be problems. Depending upon the elevation, heater design, and heater firing rate, the draft can be anywhere from 0.1 inWC (inches of water column) to 1.0 inWC. Within the industry, sometimes the draft measurement is expressed as a differential pressure ($DP = P_{firebox} - P_{atmosphere}$). Under this convention, the draft value is negative during normal combustion, and goes positive only if there are combustion problems (e.g. flame instability). Throughout the paper, we utilize this convention, showing a negative draft value in all of the trends.

As described previously, substoichiometric combustion may lead to instability in the burner flames. If someone is standing near the heater, flame instability can be heard audibly, as a "huffing" or "panting" in the heater. It is known in the industry that flame instability causes pulsations (or oscillations) in the firebox draft measurement. Sometimes flame instability is confirmed by an operator visually observing a draft gauge swinging between the positive and negative. Thermo acoustic coupling [1, 2, 3, and 4] is one mechanism that may contribute to flame instability and draft oscillations. Industry has investigated utilizing patterns in the draft oscillations for detecting flame instability [5].

Figure 4 shows an example of draft oscillations that occur under flame instability. Under normal heater combustion, the draft measurement is negative and relatively steady. However, when the flames go unstable, the draft starts oscillating, and may periodically go into positive. In the example below, during Healthy Combustion, the draft is approximately -0.2 inWC. But during Flame Instability the draft starts oscillating at a frequency of about 3 Hz, with a peak-to-peak amplitude of about 0.6 inWC. The draft even periodically goes as high as +0.2 inWC.

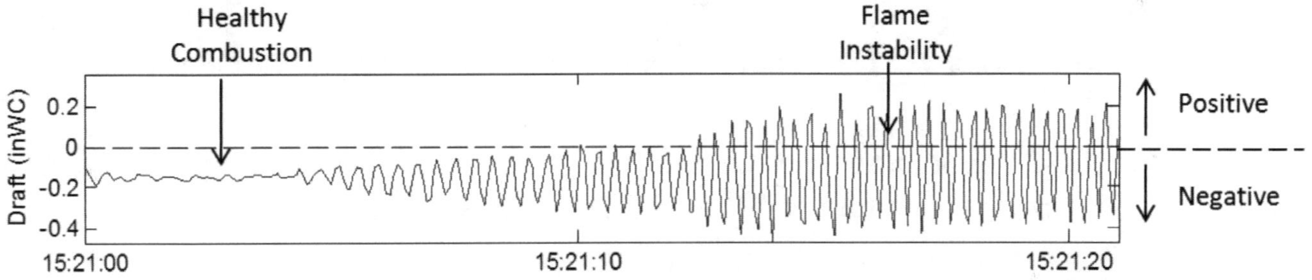

Figure 4 – Example Draft Measurement Signal during Flame Instability

However, the draft oscillations associated with flame instability are not normally seen in the draft measurement because of typical damping, filtering, or sampling, which may be implemented in the draft transmitter and/or in the Distributed Control System (DCS). Figure 5 shows the effect of transmitter damping (PV damping) on the flame instability draft oscillations. Many pressure transmitters are pre-configured with a PV damping of approximately 0.4 seconds. As shown in the middle trend, this damping flattens out the draft oscillations significantly. Sometimes a draft measurement is configured with a much higher damping (e.g. 5 seconds as shown in bottom trend) to smooth out the impact of combustion rumble on the signal, and make the draft steady when a heater is operating normally. A damping this high makes the flame instability draft pulsations impossible to see. With either the 0.4 second or 5 second PV damping, neither an operator nor an automatic control system can see the pulsations and positive differential pressure associated with flame instability.

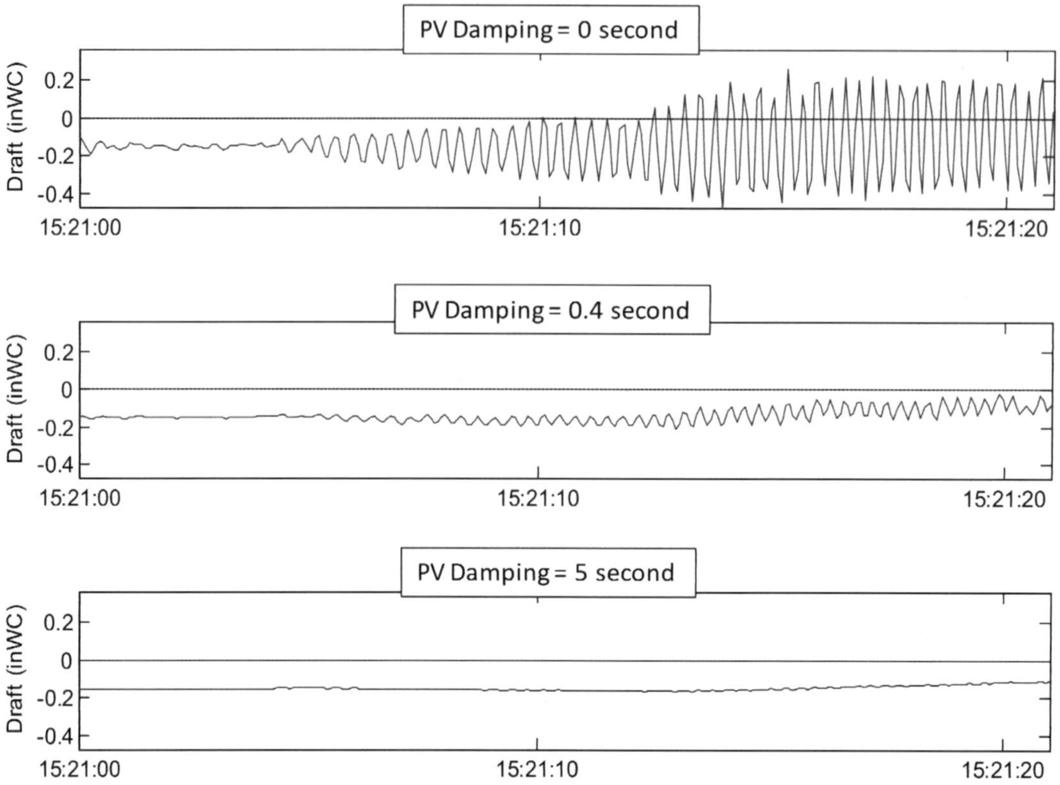

Figure 5 – Effect of Draft Transmitter Damping on Flame Instability Oscillations

The other limitation to observing the draft pulsations associated with flame instability is the damping and slow sampling rate of a typical DCS. A DCS may contain additional damping or filtering to further smooth out the draft measurement. A DCS also typically has a sampling rate too slow to see these draft oscillations. For example, a DCS may sample the pressure only once per second. Figure 6 illustrates a 1-second sampling of the draft measurement during flame instability. Because this is much slower than the 3 Hz oscillations of the draft, none of the positive pressure peaks are detected.

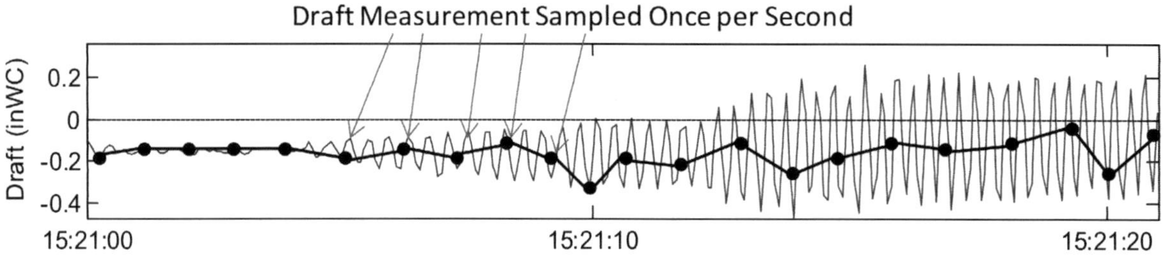

Figure 6 – Slower Sampling in the Distributed Control System (DCS)

Thus, because of limitations of transmitter PV damping, and DCS sampling rate, the draft oscillations associated with flame instability are normally not seen by a heater operator.

Advanced Pressure Diagnostics

Over the last twenty years, smart field devices have become increasingly more prevalent in the process industry. Smart field devices provide not just the basic process measurement (e.g. pressure, differential pressure, draft, temperature, or flow) but they provide a wide range of additional information, such as second and third process variables, device identification and health, and device diagnostics information. One example of such smart field devices is the Rosemount 3051S Advanced Diagnostics Pressure Transmitter with Statistical Process Monitoring (SPM) capability. SPM is available on most types of pressure measurements within the 3051S family, including absolute and gauge pressure, differential pressure (DP), DP flow, DP level, and draft. As used herein, the term "pressure transmitter" encompasses all of these pressure measurement types including fired heater draft measurement.

SPM technology allows a plant engineer or operator to see the variation (or noise) in a process that is normally filtered out when the variable is measured and displayed in the DCS. Because the pressure transmitter is closer to the process than the DCS, it can sample the pressure at a much faster rate, and turn any process variation into valuable information. The Rosemount 3051S pressure transmitter samples the pressure at 22 Hz, much faster than the typical DCS sampling rate of once every 1-2 seconds. The diagnostics transmitter calculates mean and standard deviation of the pressure and sends this information to the host system via digital communication protocol, such as HART or Foundation Fieldbus.

Figure 7 illustrates the architecture and implementation of a pressure transmitter with Advanced Diagnostics SPM capability.

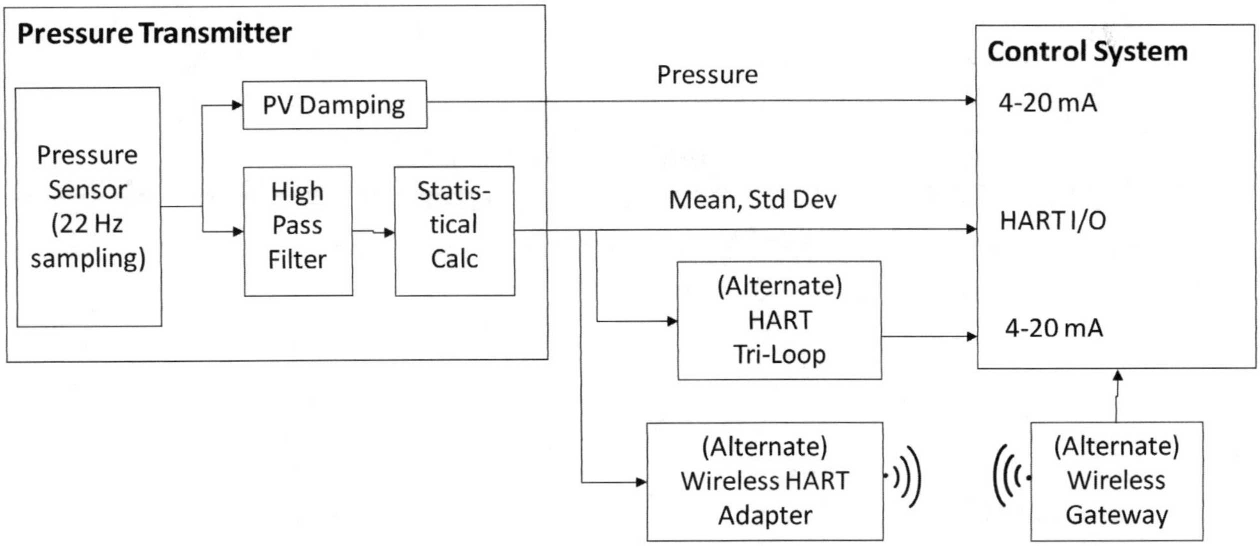

Figure 7 – Architecture and Implementation of Advanced Pressure Diagnostics

The pressure sensor measures the pressure at 22 Hz, applies the normal PV Damping, and then sends the measured pressure to the Control System via 4-20 mA. On a parallel path, a High Pass Filter and Statistical Calculations module takes the un-damped pressure and calculates mean and standard deviation. The statistical values are made available to the control system as secondary variables via HART I/O.

If a control system does not have HART I/O ability, the statistical values can still be accessed via a HART Tri-Loop (e.g. Rosemount 333) which converts the HART data to a 4-20 mA. Alternatively, a wireless HART adapter (e.g. 775 Smart Wireless THUM adapter) can be used to access the advanced diagnostics information, and transmit the information wirelessly to a gateway (e.g. 1420 Smart Wireless Gateway), which then integrates the diagnostics information with the Control System.

Figure 8 further illustrates Statistical Process Monitoring technology. On the left, is the process variable as measured by the pressure transmitter, with different amount of process variation (Normal, Noisy, and Quiet). On the right are the variables that the advanced diagnostics pressure transmitter sends to the control system.

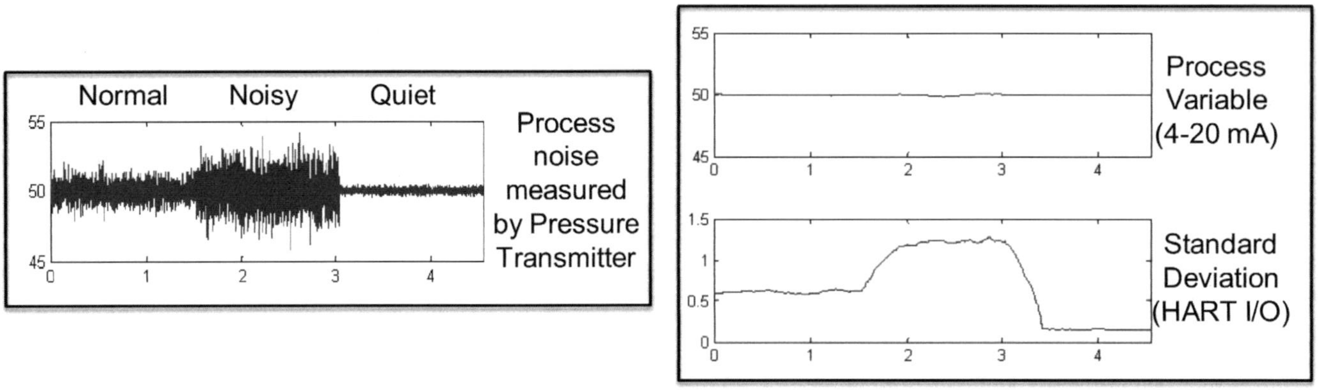

Figure 8 – Statistical Process Monitoring in a Pressure Transmitter

Note that the process variable (4-20 mA) is relatively flat, because the PV damping has removed all of the process variation. However, the standard deviation is a measure of how much process variation is in the original signal. The standard deviation is higher when the process is noisier, and it is smaller when the process is quieter.

As shown in Figure 7, a high-pass filter is used prior to the Statistical Calculations module. The purpose of the high-pass filter is to ensure that the SPM analyzes only the process variation at higher frequencies (approximately 1 Hz and higher). When there is just a slow change in the pressure measurement (such as the heater draft), this should not be interpreted as an increase in the process variation. Figure 9 illustrates the effect of using the high-pass filter. The top trend (a) represents the actual process behavior, including both a nominal (average) pressure, and random process noise. The middle trend (b) represents the pressure reading that is transmitted to the DCS via the 4-20 mA. The PV damping has smoothed out all the noise, and preserved the nominal pressure value. The bottom trend (c) represents the high-frequency noise that is analyzed by the SPM. The high-pass filter allows the SPM to analyze the pressure variation that is removed by the PV damping.

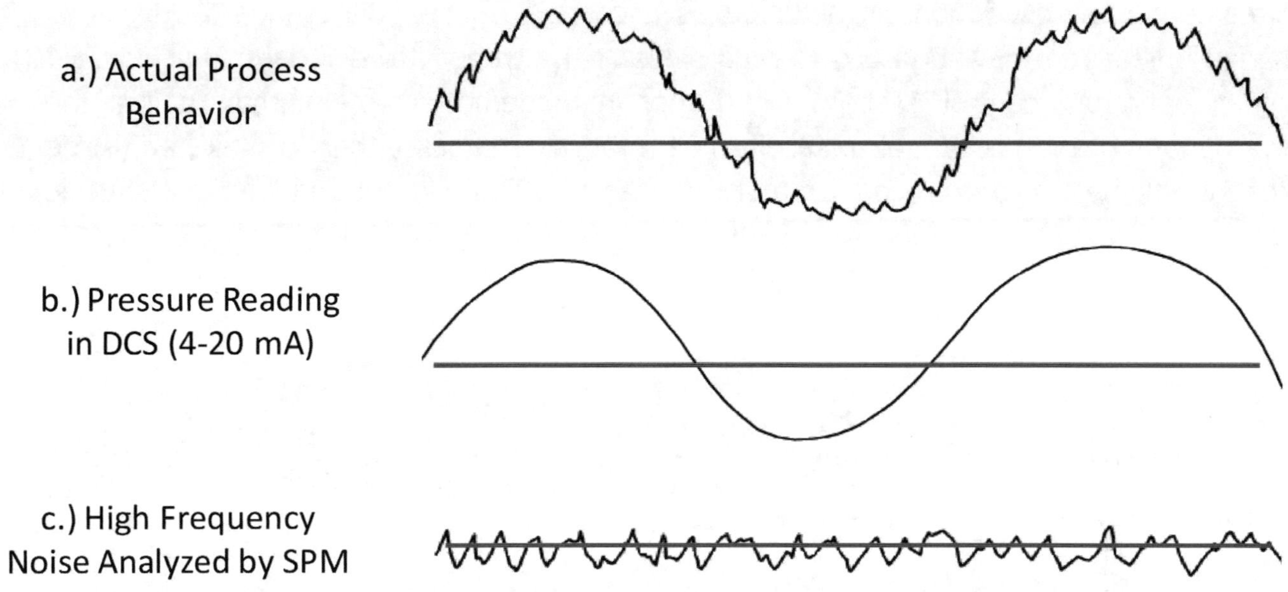

Figure 9 – Effect of High-Pass Filter on the SPM Algorithm

When the advanced diagnostics standard deviation is viewed in the control system, it is possible to detect abnormal conditions that would otherwise be unobservable using only the standard 4-20 mA pressure signal. Many different abnormal conditions have been shown to be detectable using advanced pressure diagnostics SPM technology, including plugged impulse lines [6, 7], entrained gas in liquid flows [8], increased liquid loading in natural gas wells [9], and distillation column flooding [10, 11].

Flame Instability Detection with SPM

As published previously [12, 13, 14, 15, 16], SPM technology can also be utilized to detect the draft oscillations associated with flame instability. Referring back to Figure 1, the increase in draft oscillations is detected as a change in the SPM standard deviation. This change in standard deviation occurs regardless of the transmitter PV damping (Figure 10). A heater operator, control system, or burner management system may utilize the SPM information as an indicator of instability in the burner flames either to make adjustments to prevent a flame out, or to recognize that the heater has flamed out and initiate a shut-down.

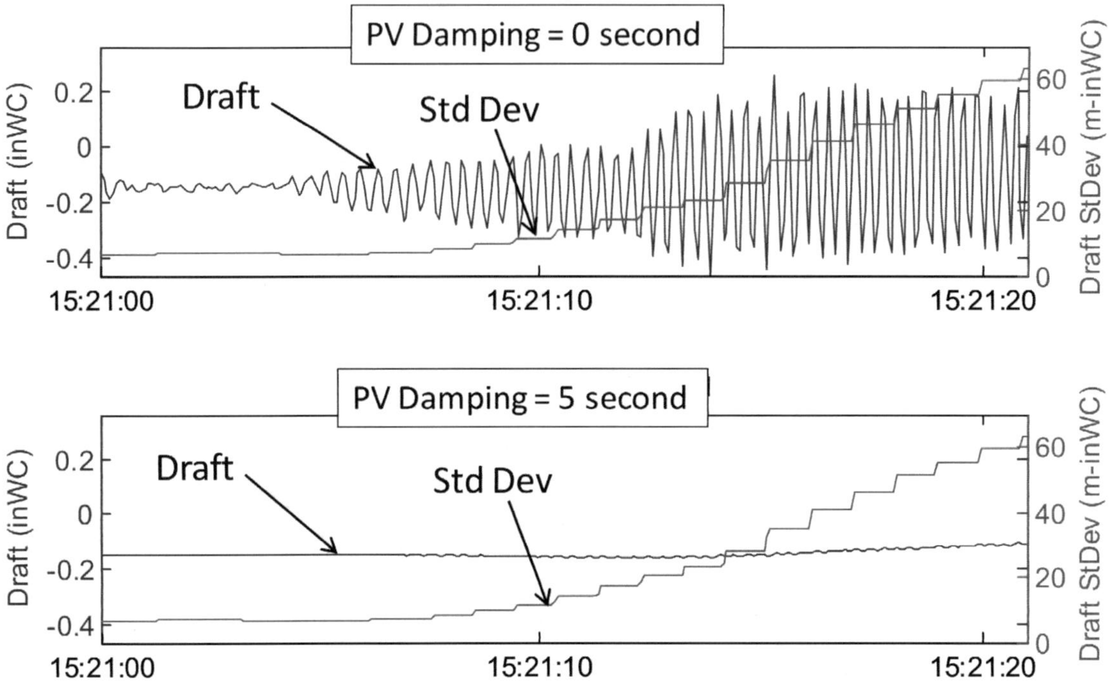

Figure 10 – SPM Standard Deviation is not affected by PV Damping

Implementing SPM technology to monitor an operating fired heater requires the following:

1. Measure draft using a draft transmitter with SPM
2. Trend SPM data into DCS process historian
3. Configure flame instability alarm

Item 1 requires simply replacing or upgrading the existing draft measurement with an advanced diagnostics draft transmitter with SPM. Referring back to Figure 3, the draft can be measured at any elevation on the heater. Previous work [13, 14, and 17] has shown that the SPM can detect flame instability regardless of the elevation of the draft measurement.

Item 2 requires first bringing the SPM data into the control system I/O via any of the interfaces shown in Figure 7, and then configuring the process historian to be trending the SPM data into the database. The SPM data may also be displayed to the heater operator so that they can use it together with other heater operational data.

Item 3 may take some additional effort. The flame instability alarm limit is the minimum value of the SPM standard deviation at which to notify a heater operator of combustion problems and/or flame instability. Obviously, in an operating fired heater, one cannot deliberately make the heater flame out, just to determine how much the standard deviation changes, in order to set the alarm limit. However, manufacturers of burners have test facilities that enable burners to be tested in a wide variety of conditions, including substoichiometric and flame instability conditions. It is possible that results from these tests at burner vendors may be utilized to gauge how much the SPM standard deviation changes under flame instability. The purpose of this joint effort between Rosemount and Valero was to evaluate the use of SPM technology for flame-instability detection, and to determine a method for setting the alarm limit on operating fired heaters.

Testing at Burner Vendors

Over two weeks in December of 2011, Valero tested five different burners at three different burner vendors. These tests were part of the standard Valero procedure for validating the suitability of each burner for various heater design or retrofit projects across the company. During these tests, each burner was evaluated on typical burner performance factors, such as range of firing rate and turn-down, flame size (length and diameter), emissions (CO, NOx, etc.), and acoustic noise levels. All five burners were the newest Ultra-low NOx burners made by each of the respective burner vendors. Four of the burners were round-flame burners, while one was flat-flame burner geometry.

Rosemount was also invited to participate in these burner tests to help evaluate the use of Advanced Pressure Diagnostics Statistical Process Monitoring technology for flame instability detection. On each test furnace, a Rosemount 3051S draft transmitter, with the Advanced Diagnostics (DA2 option), was installed. On all tests, the draft transmitter was installed at approximately the burner level (location 6 or 7 in Figure 3). Throughout the burner tests, both the un-damped draft measurement, and the draft standard deviation were recorded.

In addition to the regular burner tests, each burner was driven to various levels of substoichiometric combustion. Although burner vendors may have the capability of testing substoichiometric operation, this testing is usually not part of the standard burner test protocol, because heaters are not typically operated below stoichiometry. However, the refining industry recognizes that although they would never purposely operate a heater sub-stoked, unexpected events (e.g. a surge in the fuel gas composition) may occur periodically, and they want to be certain that their heaters will not flame out under such an event. Thus, substoichiometric testing at burner vendors is becoming more prevalent.

A typical substoichiometric / flame instability test would consist of the following:

1. Set the burner at a desired firing rate (e.g. minimum, normal, or design) and normal excess oxygen (e.g. 3% O_2, wet basis, in the flue gas)
2. While keeping the stack and burner air dampers at a constant position, gradually increase the firing rate, until the stoichiometric point (0% O_2 in flue gas) is reached. This normally coincides with the CO Break (the point where the concentration of carbon-monoxide in the flue gas exceeds 1000 parts per million).
3. While still keeping air dampers at a constant position, continue increasing the firing rate into the substoichiometric range. Because the % of stoichiometry cannot be measured directly, it must be estimated based on how much the fuel gas flow is increased.
4. Continue increasing the firing rate until either flame instability is observed (on a camera or through a test furnace viewport) or until the test furnace operators feel they cannot safety operate any further into the substoichiometric range.

Figure 11 shows an example of one of the flame-instability tests at a burner vendor. The top trend shows the un-damped draft measurement, while the bottom trend shows the draft standard deviation. Note that throughout this paper, the draft standard deviation is shown in units of thousandths inch of water column (m-inWC). For example 6 m-inWC = 0.006 inWC.

In this burner test, operation started at a baseline firing rate (3.6% O_2), and then the firing rate was gradually increased. The burner was still stable at both the stoichiometric point (3.15 MMBtu/hr) and at 90% of stoichiometry (3.47 MMBtu/hr). However, when trying to hold the burner at 85% of stoichiometry (3.6 MMBtu/hr) the burner flames would suddenly start becoming unstable. The flame instability could be seen both on a camera, visible in the control room, and in observing directly through a viewport on the test furnace. During flame instability, the sides of the burner flames could be described as "clapping hands together", and an audible "huffing" sound near the burner air intake.

Prior to the point of flame instability, the draft standard was approximately 4 to 7 m-inWC. However, at the point of flame instability, the standard deviation went up to between 60 and 70 m-inWC (about a 10 times increase). In all of the observed instances of flame instability, there was a similar large increase in standard deviation.

An alarm limit on the draft standard deviation could be defined for detecting flame instability. In this example, 30 m-inWC may be an appropriate alarm limit.

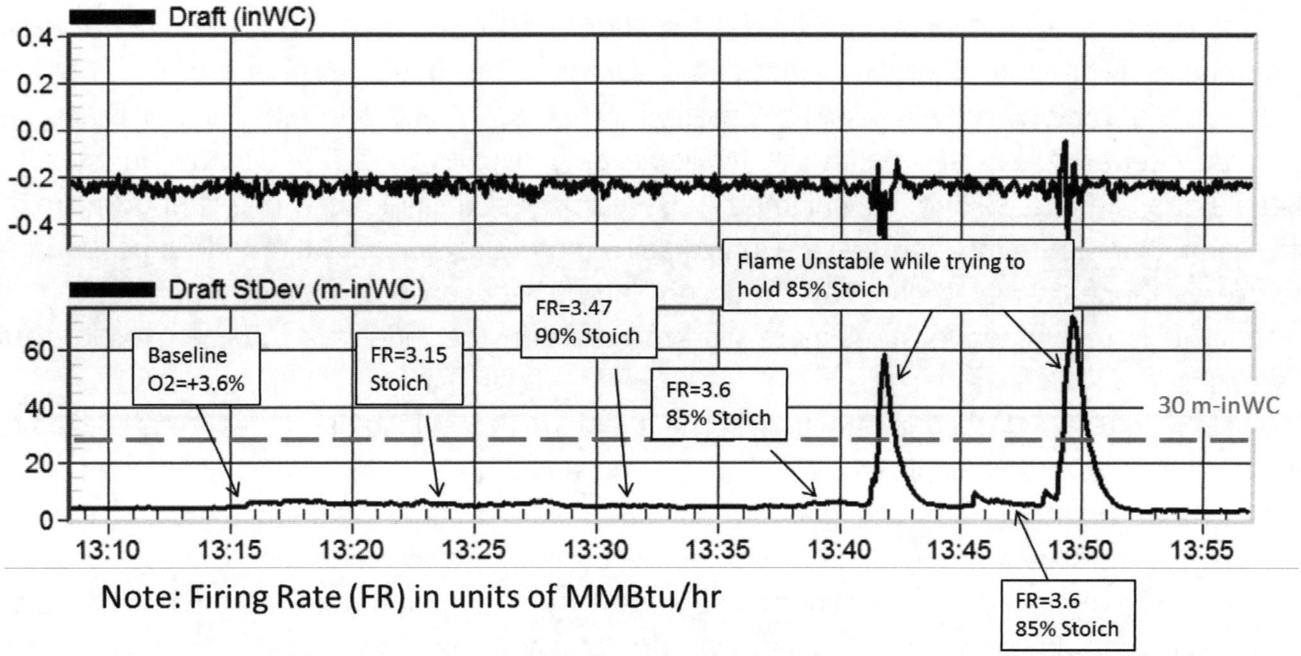

Figure 11 - Example Substoichiometric and Flame Instability Test at a Burner Vendor (Burner D-2 at Normal Firing Rate)

Table 1 shows a summary of all of the flame instability tests conducted at the burner vendors. Only burner vendor D was able to run the burners far enough into the sub-stoichiometric operating region to create flame instability. Burner vendors E and F showed their burners were stable down to 71% and 90% of stoichiometry respectively, but neither one had a test protocol that could take their burners to the point of flame instability.

Table 1 – Flame Instability Tests at Burner Vendors

Vendor	Burner	Geometry	Burner Firing Rate Conditions (MMBtu/hr)			Lowest % of Stoichiometry	Flame Instability in Burner	Flame Instability Detected with SPM
			Min	Norm	Max			
D	D-1	Round	1.12	5.05	6.72	57% (Norm FR) 70% (Max FR)	Yes *	Yes
	D-2	Flat	0.59	2.65	3.52	85% (Norm FR) 92% (Max FR)	Yes **	Yes
E	E-3	Round	2.29	7.96	9.15	71%	No	No
F	F-4	Round	1.96	6.26	7.83	90%	No	No
	F-5	Round	1.31	4.2	5.25	90%	No	No

* Burner D-1: When starting at the Normal Firing Rate, burner was stable down to 57% of Stoichiometry. Starting at the maximum firing rate, flame instability observed at 70% of Stoichiometry

** Burner D-2: Flame instability observed both when starting at the normal firing rate, and when starting at the maximum firing rate

It is interesting to note that of the two burners from vendor D, one of them was a round burner, while the other was a flat flame burner. The round flame burner (D-1) proved to be stable at as low as 57% of stoichiometry. (That is a, firing rate 43% higher than what would be needed to exactly match the available air.) However, the flat flame burner went unstable at as little as 92% of stoichiometry. Based on just a single test, one cannot make the conclusion that flat-flame burners are universally more unstable than round-flame burners. However, it appears that there may be a correlation between the geometry and stability of burners. It is recommended that more work be done in the industry to further explore this relationship.

Note in Table 1 that in the two burners where flame instability was created, it was detected by the draft transmitter with SPM. In the three burners that were not driven to the point of flame instability, the SPM standard deviation did not show the same large increase.

Figure 12 shows an example trend from a burner test at vendor F. This figure shows the values of draft and draft standard deviation that were observed over the full day of burner tests. Note that although there were a few periods when the burner operation was taken to the point of 90% of stoichiometry, the burner flames never went unstable. There were small changes in the draft standard deviation over the day of the tests, but it never went higher than 9 to 10 m-inWC. Using the same flame instability alarm limit of 30 m-inWC, we can see that this limit was not exceeded during the full day of the burner tests. Thus, the SPM did not give any false alarms.

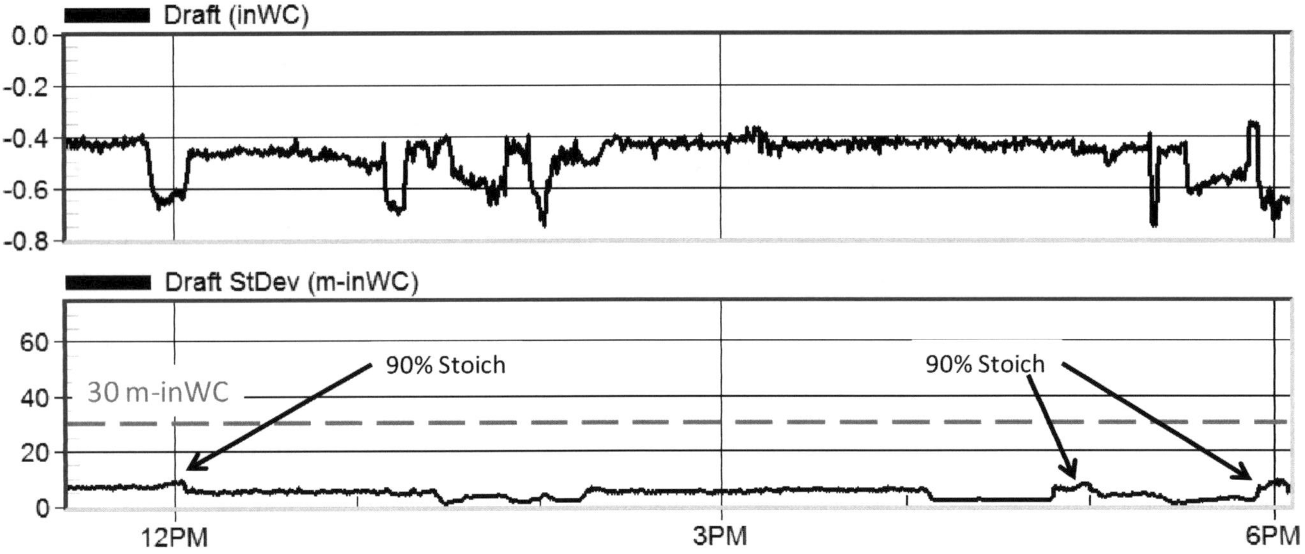

Figure 12 - Example Sub-stoichiometric test - Flame Instability Not Observed (Burner F-4)

Figure 13 shows one data point from a later test run on the same burner F-4. This data point was found rather opportunistically, as it occurred under a circumstance that typically doesn't happen in burner tests. In this case, the test furnace operators opened the stack damper to increase draft in the test furnace instead of keeping the damper partially closed and using steam injection to the base of the stack to increase draft in the firebox. Once the firebox was given line-of-sight to the sky, the burner flames started pulsating, and soon flamed out. A view of the trend of the draft standard deviation (Figure 13) shows that just prior to the flame out, there was a large increase in the standard deviation. In this case, the alarm limit of 30 m-inWC would be appropriate for detecting flame instability. There was also a point during light-off where the standard deviation exceeded 30 m-inWC. However, an implementation in an operating firing heater (such as in the Burner Management System) could have logic to ensure that a flame instability alarm is not generated during light-off.

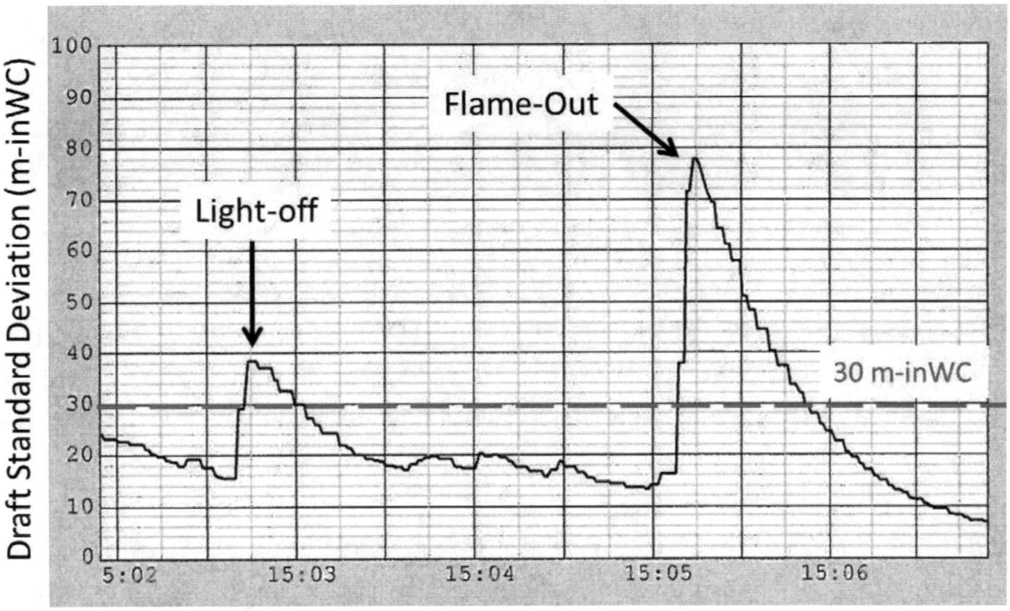

Figure 13 - Flame out Event Observed during a Later Test of Burner F-4

Refer to Appendix A for additional tends of data from the tests at the burner vendors.

Test on Refinery Fired Heater

As a follow-up to the burner vendor tests, in May of 2012, heater tests were conducted on the H-31 fired heater at the Valero Texas City Refinery. This is a radiant/convective vertical cylindrical heater, with a single row of tubes along the radiant axis. The heater has four burners, two on each side of the radiant tube bank. The burners in this heater were raw gas burners. This is an older model burner, not designed for the same low NOx levels as the burners tested at the vendors. Figure 14 shows a view looking from the bottom up into the radiant section of heater H-31.

Figure 14 – Inside the Radiant Section of Heater H-31

This test was a follow-up to the December 2011 burner vendor tests. The purpose of this test was to characterize changes in the draft standard deviation and to compare with the data that was taken at the burner vendors. Since this is an operating heater, the test team could not purposely flame out the heater, nor could the team create the same extreme substoichiometric conditions as were simulated in the test furnaces. However, it was determined that limited manipulation of firing rates, fuel pressures, and individual burners could be performed safely. During the tests, the team would be looking for characteristics of flame instability in the burner flames, and monitoring for any corresponding change in the SPM standard deviation.

Specifically, four tests were performed on this heater:

1. Low Burner Pressure on a single burner – Gradually close off the fuel supply valve to a single burner. See if as the burner is starved of fuel, it becomes "lazy" and affects the draft standard deviation.
2. High Fuel Gas Pressure – Increase the fuel gas pressure to feed more fuel to the burners (and reducing the O_2). See if burner flames go unstable.
3. Low O_2 – Increase fuel flow to drive the O_2 lower. Keep going until excess O_2 reaches 0.5%. (Typical O_2 on this heater is 3-4%). See if burner flames become unstable at low O_2.

4. Starve Single Burner – Close off the air register of a single burner to starve it of air. See if the burner flame goes unstable and affects the draft standard deviation.

Table 2 below summarizes the results from these heater tests. Flame instability could not be created in any of the tests. On the Low Burner Pressure test, the flame was completely extinguished before any change in draft standard deviation was observed. On both the High Fuel Gas Pressure and Low O_2 tests, there were small changes in standard deviation as the firing rate changed. However, these changes were significantly smaller compared to what is typically seen under flame instability. As seen looking through the furnace viewport, the burner flames remained in a stable, healthy combustion in both tests.

Table 2 - Heater Tests performed at Valero Texas City

Test	Lowest Recorded O_2	Flame Instability in Heater	Flame Instability Detected with SPM	Note
Low Burner Pressure on Single Burner	3.8%	No	No	Burner Completely Extinguished before any Change in Std Dev
High Fuel Gas Pressure	1.6%	No	No	Small increase in Std Dev with Firing Rate. Burner flames never went unstable.
Low O_2	0.3%	No	No	Small increase in Std Dev with Firing Rate. Lowest safe O_2 level reached. Burner flames never went unstable.
Starved Single Burner	2.1%	No	No	Single burner starved of air. Substoichiometric combustion observed through viewport.

The last test involved starving a single burner of air. There were small changes in the SPM standard deviation, but not of the magnitude that would indicate flame instability. A view of a starved burner through the heater viewport (Figure 15) showed very bad combustion (a bright orange flame that would jump up and down, as opposed to the stable blue flame of the normal operating burner). However, it was not the same as flame instability in the whole heater. Consequently, there was a small change in the draft standard deviation, but within the range of standard deviation changes that would be seen under normal heater operation.

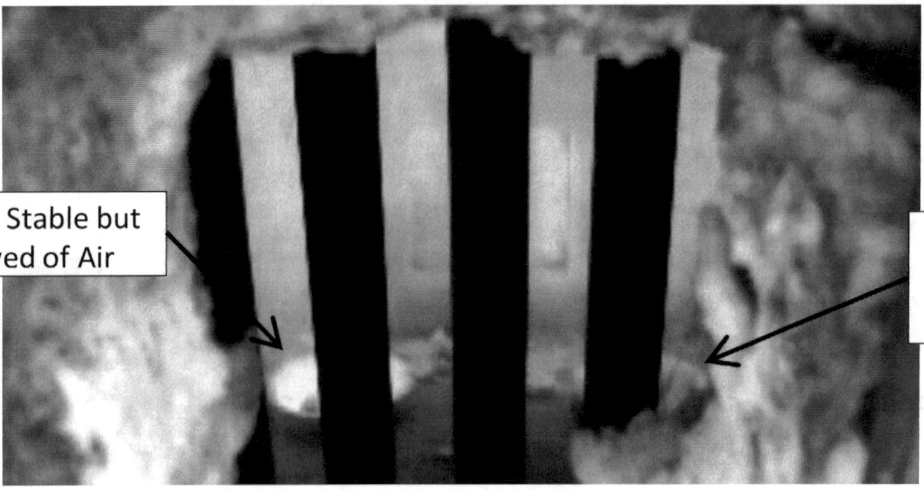

Figure 15 - Starved Burner seen in Heater Viewport

Figure 16 shows the trend of SPM Standard Deviation over the day of the heater tests. Although there were small changes in standard deviation with the different heater tests, these changes were all substantially less than what would be expected under flame instability. The alarm limit of 30 m-inWC was not tripped during the test.

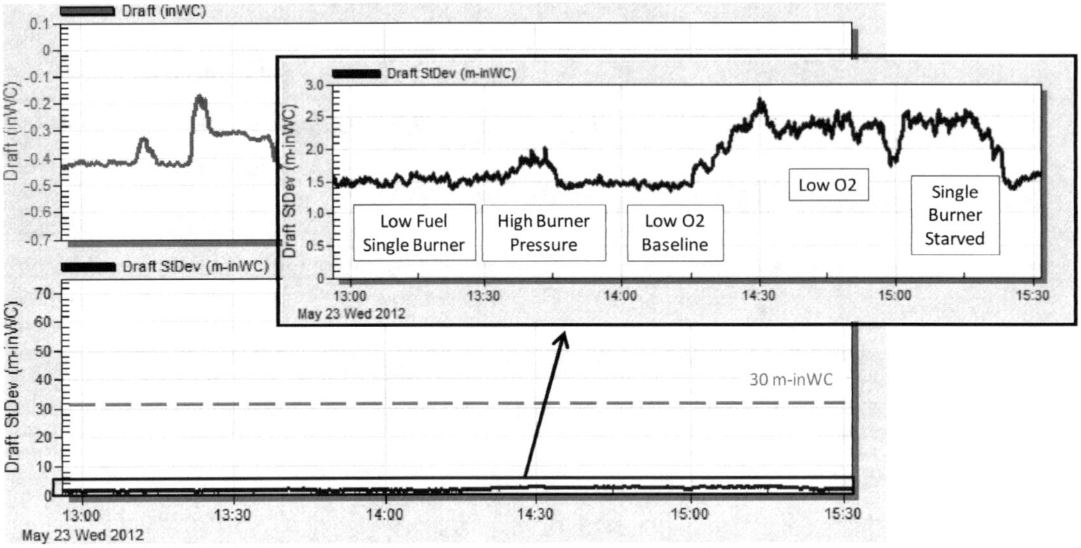

Figure 16 - Trend of Draft Standard Deviation during H-31 Heater Tests

Following the H-31 heater tests, the SPM standard deviation was monitored over five weeks of normal heater operation. During that time, the standard deviation did not go any higher than 2 m-inWC, always significantly less than the alarm detection limit of 30 m-inWC. Although no false alarm occurred during this time, we still don't know how high the standard deviation would go if flame instability and/or flame out actually did occur on this heater. It is possible that a much lower alarm limit (anywhere from 5 to 10 m-inWC) could be selected, which would not be tripped during normal heater operation. Additional heater operational experience is needed to confirm this.

Effect of Firing Rate and Comparison of Among Different Burners

During the burner and heater tests, it was observed that there was a correspondence between firing rate and draft standard deviation. In general, a higher firing rate of the burner(s) leads to a higher draft standard deviation. This makes sense because a higher firing rates means that there is more fuel and more combustion, thus more turbulence or combustion rumble. Figure 17 shows a comparison of the standard deviation values observed under normal operating conditions (e.g. minimum, normal, or maximum firing rate at 3% O_2), on each of the five burners tested at the vendors. The figure also shows standard deviation values observed during the H-31 heater tests.

For every burner, one can see that the standard deviation is the highest at the maximum firing rate, and the lowest at the minimum firing rate. This should be taken into account when setting the standard deviation alarm limit for detecting flame instability. The alarm limit must be chosen such that it would not be exceeded when the heater/burners are at the maximum firing rate. Note that the standard deviation values when burners D-1, D-2, and F-4 were unstable were in the range of 60 to 100 m-inWC and would be off this chart.

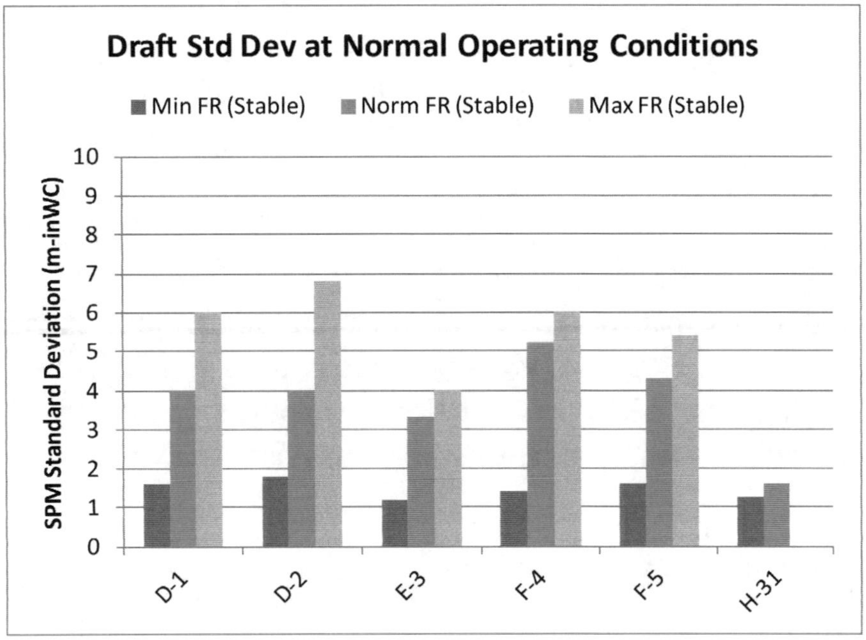

Figure 17 – Comparison of Draft Standard Deviation at Nominal Operating Conditions

Figure 18 shows a comparison of Total Firing Rate (MMBtu/hr) versus standard deviation for all five test furnace burners, and the four-burner heater H-31. All of these data points are from normal (stable) burner operation. For any single burner, there is the general relationship that standard deviation increases with firing rate. However, there is a large degree of separation between the different burners. In particular, note that on H-31, at higher firing rates, the draft standard deviation was substantially lower than that observed in all of the test furnaces.

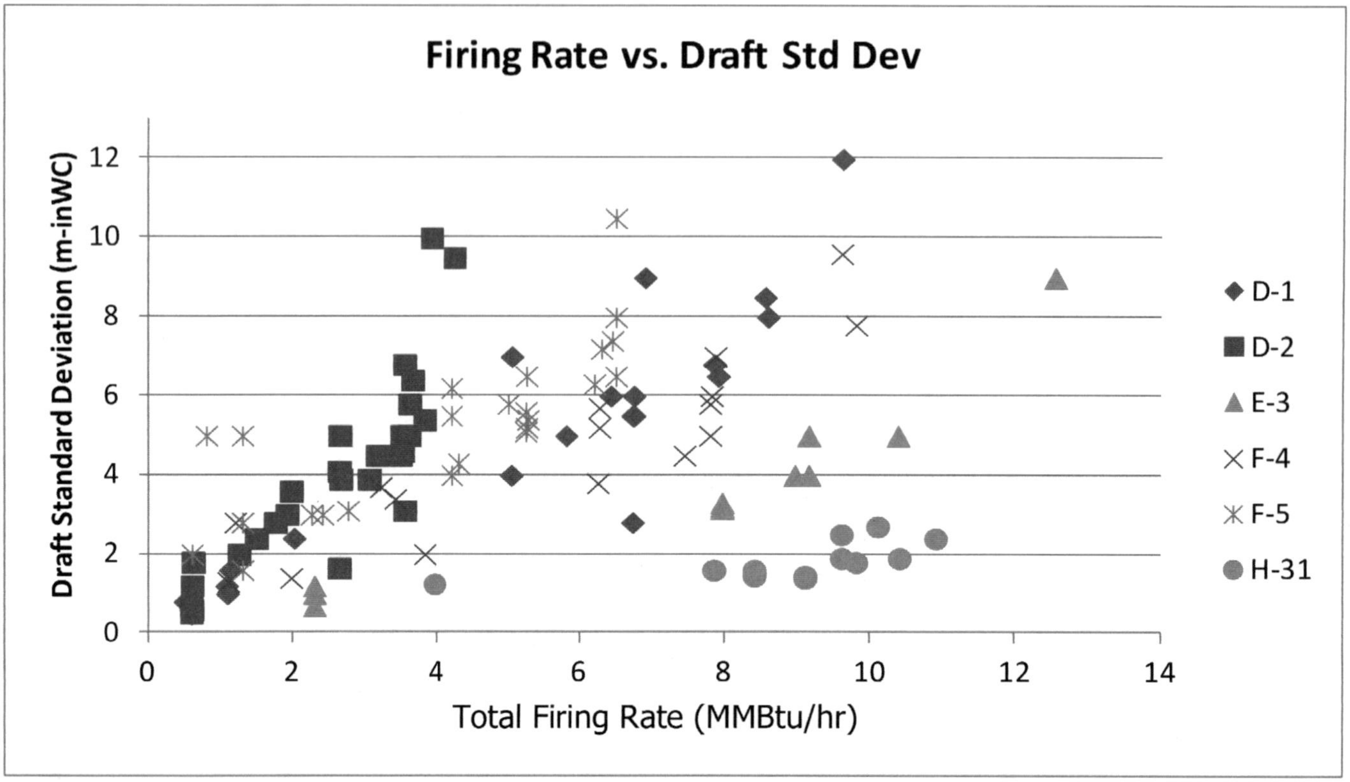

Figure 18 – Correlation of Firing Rate and Draft Standard Deviation

One factor that may explain the difference in draft standard deviation values among the different tests is the volumes of the fireboxes of the test furnaces and heater. A larger volume of the firebox means that the energy and turbulence of combustion is spread out more widely, and therefore the pulsations measured by the draft transmitter may not be as large.

Table 3 shows the approximate firebox volume on each of the test furnaces, and the calculated volumetric heat release at the Minimum, Normal, and Maximum firing rates.

Table 3 – Volumetric Heat Release for Each Test Furnace or Heater

Burner or Heater	# of Burners	Firebox Volume (ft^3)	Total Firing Rate (MMBtu/hr)			Volumetric Heat Release (Btu/hr/ft^3)		
			Min	Norm	Max	Min	Norm	Max
D-1	1	1215	1.12	5.05	6.72	922	4156	5531
D-2	1	792	0.59	2.65	3.52	745	3346	4444
E-3	1	1440	2.29	7.96	9.15	1590	5528	6354
F-4	1	2500	1.96	6.26	7.83	784	2504	3130
F-5	1	1000	1.31	4.2	5.25	1310	4200	5250
H-31	4	3049			12.5			3411

Figure 19 shows the same draft standard deviation values plotted against volumetric heat release. There appears to be a very definite correlation between draft standard deviation and volumetric heat release that is even consistent across different burners and fireboxes. Using volumetric heat release may be a way to predict the draft standard deviation values in a heater, using test data from a different furnace or heater. However, more work is required to validate this.

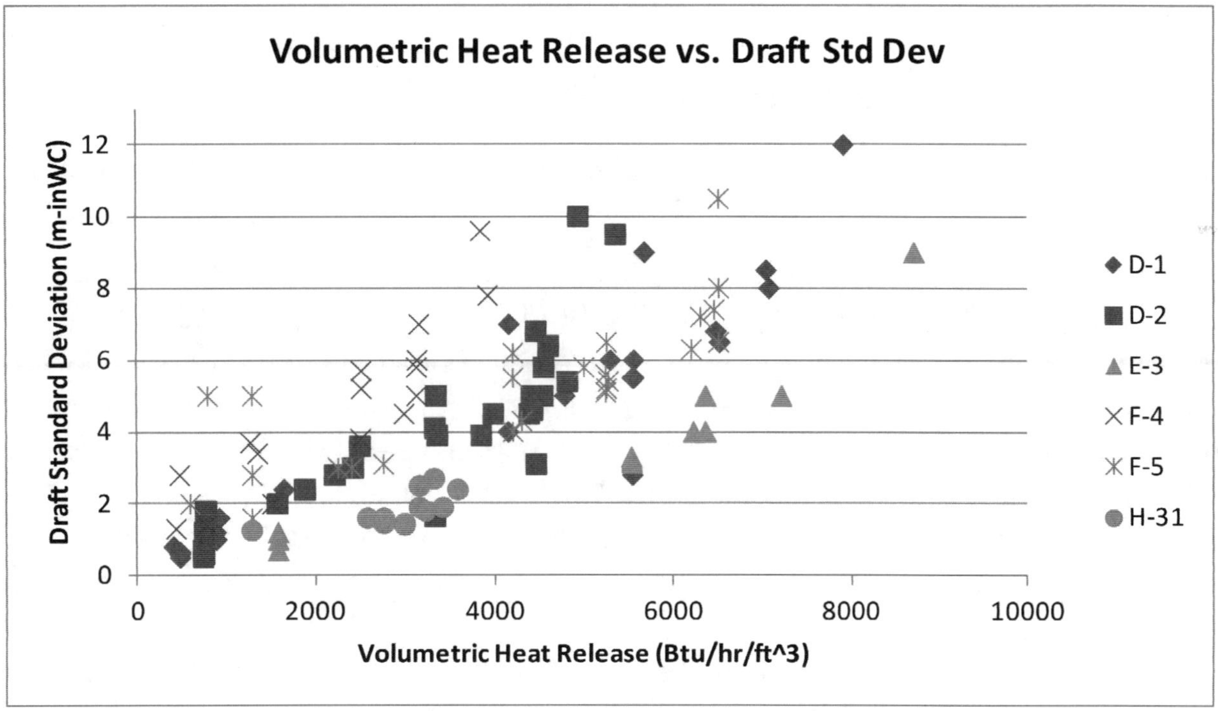

Figure 19 – Correlation of Volumetric Heat Release and Draft Standard Deviation

It may be possible to use operational data from a heater running in normal operation to set an alarm limit for flame instability detection. Figure 20 shows the draft standard deviation versus volumetric heat release for just the H-31 heater at the Texas City Refinery. Remember that during the tests on this heater, and in the weeks following, flame instability never occurred.

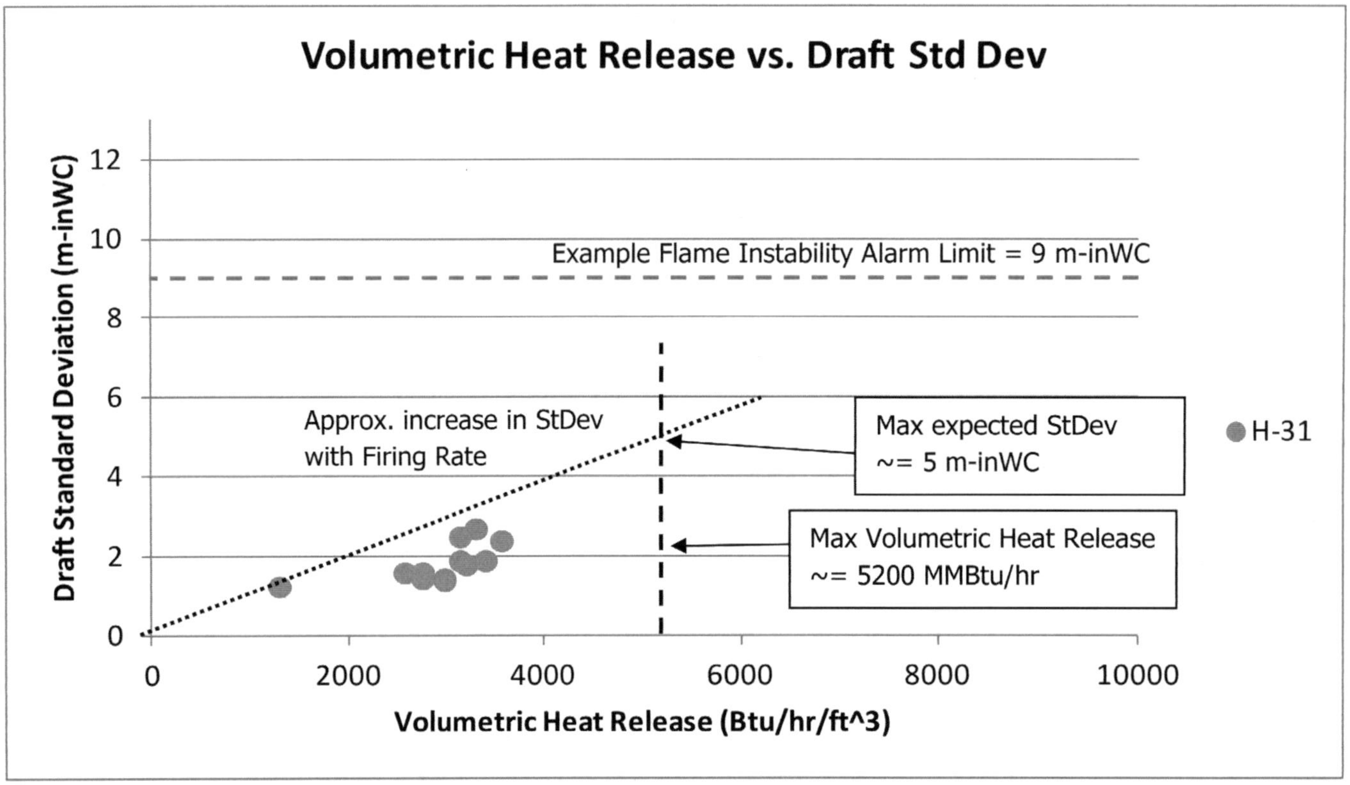

Figure 20 – Using Normal Heater Operational Data for Determining Flame Instability Alarm Limit

The H-31 heater has a maximum firing rate of 16 MMBtu/hr, corresponding to a maximum volumetric heat release of about 5200 Btu/hr/ft^3. This is marked with the vertical dashed line. The approximate increase in standard deviation with firing rate is marked by the diagonal dotted line. The point where these lines cross may correspond to the largest standard deviation value that would be expected in normal heater operation (approximately 5 m-inWC in this example). One could choose the flame instability alarm limit to be some multiple of this value. In the example, we have selected an alarm limit at 9 m-inWC. There would have to be substantial changes in the combustion conditions to make the draft standard deviation go higher than this limit.

The advantage to choosing a smaller alarm limit (e.g. as opposed to the 30 m-inWC from the burner vendor test data), is that, in this heater, we really don't know how high the standard deviation will go during substoichiometric combustion or flame instability. A smaller detection limit gives greater possibility that the alarm will occur under some future, but unknown magnitude, increase in draft oscillations.

The method shown in Figure 20 could be used to set the flame instability alarm limit on any operating fired heater. However, this method is based on only a limited set of test data. A linear relationship between standard deviation and firing rate may not necessarily exist on every burner/heater/furnace system. Furthermore, the actual magnitude of increase of the standard deviation during flame instability may vary between different heaters or burners. In addition to substoichiometric combustion, there may be other mechanisms that cause flame instability and/or flame out. As additional industry experience is gained using SPM technology for heater monitoring, it is expected these other factors will be better understood, and that new guidelines and best practices may be determined for setting the alarm limit, and integrating SPM technology with fired heater operation.

Considerations for Industry

Statistical Process Monitoring technology has been shown to have significant promise for monitoring heaters and detecting flame instability and flame out. However, more data is needed to fully evaluate the potential of this approach. Whenever adding a new draft transmitter to an oil refinery fired heater, owner/operators may wish to purchase a pressure transmitter with SPM technology. The SPM Standard Deviation could be integrated with the other heater operational data either via direct HART I/O, with a HART to analog converter, or with a wireless interface. Heater operators could monitor the Standard Deviation along with other heater measurement points, and use it as an indicator of combustion problems or flame out. An initial flame instability alarm limit could be set based on the standard deviation during normal heater operation and/or standard deviation values observed under flame instability in a test furnace. This alarm limit could be adjusted as experience is gained with a given heater.

SPM is a relatively new technology for monitoring fired heaters. If one chooses to implement SPM for monitoring heaters, the technology should be utilized in addition to (not as a replacement for) industry recommended best practices. Refer to API Recommended Practice 556 [18] for best practices in the instrumentation, control, and protective systems for fired heaters.

It is recommended that purchasers request burner vendors to monitor and trend the SPM standard deviation during burner tests. This data can be utilized during substoichiometric testing as an indicator of the stability of the burners. The SPM data can be used as a means to compare the stability of different burners at various levels of sub-stoichiometry. The SPM data can also be given to a heater designer or an end-user owner-operator, and used by the industry to further understand the application of SPM technology for heater monitoring.

Summary and Conclusion

Advanced pressure diagnostics Statistical Process Monitoring is a promising technology for detecting flame instability and/or flame out in fired heaters. The draft pulsations associated with flame instability can be detected by a draft transmitter with SPM. An increase in the draft standard deviation may be indicative of flame instability. A heater operator may be able to use the draft standard deviation together with other heater operational data to gain insight into the health of the burner combustion. An alarm may be configured to generate an indicator of flame instability when the draft standard deviation exceeds the alarm limit. Initial results show that for some heater/furnace/burner systems, an alarm limit of 30 m-inWC may be suitable for detecting flame instability. On other systems, a limit of 5 to 10 m-inWC may be appropriate. There is a general relationship between volumetric heat release and draft standard deviation. This relationship may be exploited to select an initial flame instability alarm limit on an operating fired heater. Although additional work may allow more precise determination of the optimal alarm limit for a given heater, current experience suggests that all fired heaters would benefit by utilizing pressure transmitters with SPM technology for the draft measurement.

References

1. F. Eisinger, R. Sullivan. *Evaluation of Non-Vibrating Utility Burner/Furnace Systems for Resistance to Thermoacoustic Oscillations*. PVP2006-ICPVT-11-93054. Proceedings of 2006 ASME Pressure Vessels and Piping Division Conference July 23-27, 2006, Vancoucer, BC, Canada
2. F. Eisinger, R. Sullivan. *Elimination of Thermoacoustic Furnace Vibration in a Gas-Fired Package Boiler--A Case Study.* PVP2008-61209. Proceedings of 2008 ASME Pressure Vessels and Piping Division Conference July 27-31, 2008, Chicago, Illinois, USA.
3. F. Eisinger, R. Sullivan. *Avoiding Thermoacoustic Vibration in Burner/Furnace Systems.* Transactions of the ASME, Journal of Pressure Vessel Technology, Vol. 124, November 2002, pp 418-424.
4. F. Eisinger, R. Sullivan. *Prediction of Thermoacoustic Vibration of Burner/Furnace Systems in Utility Boilers.* Paper No. PVP 2007-26090. ASME PVP 2007 Conference, July 22-26, 2007, San Antonio, TX.
5. Patent #7,950,919 – *Method and apparatus for monitoring and controlling the stability of a burner of a fired heater.* (G. Johnson, B. Hamilton). Filed: October 12, 2005. Issued: May 31, 2011.

6. D. Wehrs . *Detection of Plugged Impulse Lines using Statistical Process Monitoring Technology*. Whitepaper, Rosemount.com, December 2006.
7. R. Szanyi, M. Raterman, E. Eryurek. Diagnostics capabilities of FOUNDATION fieldbus pressure transmitters. Hydrocarbon Processing, April 2003, pp. 53-59.
8. D. Wehrs, A. Klosinski. *Entrained Gas Diagnostic with Intelligent Differential Pressure Transmitter*. Whitepaper, Rosemount.com, January 2008.
9. D. Wehrs, A. Klosinski. *Wet Gas Diagnostic with Intelligent Differential Pressure Transmitter*. Whitepaper, Rosemount.com, January 2008.
10. Rosemount Technical Note – *Distillation Column Flooding Diagnostics with Intelligent Differential Pressure Transmitter* – www.rosemount.com
11. R. Pihlaja. *Detecting Incipient Flooding in a Distillation Column* – Presented at 2010 AIChE Spring Meeting and 6th Global Congress on Process Safety – March 21-25, 2010 – San Antonio, TX, USA.
12. E. Eryurek, P. Sharpe, D. White. Abnormal Situation Prevention through Smart Field Devices. NPRA Hydrocarbon Processing, March 2006, pp. 41-48.
13. J. Miller - *Furnace Flame Instability Detection Using Pressure Measurement with Advanced Diagnostics* – Session 57 – Advances in Fired Heater Safety and Operations – 2010 AIChE Spring Meeting and 6th Global Congress on Process Safety – March 21-25, 2010 – San Antonio, TX, USA
14. Rosemount Technical Note 00840-0500-4801 – Furnace Flame Instability Detection with Advanced Pressure Diagnostics – J. Miller – July 2010
15. R. Stier, J. Miller – Session 6B-2216 *Improving Refinery Fired Heater Safety using Advanced Pressure Diagnostics* – presented at Emerson Global Users Exchange, October 8-12, 2012, Anaheim, CA.
16. US Patent #7,536,274 – System and method for detecting an abnormal situation associated with a heater (L. Heavner, J. Sharpe, E. Eryurek, K. Kavaklioglu) Filed: March 14, 2005. Issued: May 19, 2009
17. Patent Publication No: 2010/0151397 – System and Method for Controlling Fired Heater Operations. (J. Farrell, S. Chhotray, G. Dobbs, M. Alvarez, P. Schweitzer). Filed: December 14, 2009
18. API Recommend Practice 556 – Instrumentation, Control, and Protective Systems for Gas Fired Heaters, Second Edition, April 2011

Appendix A

This appendix shows the trends of draft and SPM Standard Deviation for all of the burner vendor tests that were performed in December 2011. All labels of Firing Rate (FR) are in units of MMBtu/hr.

Burner E-3

Vendor	Burner	Geometry	Burner Firing Rate Conditions (MMBtu/hr)			Lowest % of Stoichiometry	Flame Instability Observed
			Min	Norm	Max		
E	E-3	Round	2.29	7.96	9.15	71%	No

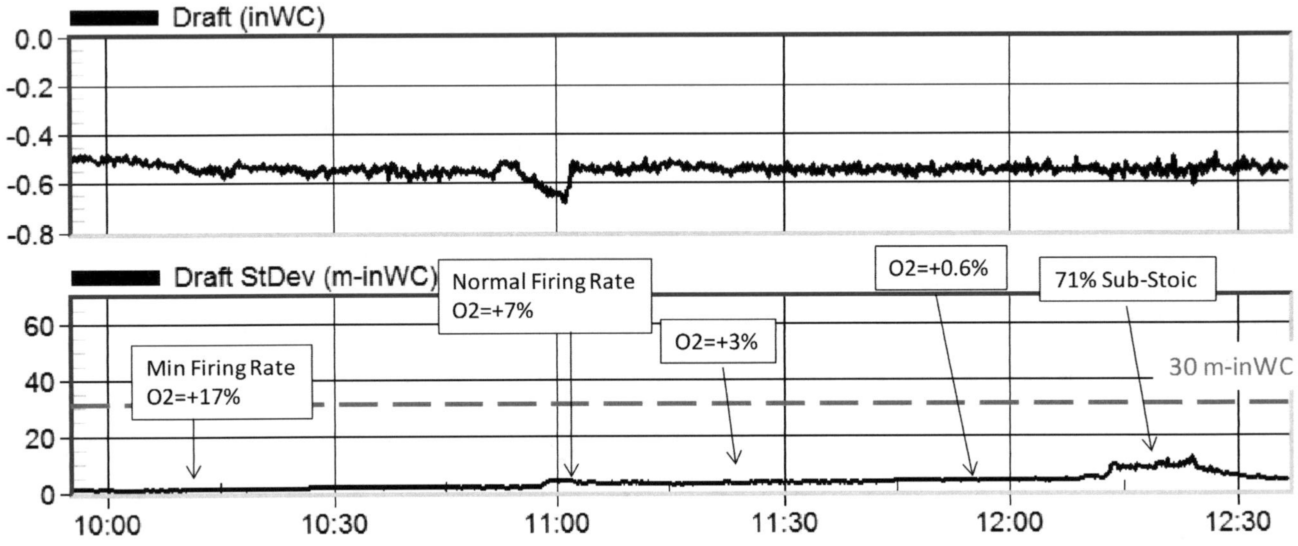

Burner D-1

Vendor	Burner	Geometry	Burner Firing Rate Conditions (MMBtu/hr)			Lowest % of Stoichiometry	Flame Instability Observed
			Min	Norm	Max		
D	D-1	Round	1.12	5.05	6.72	57% (Norm FR) 70% (Max FR)	Yes (Max FR only)

Burner D-1: Normal Firing Rate

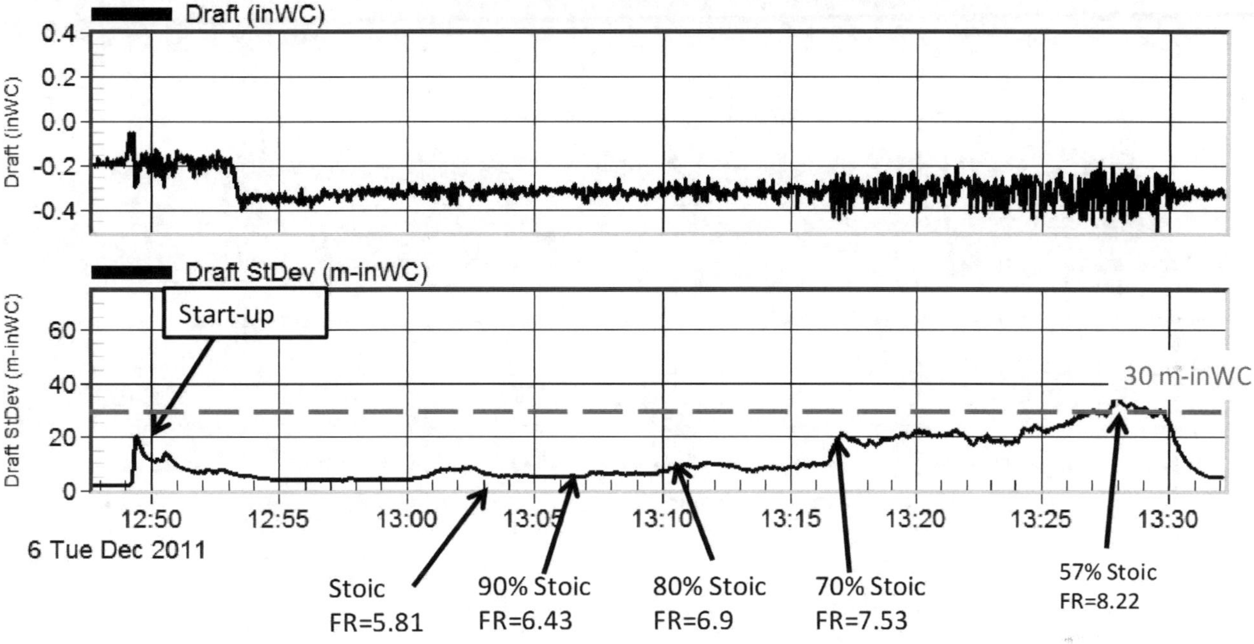

Burner D-1: Maximum Firing Rate

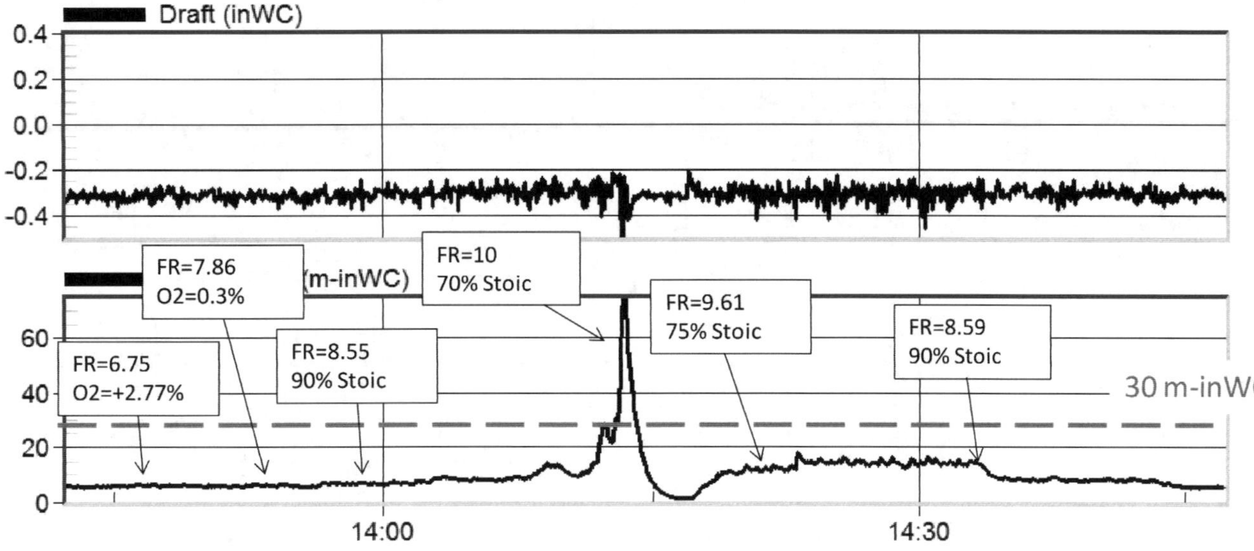

Burner D-2

Vendor	Burner	Geometry	Burner Firing Rate Conditions (MMBtu/hr)			Lowest % of Stoichiometry	Flame Instability Observed
			Min	Norm	Max		
D	D-2	Flat	0.59	2.65	3.52	85% (Norm FR) 92% (Max FR)	Yes

Burner D-2: Normal Firing Rate

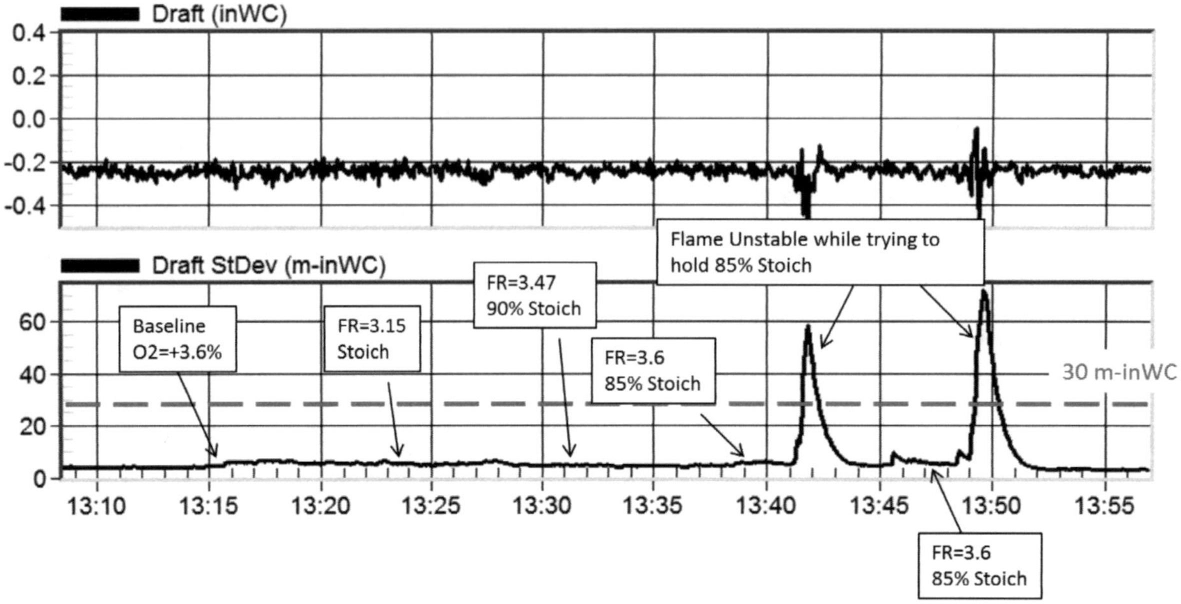

Burner D-2: Maximum Firing Rate

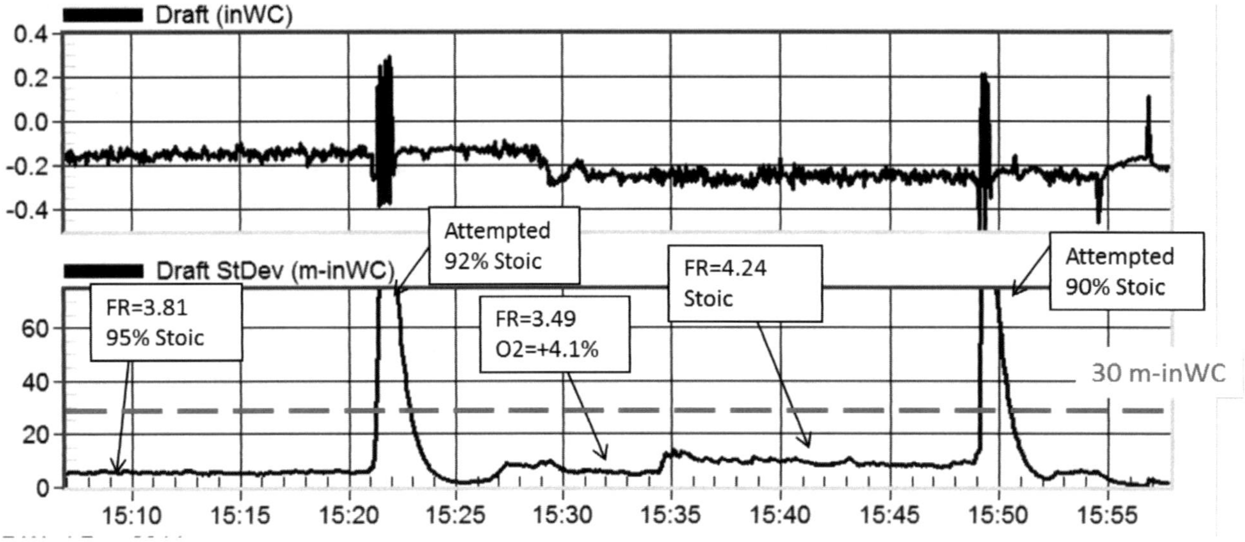

Burner F-4

Vendor	Burner	Geometry	Burner Firing Rate Conditions (MMBtu/hr)			Lowest % of Stoichiometry	Flame Instability Observed
			Min	Norm	Max		
F	F-4	Round	1.96	6.26	7.83	90%	No

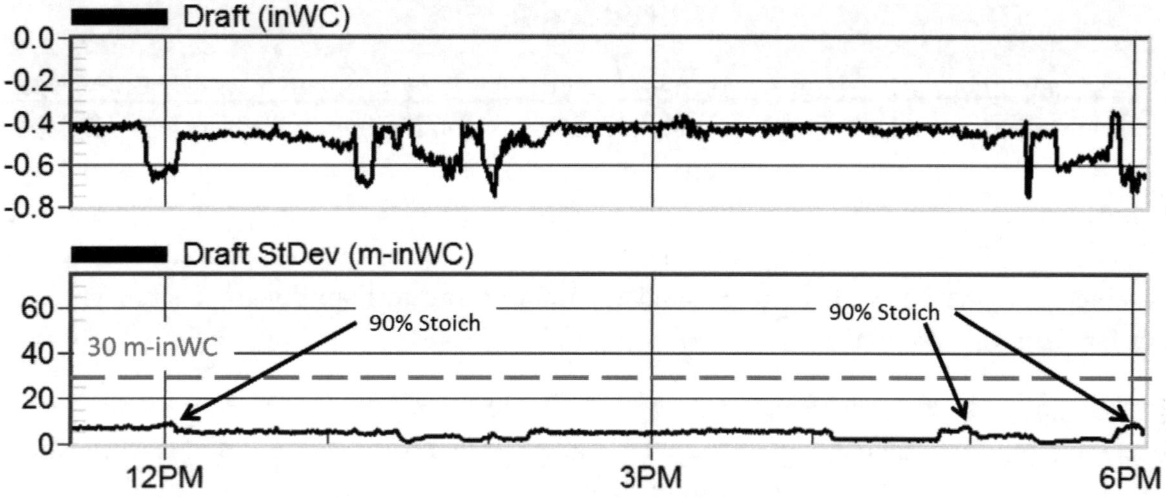

Burner F-5

Vendor	Burner	Geometry	Burner Firing Rate Conditions (MMBtu/hr)			Lowest % of Stoichiometry	Flame Instability Observed
			Min	Norm	Max		
F	F-5	Round	1.31	4.2	5.25	90%	No

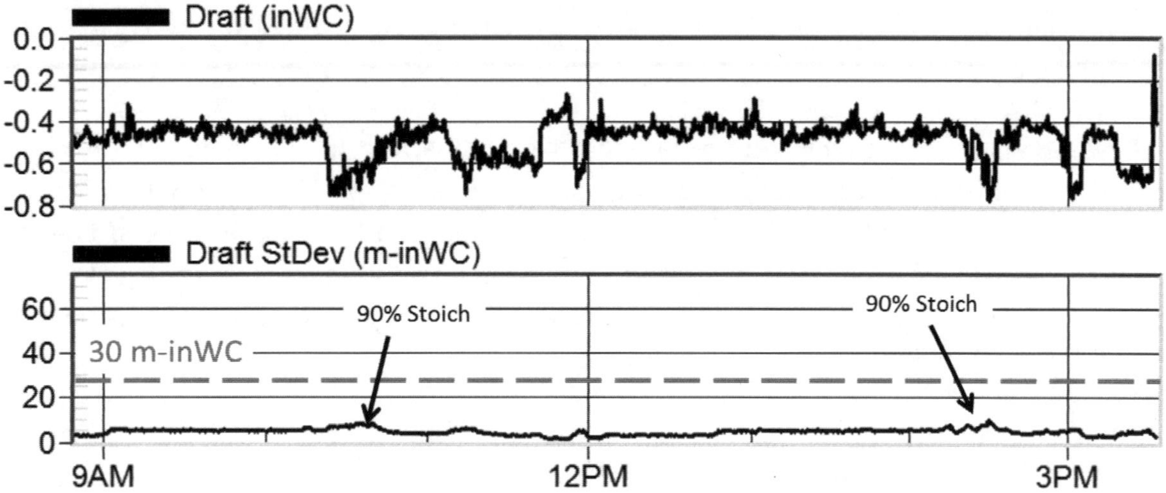

Abqaiq Plant Virtual Plant RTO Implementation

Dr. Kuangyao Brian Peng and Saleh H. AL-Musallam,

Saudi Aramco, Abqaiq, Saudi Arabia

Introduction

Abqaiq Plant is the largest worldwide crude stabilization plant with $200MM energy consumption per year. Abqaiq NGL plant shares 30% of total energy consumption. High NGL recovery rate has been the main objective because of undesired loss of hydrocarbon liquid condensate in the offgas pipeline. Recent efforts for process improvements including the commissioning of TEG dehydration unit and the compression train upgrades have been implemented to enhance NGL recovery and to improve the availability of desired cryogenic cooling for C2/C3 separation.

Abqaiq NGL Plant includes three major processes: compressing, high pressure stripping and deethanizing. Simultaneous optimization for all three major processes under the constraints of NGL product spec. requires fundamental understanding and a robust simulation tool. It has been a dream and a challenge for process and control engineers to develop such a complete model which can map the plant operating conditions in situ and perform a complete plant optimization online.

ROMeo (Rigorous Online Modeling with Equation-based Optimization), a widely used real time optimization (RTO) platform, can be used to achieve our goal. A real time optimizer can take the variations in upstream units such as changes in crude feed flow combinations and stabilization depth in real time, and calculates the most profitable mode of operating condition by using the rigorous plant model with energy consumption equations and process and product constraint equations. Then the calculated optimal set points are submitted to the DCS control room to ramp the plant process.

Abqaiq Plant NGL engineering department has successfully implemented and tested an in-house developed RTO NGL plant model. A potential increase of $200K to $400K revenue increase for each adjustment has been predicted under the current operating mode and unit operation constraints. In addition to the quantifiable revenue advantages, this presentation will also highlight other substantial advantages of the system such as measurement health monitoring, next generation process KPIs, and energy optimization.

Background of NGL process optimization initiatives

To meet the future energy demand and sustainability requirements in the Kingdom of Saudi Arabia, Abqaiq Plant starts several energy efficiency programs for all process units. A rough study with Saudi Aramco P&CSD for NGL recovery enhancement has shown great potentials with online optimization. As a result of this study, by simply adjusting the operating conditions of HP-Stripper columns could increase NGL recovery rate by 3 to 5 MBD. Further study with relaxing product spec. of individual columns and maintain the final product spec has shown an increase of NGL production up to 9 MBD.

A joint effort among corporate optimization specialists and plant operation engineers took off in 2010. A road map was proposed to implement ROMeo RTO at Abqaiq NGL plant by the end of year 2013. The road map includes three major milestones: rigorous model development, PI integration and offline optimization.

Process description

Continuous feed gas from oil stabilization units is fed into the NGL plant with a flow rate of 550MMSCFD at 2psig and 140°F. Feed gas is compressed with two stage compressors to 450psig and is cooled by cooling tower for the outlet temperature at 110 -125 °F.

Feed compositions are studied with the available historical data. The major variation of feed compositions come from crude oil mix containing neat condensate, Arabia Light and Arabia Extra light. As a baseline model, average compositions over the past few years are used for model stability. Typical compositional data for hydrocarbon contains C3+ above 70%. Table 1 shows the average compositional data for a typical feed.

	N2	C1	CO2	C2	H2S	C3	IC4	NC4	IC5	NC5	C6	C7	C8
Overall	0.019706	0.0785301	0.028414	0.224106	0.043989	0.342631	0.056587	0.132501	0.032532	0.032833	0.006334	0.0018	3.4E-05

Table 1

Abqaiq NGL plant contains five major high pressure stripper columns and one deethanizer column. The average operation condition of high pressure column is 145 °F and 400 psig. Column efficiencies vary from 68% to 85% because of different tray designs. The five high pressure stripping columns were designed with 22 bubble trays for separation of C2 and C3+. Each column equips with steam kettle reboilers and temperature controller at the 7th stage of the column. The column temperature is the key handle for NGL product specification.

The deethanizer chiller is designed to operate at 65 °F and the reflux condenser is 60 °F . Currently, operation requires all columns product spec at 0.4 vol% which is a conservative approach. Overstripping of NGL product to offgas line would raise the liquid condensation in the pipeline, reduction of valuable NGL product recovery and waste of energy.

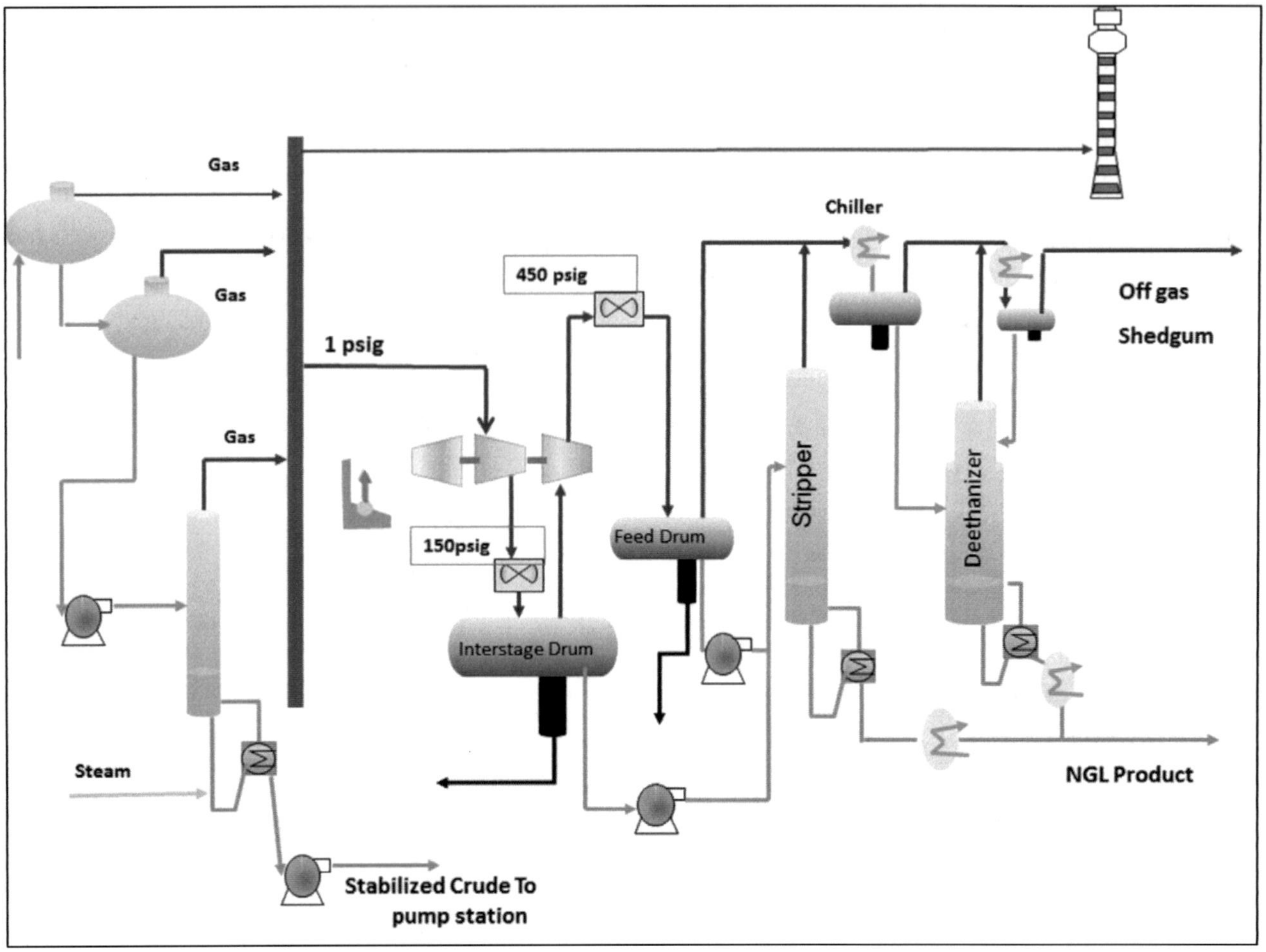

Figure 1. A simplified NGL process flow diagram.

Model development and implementation

Rigorous process model is developed based on the current process conditions in the field. Due to time constraint, the preliminary process model does not integrate with detail performance data of compressors. However, because phase I of this program is to maximize NGL recovery and to minimize the impact of liquid condensation, the performance functions are not required at this stage.

One of the challenges of the team is to identify all the piping routes of cross feed lines. The cross feed flow rates have to match the field instrumentations. The on-off setting of process units is relied on the macro programs that were developed in house with the engineering support and training from Invensys. A total number of 480 major PI-tags were linked to PI historian for data reconciliation.

Economic variables for the sources include 60# and 625# steam, propane and feed gas. For the product steams, two offgas streams and one liquid NGL stream are included. The environmental variables include the release of CO2 to atmosphere and the amount of sour water steam to downstream facility.

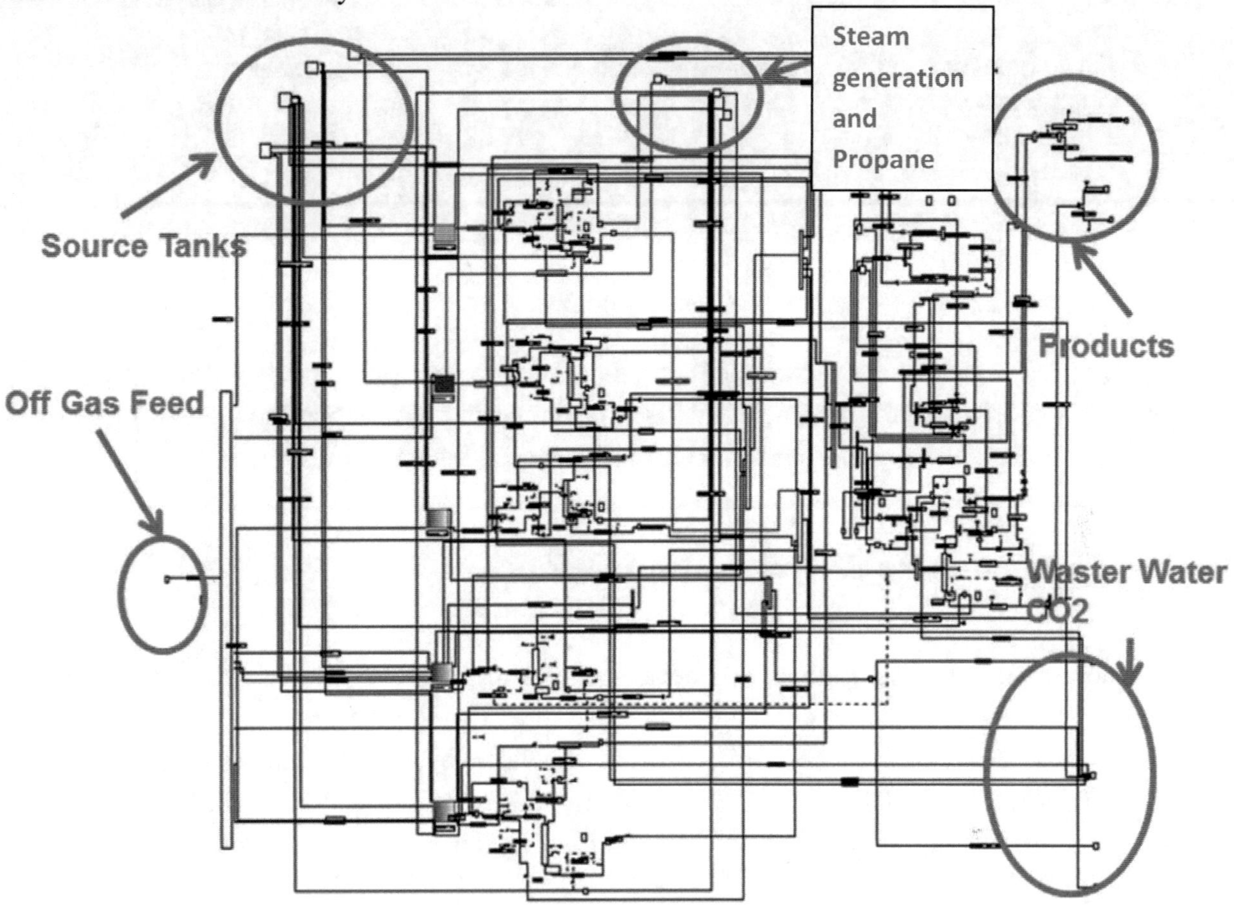

Figure 2

Figure 2 demonstrates the process models with eight compression trains, five high pressure stripping columns, one deethanizer column and a propane refrigeration system. All process units in this project are simulated with rigorous thermodynamic models.

Data Reconciliation

Process model has been integrated with PI historian for data reconciliation purpose. Macro routines with TCL scripts are developed to turn on and turn down major equipment under different operation modes. Data reconciliation results deliver the report for measurement system analysis that indicates major instrumentation issues. Operation, instrumentation engineering and maintenance are able to utilize this report to identify the issue and improve the situation.

Furthermore, accuracies of some major KPIs are improved after the improvement of measurement system.

Tags	Unit	Scan	Model	Bias	% ABS(Bias)	Comments
462-FI-310.PV	MMft3/day	114.851	91.047	-23.804	26.14474	
340-Total-Off-Gas.CAL	MMft3/day	393.3	297.296	-96.004	32.2924	
461-FIC-182.PV	MMft3/day	18.471	14.189	-4.282	30.17831	
499-TI-764AI1PV.CV	F	159.65	125.356	-34.294	27.35729	
462-FI-1147.PV	MMft3/day	79.345	61.404	-17.941	29.21797	
334-FI-1005.PV	lb/hr	38297.5	25444.8	-12852.7	50.51209	
461-FIC-12.PV	lb/hr	62396.6	29909.7	-32486.9	108.6166	

Table 2

Table 2 shows an example of data reconciliation results with problematic instrumentations related to the major KPIs. Engineering and operation was able to take immediate actions with the reported warnings and resolved the issue in time.

Online Open Loop Real Time Optimization

The online process model has been up running to provide advisory for operation engineers. The new process set points for high pressure stripping columns such as the feed flow rates and individual column specs are reported on daily basis. A typical adjustment could increase a few MBD of NGL product. When feed composition shift due to variation of oil feed combination and stabilization depth, the NGL recovery can be enhanced and lined out quickly with ROMeo RTO.

Table 3 shows an example of production increase with RTO. The long term benefit is estimated at least $3MM per year.

Optimization Summary

Variable	UOM	Value	Change
Net Objective	currency/day	5.628E+006	56752.500
Positive Contributors			
+ Products	currency/day	1.119E+007	54915.900
Negative Contributors			
- Steam Feeds	currency/day	5.563E+006	-1836.620

Table 3

ROMeo has also identified a potential increase of 3MBD recovery with simple process modification. As shown in the figure 3, under the optimization mode, ROMeo tends to divert all the overhead gas from deethanizer feed drum to bypass deethanizer overhead condenser. It has been the normal practice for NGL operation since the system was designed. Similar result was concluded from an independent energy optimization study report from corporate process optimization group. The online RTO demonstrates capability of evolving process by justifying the boundaries of current process design.

Other Monitoring System Development

A couple of monitoring indexes have been proposed for the NGL plant and pipeline operations. For NGL process, a fouling index is defined to capture the early stage of fouling. Instead of relying on the steam flow meter, a fouling index is defined with fundamental equations to remove the impact of temperature gradient across the heat exchanger. The fouling index can provide information for unplanned maintenance schedule.

Table 4 shows the examples of fouling indexes for the kettle reboilers. Out temperature at the cold side is assumed to be the bubble point of the hydrocarbon flow. Hot side temperature is assumed to be the dew point of the superheated steam flow. Temperature difference is calculated with the average temperature from both sides. Based on the engineering data, the heat transfer coefficient in red shows the inconsistent dimension which requires further validation during turn around activity.

Fouling Studies

Variable	UOM	Initial	Value
/340_E3_Duty_CU/e3ua	BTU/hr-F	337104.000	317283.000
/340_E3_Duty_CU/e3u	BTU/hr-ft2-F	67.421	63.457
/461_C_101Duty_CU/e103ua	BTU/hr-F	1.900E+006	1.789E+006
/461_C_101Duty_CU/e102u	BTU/hr-ft2-F	383.799	361.470
/462_stripper_duty_cu/e203ua	BTU/hr-F	2.037E+006	1.905E+006
/462_stripper_duty_cu/e203u	BTU/hr-ft2-F	59.916	56.036
/499_E3_Duty_CU/E3UA	BTU/hr-F	1.515E+006	1.515E+006
/499_E3_Duty_CU/E3U	BTU/hr-ft2-F	69.030	69.030

Table 4

Table 5 shows another environmental variable established for carbon dioxide release. The total amount of carbon dioxide can be tracked with online optimization model. With data collection from historian, justification for carbon dioxide capture can be established because supercritical carbon dioxide injection could be the potential technology for the depleted oil fields in the Kingdom.

Overview Stream Economics: CO2 Generation

Variable	UOM	Value	Change
Flow	MMft3/day	59.916	0.000

Table 5

Figure 3 shows an important process variable, the predicted hydrocarbon liquid condensate, to be monitored. The offgas pressure will drop along the pipeline to the gas plant for C1-C2 separation and sulfur removal. When the initial temperature is higher than the ambient temperature, hydrocarbon liquid condensate is expected in the pipeline. The amount of hydrocarbon liquid condensate can be predicted with online optimization model which has been updated on daily basis. The amount of accumulated liquid can be correlated to the back pressure incidents for future optimization of pipeline scraping frequency and plant design criteria. For instance, the NGL target recovery and frequency of pipeline scrapping shall be determined based on the total cost analysis.

Tracking of hydrocarbon liquid condensate can be achieved with an online model to mitigate back pressure and flaring of valuable products. This new performance index could potentially save the company millions of dollars.

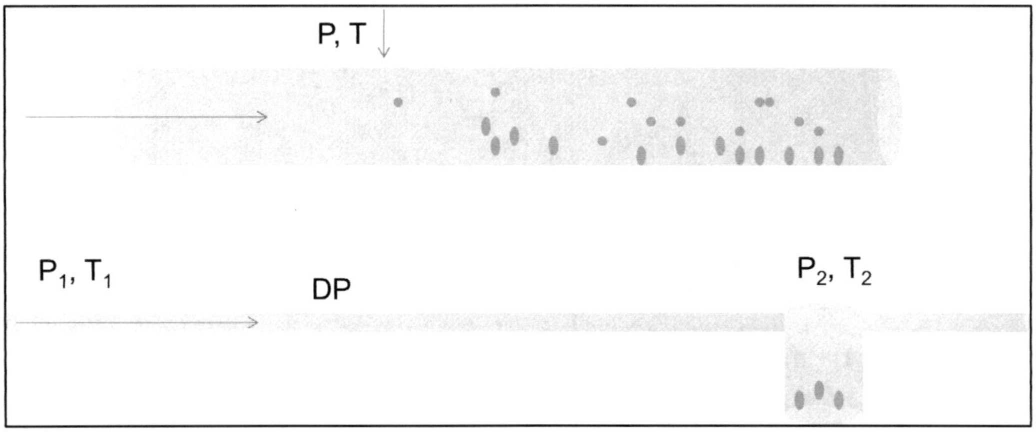

Figure 3

Conclusion

Abqaiq NGL Plant has successfully implemented and tested an in-house developed RTO NGL plant model. A potential increase of $200K to $400K revenue increase for each adjustment has been predicted under the current operating mode and unit operation constraints. In addition to the quantifiable revenue, this presentation also highlights other substantial advantages such as measurement health monitoring, next generation process KPIs, and energy optimization. Future development of this project will focus on supervisory PI Processbook for DCS operators and implementation of engineering details for major equipment.

Acknowledgement

The authors thanks for the supports from the managements of South Area Oil Production Department at Abqaiq and Process Simulation and Control System Department at Dhahran under Saudi Aramco Oil Company. This joint technology development has demonstrated a successful story for Saudi Aramco accelerated transformation program.

Economic Load Allocation Optimization of an Industrial Power Plant

Ravi Nath
Honeywell Process Solutions
Houston, TX

ABSTRACT

Operational Optimization of Industrial Power Plants (IPP) is an often ignored area. One reason for this complacency is that IPPs are usually cost centers in production plants and the cost of energy production is prorated amongst the various production units, so it suffers what is at times referred to as "the tragedy of the commons". Another reason for this complacency is that IPP structure is significantly different than other production units and traditional operations optimization techniques do not readily apply to IPP operations optimization.

This paper will discuss various aspects of IPP optimization; the solution will be illustrated with a recently completed project for a US refinery.

Unlike most refineries, this refinery purchases steam from a neighboring process facility to supplement their own steam production. The implemented solution is an on-line optimizer that optimizes purchased and produced steam to maximize the economic benefit as well as reliability. The implemented solution determines the optimum allocation of steam generators and adjustments to steam balance in the boiler house; furthermore, it also suggests changes to equipment line up in an opportunistic way. The project was completed last year and has been in operation since then, it has been instrumental in streamlining their operations and decreasing their IPP operating costs.

I. INTRODUCTION

A Process Plant comprises one or more process units and an Industrial Power Plant. The process units transform raw materials to products and in doing so consume energy (or utilities) in the form of steam, fuel, electricity, compressed air, cooling water, etc. The Industrial Power Plant (IPP), also referred to as the Plant Utility System, is responsible for providing utilities to the process units. Typically the IPP self generates some of the utilities and purchases others from external sources. Typically all of the cooling water and compressed air is self generated, much of the steam is self generated and at least some of the electrical power and much of the fuel is purchased. Electrical power is typically purchased from the local power grid. Fuel is purchased from one or more local fuel suppliers. Steam, if purchased, is usually from a neighboring process or cogeneration facility. External purchases are usually governed by legal contracts which tend to be complex with provisions for minimum "take or pay", "tier pricing", "time of use" rates etc. Figure 1 is a conceptual representation of a Process Plant showing the interactions between the Process Units and the IPP.

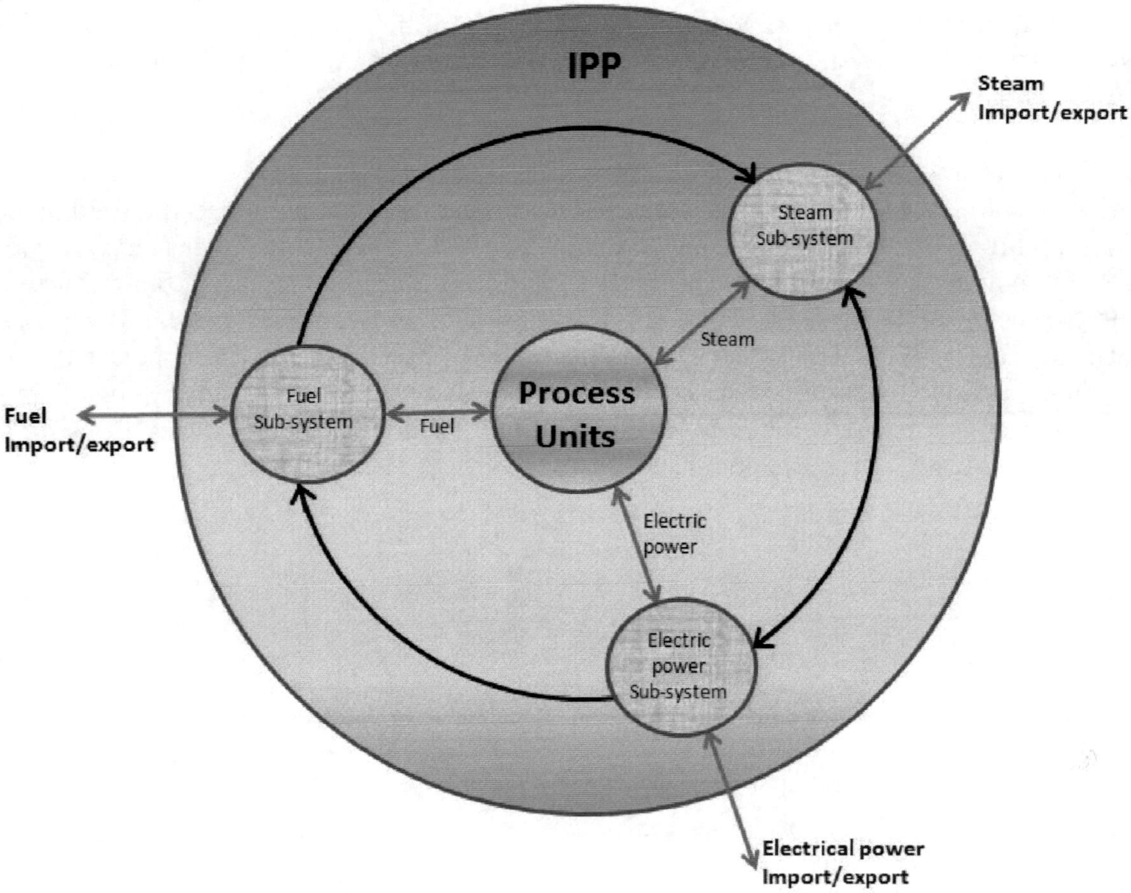

Figure 1: Conceptual Representation of an Industrial Process Plant

Depending on the process units' utilities needs and the location of the production site, there is a large variation in the structure of Industrial Power Plants; however they all share some common characteristics that will be discussed in section III.

As discussed in section II, there are many alternate ways to operate an Industrial Power Plant to satisfy the utility needs of the process units at any given instant. Each alternative however requires different mix of purchased energy and therefore has a different operating cost and different level of emissions to the environment. Because IPPs in large production sites tend to be complex, it is difficult to understand all tradeoffs; therefore there is a tendency to sub-optimize parts of the system rather than optimize the whole, doing so however could result in a significant missed opportunity.

This paper describes a methodology for operations optimization of Industrial Power Plants that has wide applicability and guarantees a global optimum. The proposed solution is based on Mixed Integer Linear Programming (MILP) formulation of the problem. An example based on a recently completed project is presented in section VI.

II. A TYPICAL INDUSTRIAL POWER PLANT

For illustrative purposes, we will consider a simplified IPP with three sub-systems: the steam subsystem, the fuel subsystem and the electric power subsystem. Each of the subsystems is briefly described below;

Steam Subsystem:

The steam subsystem comprises a network of headers that connect to the production units. As shown in Figure 2, there are three steam headers – HP, MP and LP; one boiler feed water (BFW) header and a steam condensate header. Steam is generated in multiple HP and MP boilers that supply to the HP and MP headers respectively. There are back-pressure and condensing steam turbo-generators that produce electrical power. There also are several steam turbines and electrical motors that drive various fans, pumps and compressors such as boiler fans, BFW pumps, cooling water pumps, plant air compressors etc. There are Pressure Reducing Stations that let down steam from higher to lower pressure and a steam vent to maintain header pressures. There is also a blow down flash unit that recovers LP steam from the boiler blow down streams. The BFW used in the boilers is generated in one or more deaerators that consume LP steam.

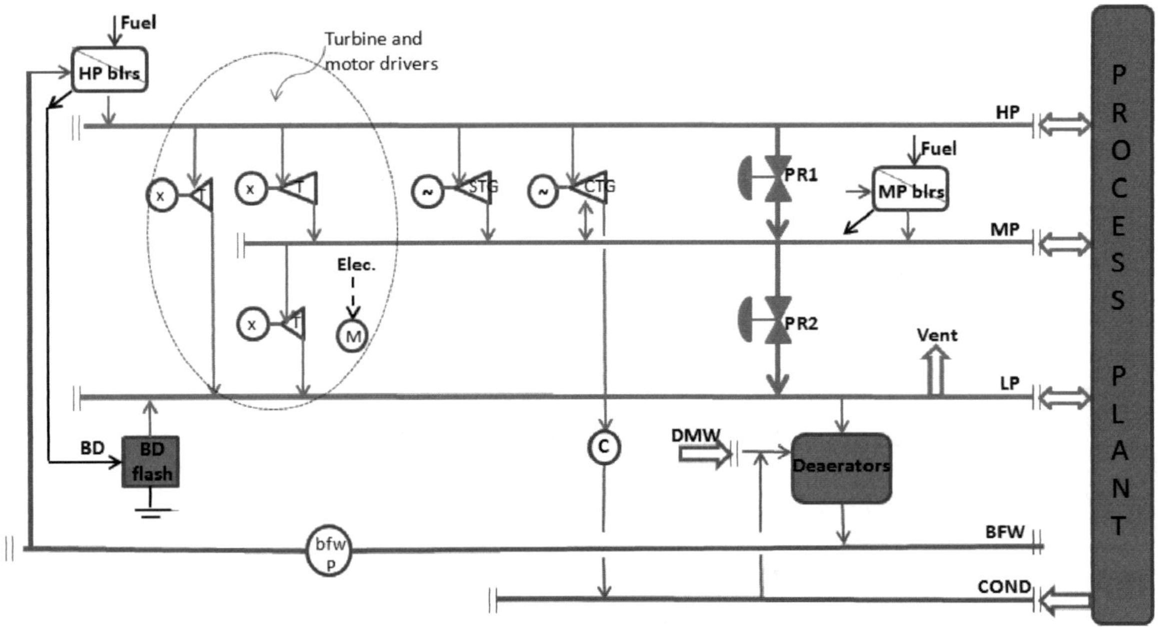

Figure 2: IPP: Steam subsystem

Fuel Subsystem:

Typically there is more than one fuel header at the site; for each fuel header there may be multiple suppliers. With each supplier there usually is a different contractual agreement; for example one may be a long term supplier with a minimum take or pay agreement, another may be short term supplier with a tiered pricing agreement and the third may be a spot supplier with price tied to a fuel price index. Fuel may be high pressure Methane, low pressure Methane or plant fuel which could be a mixture of Hydrogen and Methane. There are several consumers of fuel such as fired heaters, furnaces, steam generators, gas turbines etc. The fuel consumers may have certain restrictions, for example, some could only consume a single type of fuel, while others could consume any mix of fuels. Sometimes there are fuel compressors to uplift the fuel pressure; the fuel compressors could be driven by electrical motors, steam turbines or gas turbines. There are letdown stations and a flare to maintain fuel header pressures. One such fuel subsystem schematic is shown in Figure 3.

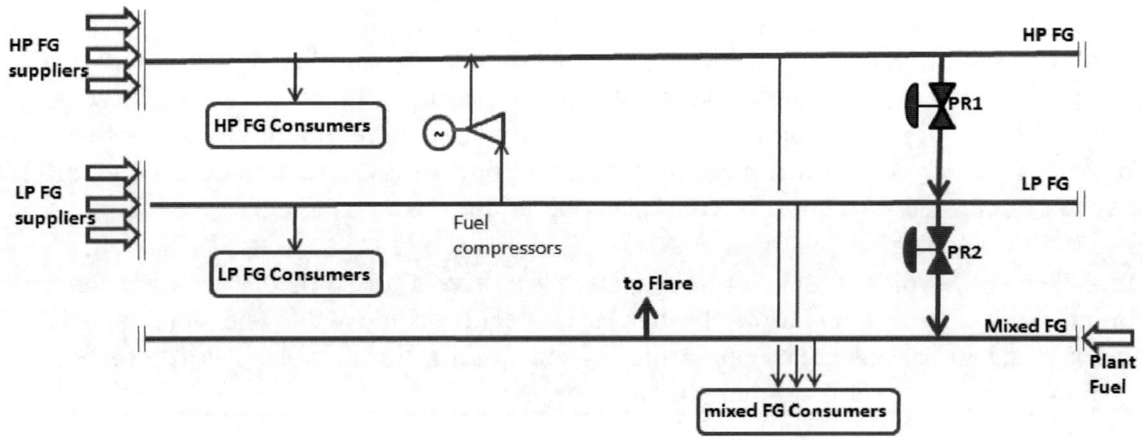

Figure 3: IPP: Fuel subsystem

Electrical Power Subsystem:

The electrical power subsystem is the simplest to conceptualize. Usually a simple balance model (IN = OUT) is sufficient, and it is usually not necessary to model the transformer stations and various voltage sub-headers unless they are constraining. One such electrical power subsystem schematic is shown in Figure 4.

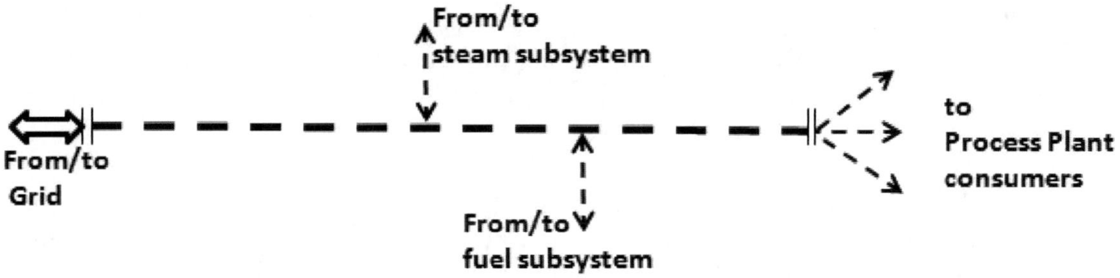

Figure 4: IPP: Electrical power subsystem

Sub-system Interactions:

The above mentioned subsystems interact with the process units by supplying the energy needs of the process. In addition, the subsystems interact amongst each other; for instance the fuel subsystem supplies the fuel required by the boilers in the steam subsystem and the electrical power subsystems accepts surplus power from the steam subsystem and supplies power to the fuel subsystem consumers. Such interactions are shown by arrowed lines in Figure 1.

III. CHARACTERISTICS OF INDUSTRIAL POWER PLANTS

There are several characteristics unique to IPP operations that are noteworthy;

- **Surplus generation capacity:** IPP typically has more generation capacity than required for normal operation. This is by design so that the IPP could accommodate the extremes in demand that are usually encountered during the process unit start ups and shut downs. The surplus generation capacity is manifested in terms of surplus boilers, pumps and compressors. For example, there may be four boilers that are required during start ups, but only two may be needed during normal operation and a third one may be in hot standby to provide additional steam on a short notice. During normal operation, the surplus boiler equipment represents additional degrees of freedom and creates an opportunity to select the most economic set of boilers that could reduce the operating cost. Similarly there would be more BFW pumps than normally needed. Since each pump has different operating characteristics, an opportunity exists to select a set of BFW pumps to reduce the operations cost.

- **Dual Drives:** Typically a boiler house has spare drivers for some of the larger fans, pumps and compressors. Typically one of the drives is steam turbine while the other one is an electric motor. Such situations create opportunities for optimization; this is, at times use of a particular drive may be more beneficial than the other drive.

- **Complex contracts with external suppliers**: Most plants are not self sufficient in their energy needs and have to purchase energy from external suppliers. For electric power, almost all plants are tied to the electrical grid for reliability reasons. And for fuel, most plants have agreements with one or more fuel supplier.

  For purchase of electrical power, there usually are charges both for the use of electrical energy as well for what is referred to as "electrical demand". Electrical demand is the instantaneous consumption peak that occurred over a specified period of time, usually the previous 12 to 24 months. This means creating a new peak in electrical demand will entail penalty for the next 12 to 24 months. Some IPPs also have "Time of Use" contracts in which case the price of electric power varies with the time of use, usually significantly higher during the "peak hours" in a day and during the "peak season" in a year.

  For purchase of fuel, typically there are multiple suppliers. Some with long term contracts to have a reliable supply for which one would usually pay a premium; for such supplier there usually would be a minimum "take or pay" type of agreement. For short term suppliers, depending on the local natural gas market, there could be volume discount or penalty for excess purchase.

- **IPP as Cost Center:** In an industrial setting, the IPP is usually a cost center and their cost is prorated amongst the various process units. Such an arrangement, many a times leads to short term thinking and discourages investment to improve the efficiency of IPP, which is unfortunate since energy costs in some industries are second only to raw material cost [1].

- **Flowsheet simplicity**: Compared to typical production processes, an IPP has a considerably simpler flowsheet. Other than steam and power generation, other unit operations could be modeled as simple mixers and splitters, there are no chemical reactors and distillation columns. In terms of number of components, we have just one in the steam subsystem.

- **Equipment performance curves**: Detailed, first principles modeling of steam and power generators (boilers, turbro-generators, GT/HRSG etc) could be complicated. However, for operations optimization of existing system, detailed modeling is not justified and instead the usual practice is to use equipment performance curves, such as boiler efficiency curves, turbine water ratios and Gas Turbine iso-charts to predict their behavior in the normal operating range.

In summary, characteristics of IPP are simplicity in flowsheet but complexity in purchased energy contracts and opportunities in equipment selection. These characteristics have profound implications on the design of an operations optimizer for an IPP.

IV. THE OPTMIZATION PROBLEM

The optimization problem comprises determining the most economical (usually least cost) operations policy that satisfies the energy demands of the process units and is consistent with external supplier contracts.

From an implementation perspective, the optimizer determines the set of desired values for the decision variables that gives the most economic operation. There are several decision variables in an Industrial Power Plant. These can be broadly classified in two groups;

- **Continuous decision variables**: are variables that could operate over a range, such as boiler and turbo-generator loadings. From an optimization point of view, continuous variables are simpler to handle as the traditional optimization methods apply to problems with continuous variables.

- **Discrete decision variables**: are variables that could take only discrete values such as the number of boilers in service. The simplest of the discrete variables are the "binary" variables that could be one of possible two values. Examples include ON/OFF status of a boiler and selection of turbine versus motor drive for a dual drive pump. From an optimization point of view, discrete decision variables are computationally demanding. Although systematic procedures for discrete optimization exist, each additional binary decision has the potential to double the computation time.

In most cases there will be many alternate ways to operate the Industrial Power Plant that will all satisfy the process energy demands and be consistent with other operating constraints. The optimization problem then is to find the operations policy that is the best amongst all such feasible alternatives.

The best operating mode may not always be obvious. Past experience could help but not always. To gain an appreciation of the problem, let us consider a simple scenario – with a steam subsystem that has only three steam headers, a HP boiler, a condensing turbo-generator CTG, two pressure reducing stations – PR1 and PR2 between steam headers, and a deaerator as shown in Figure 5.

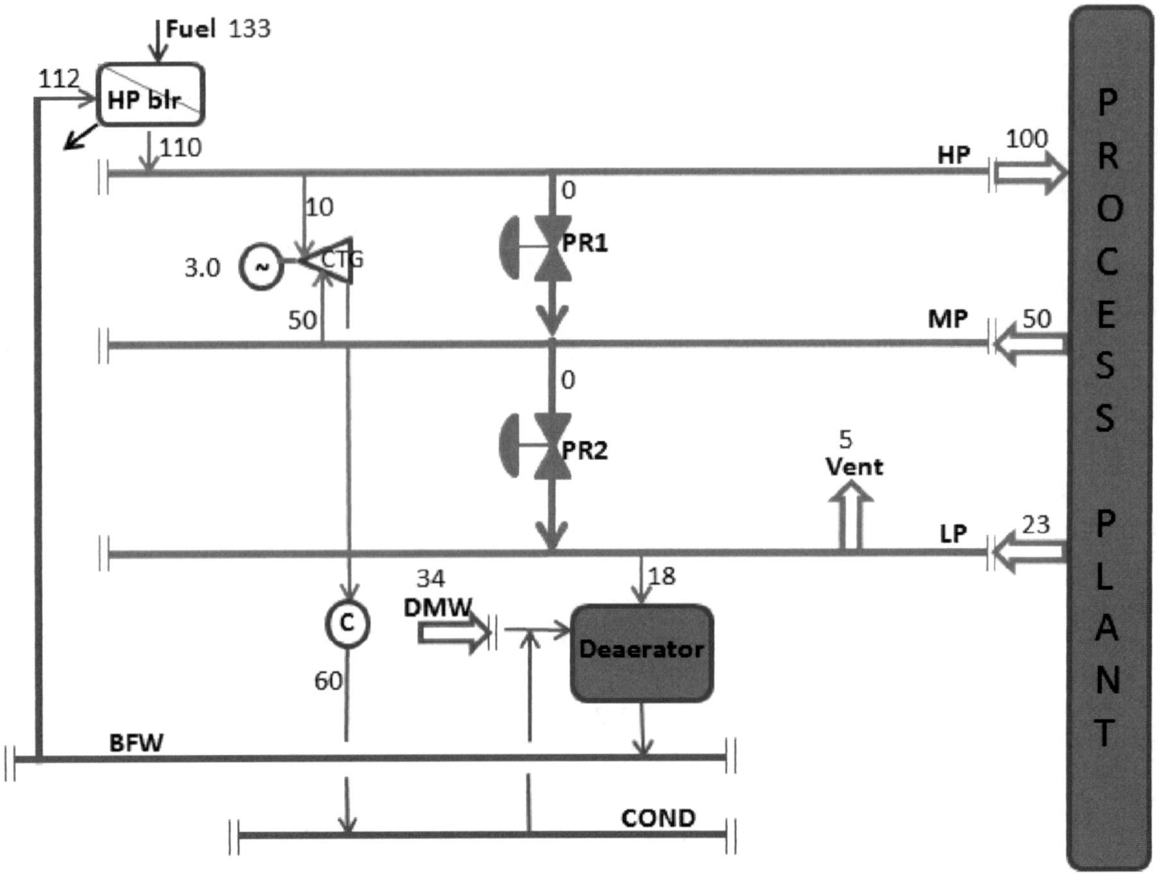

Figure 5: Simple steam subsystem operation: Base case

Note the following equipment constraints;

	Min	Max	EU
HP Boiler	100	300	MPPH
STG – HP inlet	10	85	MPPH
STG – MP inlet	20	50	MPPH
STG – generation	2	10	MW

And the process demands are as follows;

	Demand	EU
HP	100	MPPH
MP	-50	MPPH
LP	-23	MPPH

Base case

Figure 5 shows the base case operating conditions, the boiler produces 110 MPPH HP steam and the turbine throttles 10 MPPH HP steam and 50 MPPH MP steam from the process. Note that the boiler fuel consumption is 133 MMBtu/hr. and power generation is 3 MW. Note that there is steam vent of 5 MPPH, which is an obvious inefficiency and one may seek alternatives to prevent it.

Alternative A

Figure 6 shows an alternative to prevent steam venting by using the surplus LP steam to generate additional boiler feed water that is used to generate additional steam which is expanded through the condensing turbo-generator. Note that the boiler fuel consumption increases to 172 MMBtu/hr. and power generation increases to 6 MW. The venting of steam is stopped. Now, we are consuming little more fuel and producing little more power; this could be profitable if electrical power is a higher value commodity such as during peak hours. In such a case, one may wish to increase power generation further.

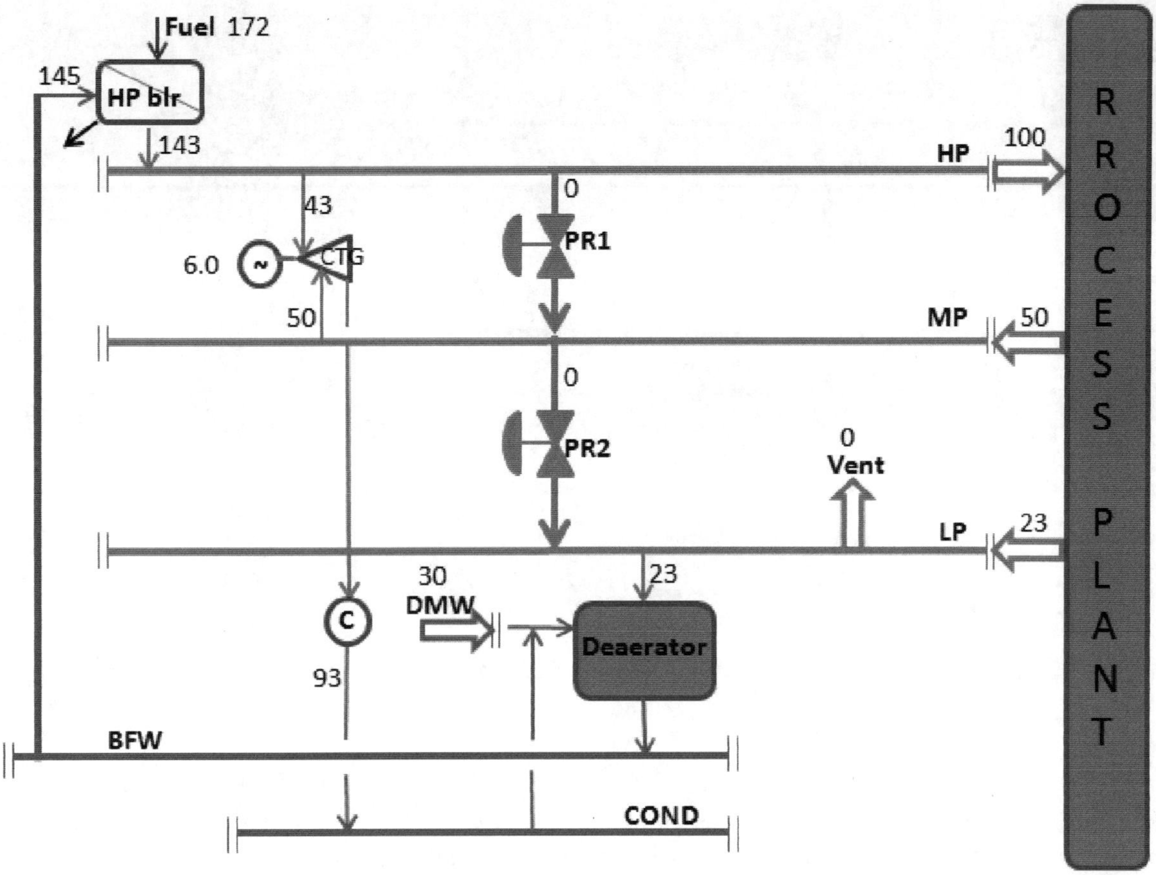

Figure 6: Simple steam subsystem operation: Alternative A

Alternative B

Figure 7 shows an alternative operation to increase power generation by generating additional steam and letting it down through the turbo-generator. Obviously the BFW demand and the deaerator LP steam demand will also increase which could be met by letting down 5 MPPH through PR2. Note that the boiler fuel consumption increases to 211 MMBtu/hr. and power generation increases to 8.6 MW. Depending on the purchased energy costs, one may wish to increase power generation further to the limits of the turbo-generator.

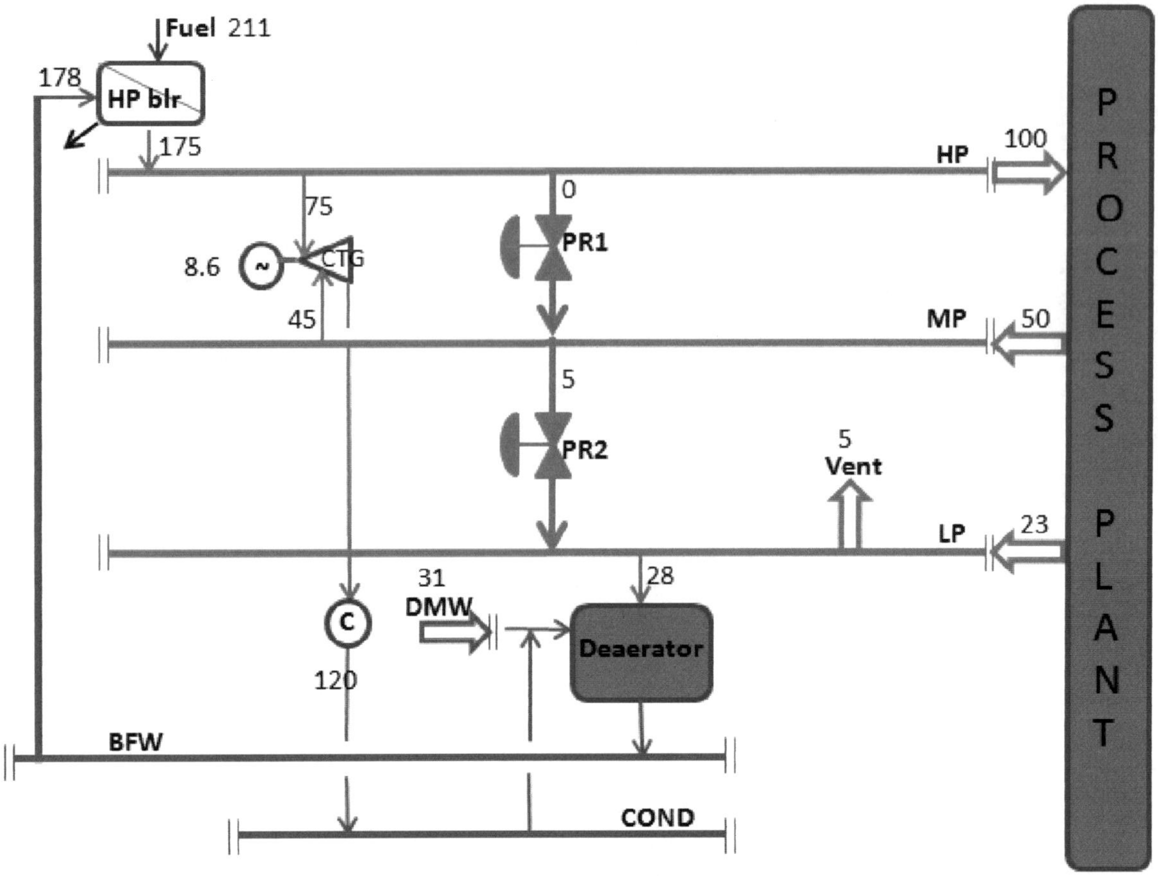

Figure 7: Simple steam subsystem operation: Alternative B

Alternative C

Figure 8 shows an alternative operation that maximizes power generation by generating additional steam and admitting it to the turbo-generator. Obviously the BFW demand and the deaerator steam LP steam demand will also increase which will now come by letting down 8 MPPH HP steam all the way down to LP. Note that the boiler fuel consumption increases to 233 MMBtu/hr. and power generation increases to 10 MW which is the power generation limit of the condensing turbo-generator.

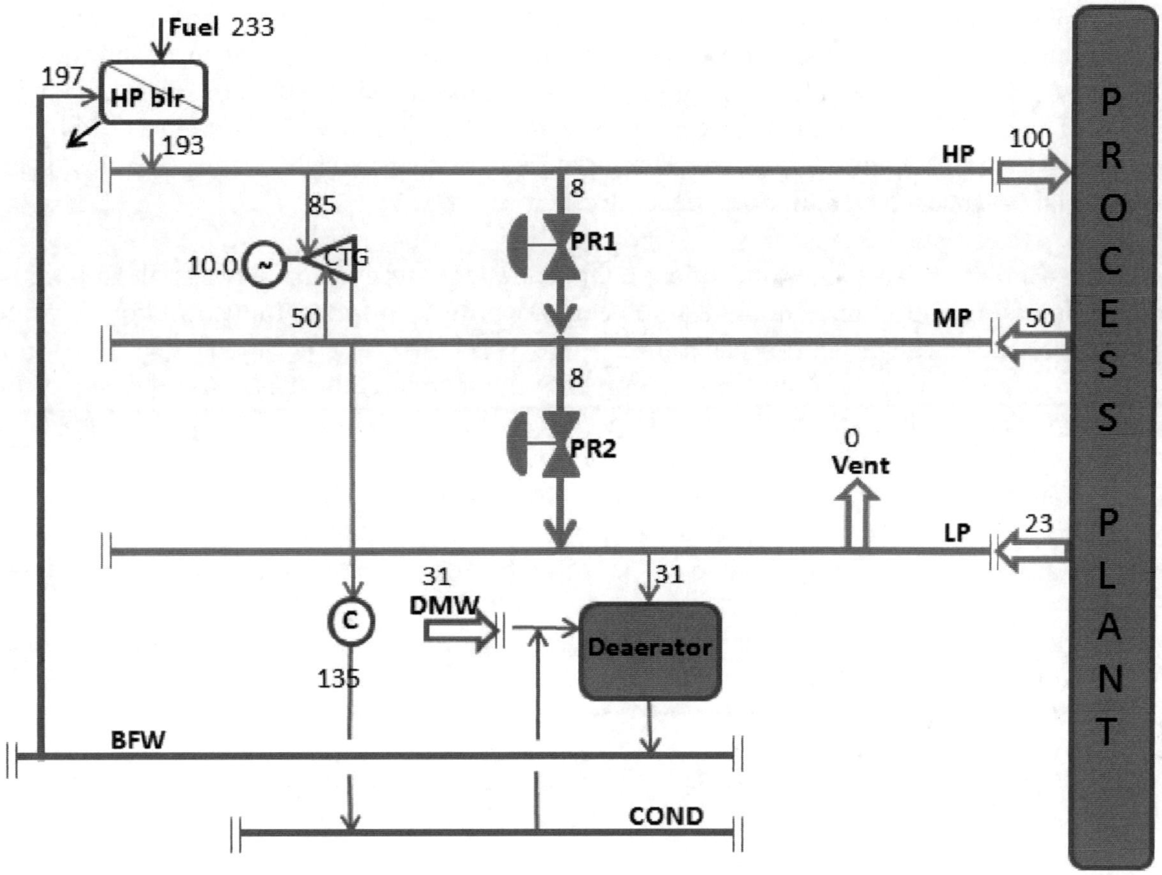

Figure 8: Simple steam subsystem operation: Alternative C

Note that each of these alternatives is feasible because the steam demand of process units is satisfied and the turbo-generator and the boiler are operating within their respective operating limits.

One may ask, "Which of these four operations policy is the best?" Answer to this question, is that any of these alternatives may be the best, it all depends on the relative cost of fuel and power. For relatively inexpensive power, the base case would be the best. As the relative cost of power increases (or fuel cost decreases), the alternative A will become the best. As the power cost increases further (or fuel cost decreases further) alternative B will become the optimum. And for even higher power cost (or lower fuel cost) the alternative C will become the optimum.

The example shown here has been simplified for clarity; it only has 2 degrees of freedom, has simple pricing for fuel and power and no surplus equipment. And even here the optimum is not obvious. For an industrial system, with multiple continuous degrees of freedom, complex contracts and multiple discrete degrees of freedom corresponding to surplus equipment, the problem can quickly become overwhelming and beyond manual optimization; a computerized solution will become a necessity to obtain a definitive solution in a reasonable amount of time.

V. SOLUTION TECHNOLOGY

In simple terms, optimization, is a search task. It is a search for the most desirable feasible operation amongst all feasible operations. Most desirable is usually defined in economic terms, for example for an Industrial Power Plant operation, the best is usually defined as the least expensive mode of operation. The constraints define as to what is operationally possible. Constraints could be as simple as the operating range of equipment or complex relationships that exist between variables, such as the heat and material balances for a boiler or a deaerator.

For most practical problems of modest complexity, the number of all possible solutions is still very large. For example, optimization with fixed equipment line up for an Industrial Power Plant with 4 boilers and two back-pressure turbo-generators there would be 6 continuous decision variables – corresponding to the loading of individual equipment. Mathematically it is a 6 dimensional problem. So optimization is a search in a 6 dimensional space, it is like finding a needle in a 6 dimensional hay stack – certainly a daunting task. However, if we now consider the flexibility in equipment lineup changes we would add another dozen or so discrete decision variables corresponding to 4 boilers, 2 turbo-generators, and perhaps half a dozen dual drives for various pumps and compressors in the IPP; thus creating a search for the needle in 4096 (2^{12}), 6-dimensional hay stacks, by no means a trivial task.

Fortunately, the mathematicians over the years have developed some very efficient search methods, called optimization algorithms [2]; that computer scientists and engineers have coded in computers for use by non-mathematicians.

Mathematically, an optimization problem is;

$$\text{MINIMIZE } F(X, Y) \qquad (1)$$

Subject to

$$G(X, Y) \gtreqless RHS \qquad (2)$$

$$Xmin \le X \le Xmax \qquad (3)$$

$$0 \le Y \le 1 \qquad (4)$$

Equation 1 is the objective function, F(X, Y), which specifies the cost function to be minimized; for example, it could be the total cost of purchased fuel and electrical power by the Industrial Power Plant. Equations 2 are the constraints, G(X, Y), that define all relevant relationships between the variables; for example, boiler and turbine performance curves, material balances on the steam and fuel headers, energy balances on the deaerators etc. Note that these relationships usually are equalities. Lastly, equations 3 and 4 are the bounds on variables, for example, the operating range of equipment, flow limits through pressure reducing stations etc.

Basically, there are two types of optimization problems

Linear Programming

An optimization model is linear if the objective function F and constraints G are all linear expressions, that is, all equations describing the various relationships are linear.

An expression is linear if each term in the expression is a constant times a variable. The following is an example of linear expression,

$$A1 * X1 + A2 * X2 + A3 * X3 + A4 * Y1 + \ldots$$

Linear optimization problems are also called Linear Programs (LP). The technology of solving LPs is very advanced and LPs can be solved very efficiently with guaranteed globally optimum solutions. Global optimum means that a better solution does not exist which is very reassuring. A variety of commercial solvers are available [3]. Such solvers are robust and many are fairly inexpensive.

Non Linear Programming

An optimization model that has at least one non-linear term in the objective function F or one or more of the constraints G is a non-linear optimization problem, also called a Non Linear Program (NLP).

An expression is non-linear if any term in the expression is anything other than a constant times a variable. Following are two examples of non-linear expressions;

$$A1 * \sqrt{X1} + A2 * X2 + \ldots$$
$$A1 * X12 + A2 * X1 * X2 + A3 * \log(X3) + \ldots$$

The technology of solving NLPs although has advanced significantly, is not as robust as that for LP. In general. Solution to NLPs is not as efficient and the solution is only a local optimum. Local Optimum means that is there is no guarantee that a better solution does not exist, which is not very reassuring. Although a variety of commercial solvers are available, they all suffer from the aforementioned difficulties.

Integer Variables

So far the optimization discussion in this section has been limited to continuous variables; that is the variables could take fractional values, such as 0.4 and 1.928. However as mentioned earlier IPPs have spare equipment so equipment selection variables need to be included in the optimization variables. Selection variables are the simplest integer variables as they could either be 0 or 1 corresponding to "on" and "off" states. Such variables are called Binary variables. General Integer variables that could take integers values over a range could be modeled using multiple Binaries. So we will limit the remainder of the discussion to binary integer variables. In the above formulation the binary variables are represented by symbol Y and are restricted to take value of either 0 or 1.

Binary variables add a new dimension to modeling and make the following possible;

- Equipment Selection (ON/OFF)
- Equipment operation with multiple states, such as ON, OFF or stand by
- Discontinuities in equipment operation, such as high range operation and low range operation.
- Minimum up time or downtime requirements, such as after start up a boiler must continue to operate for a minimum specified period
- Delay between events, such as start up and shut down time which is useful in multi-time period modeling, and
- Other operations logic

General discussion on the use of binary variables in modeling is beyond the scope of this paper. Note that the first three uses may be applicable to IPP modeling and will be discussed in the next section.

Although the binary variables add modeling capabilities that would be impossible with continuous variables alone; addition of binary variables to an optimization problem comes at a cost in terms of increased execution time. In the worst case, addition of a binary variable could potentially double the execution time as both options may have to be fully evaluated. So, one must be careful in the use of integer variables. However, properly formulated models, even those with large number of binary variables can be efficiently solved in a reasonable amount of time [4].

Use of binary variables in an otherwise LP renders the optimization problem to be a Mixed Integer Linear Program (MILP). Similarly use of integers in an otherwise NLP renders it a Mixed Integer Non Linear Program (MINLP).

Both LP and NLP solvers could be expanded to accommodate integer variables. Commercial software for both MILP and MINLP are available. Similar to their continuous counterparts, global optimum can be guaranteed for MILP, but not for an MINLP.

VI. CASE STUDY

Scope

This case study is based on a recently completed IPP optimization project for a US refinery. The scope of optimization was limited to the Boiler House and included the following;

- Energy purchase: steam, fuel and electricity
- HP and MP Boilers and auxiliaries
- Steam headers, fuel header and Electrical Bus
- Let down stations
- Blow down flash
- Deaerators
- BFW pumps, BFW pump turbines and motors
- Let down stations, vent and flare

The Boiler House includes three High Pressure (HP) and two Medium Pressure (MP) gas fired boilers. In addition, the plant purchases High Pressure steam from a neighboring Cogeneration plant. Steam from the Cogeneration plant is a little cheaper but it is also somewhat less reliable. In addition to the HP and the MP steam headers, there is also a Low Pressure (LP) header that supplies to the Process Plant. Most auxiliary pumps and fans in the Industrial Power Plant are turbine driven while a few are motor driven which offers some flexibility in balancing of steam headers. The process plant is usually a net importer of steam, however occasionally it exports LP steam to the Industrial Power Plant which can at times cause venting of LP steam in the steam subsystem.

All boilers are gas fired; the HP Boilers are tied to the purchased Natural Gas header while the MP Boilers are tied to the refinery gas and the purchased fuel headers. Overall, the site is net importer of fuel gas from a local Natural Gas Company. The site is also a net importer of electrical energy from the local grid.

Continuous blow down from the boilers is collected in two Blow Down drums that generate LP flash steam.

Steam balance is maintained by use of Pressure Reducing Stations that let down HP to MP and MP to LP steam headers.

Boiler Feed Water is prepared in LP deaerators. Process condensate return is fed to the deaerators. Make up water comes from the Demin system.

There are two groups of Boiler Feed Water pumps. Each group has an electric and two turbine driven pumps.

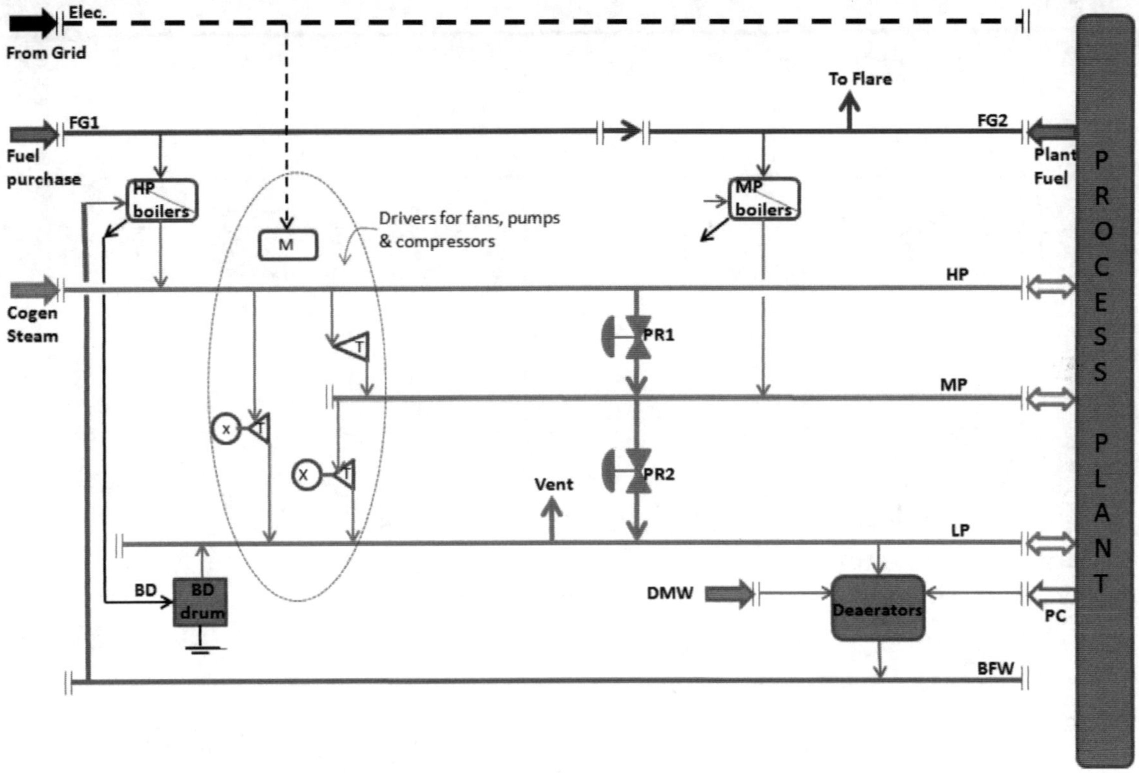

Figure gives a simplified schematic of the Refinery Boiler House.

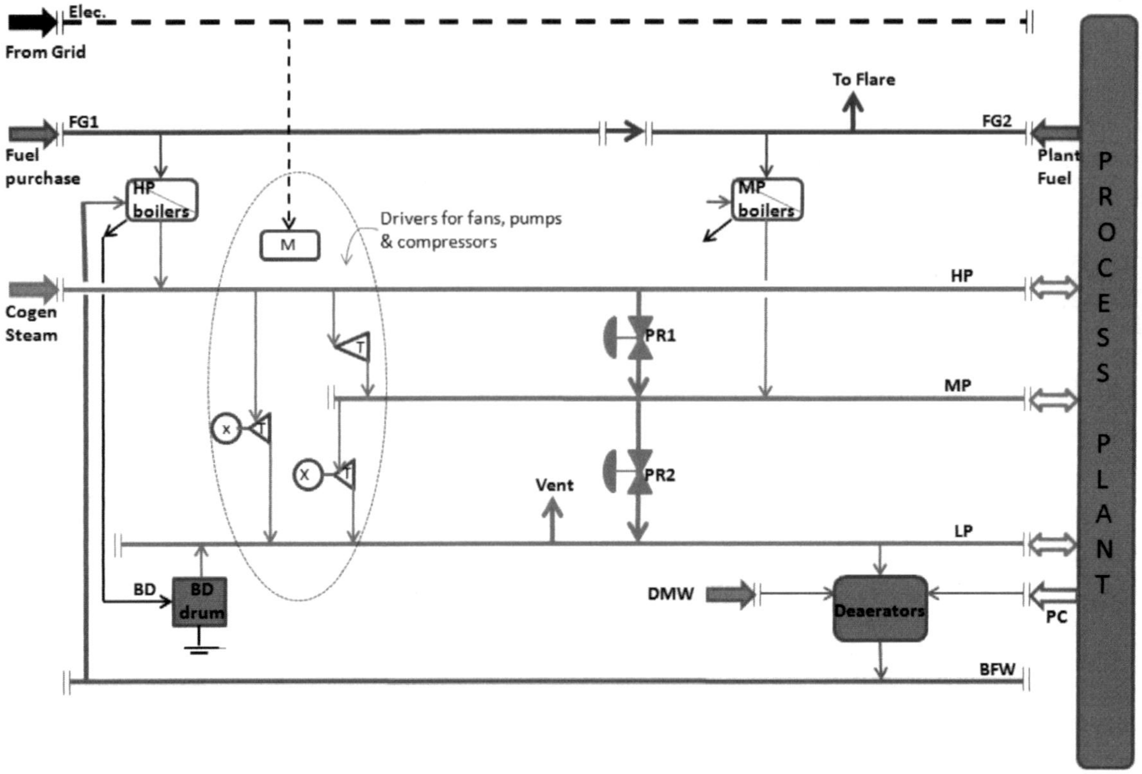

Figure 9: Case Study: IPP Boiler House – simplified schematic

The complexity and uncertainty associated with Cogeneration steam purchase, the inherent variability in process plant steam demand and flexibility in the operation of the auxiliaries, collectively create an environment where determination of minimum cost operation without additional risk is not straight forward. An automated optimization tool that could quickly analyze the situation and recommend optimal operations policy would be a valuable tool for more economical operation of the boiler house.

With this in mind, the Refinery had contracted Honeywell Process Solutions to design, configure and implement an Economic Load Allocation (ELA) Optimizer for the Boiler House. The ELA Optimizer has been designed for an on-line, open-loop execution that provides advisory information to the operations staff. Its execution is at a high frequency, once every two minutes, so the results of ELA Optimizer are always fresh. All data communication to and from the ELA application is via the Honeywell DCS. The Refinery operations staff is responsible for reviewing the ELA optimizer recommendations and making the ultimate operations changes.

The Refinery Boiler House operation usually runs with surplus steam generation capacity so as to be prepared for possible Cogeneration steam supply disruption.

Optimizer Modeling

Equipment operations data was collected and Mixed Integer Linear models for each of the equipment in the scope of the Optimizer were configured. An MILP model represented the operations quite well. In addition, several tuning constants were built in the model and were used to fit the model to actual operation in the first step of the execution cycle; this way the model maintains good fidelity.

One particular concern in this case was the reliability of steam import. This concern was mitigated by adding reliability constraints which ensured that the surplus steam production capacity was at least equal to the steam import plus a desired excess; a similar constraint was applied to boiler feed water capacity. The overall result of these constraints was that the optimum operation could gracefully handle any disruption in supply of imported steam by quickly making up for the loss by utilizing the surplus capacity in boiler feed water and steam generation.

Another aspect of optimization modeling was the recognition that making discrete changes to equipment lineup is a more deliberate endeavor which requires careful consideration of other factors. For this reason, the following three optimization problems were solved in each execution cycle;

- *COPT*: for optimization of only the continuous variables with current equipment line up fixed. This solution is suitable for immediate implementation manually by the operator or via automation.

- *POPT*: for optimization of continuous variables plus added flexibility to be able to change the pump lineup. The motivation behind this formulation is the recognition that implementing pump lineup change is easier then implementing boiler lineup changes. POPT recommendations need consideration only when the additional benefits beyond COPT are worth the additional trouble.

- DOPT: for optimization of all continuous variables as well as all discrete variables. DOPT case had the additional flexibility to change the boiler lineup. DOPT recommendations need consideration only when the additional benefits beyond POPT are worth the additional effort.

On-line Execution

In absolute terms, Industrial Power Plant optimization is a relatively small optimization problem and execution is extremely fast. The three Optimization cases mentioned earlier were solved every two minutes and the results of optimization were presented to the operators on the DCS. Figure 10 summarizes the optimization execution cycle.

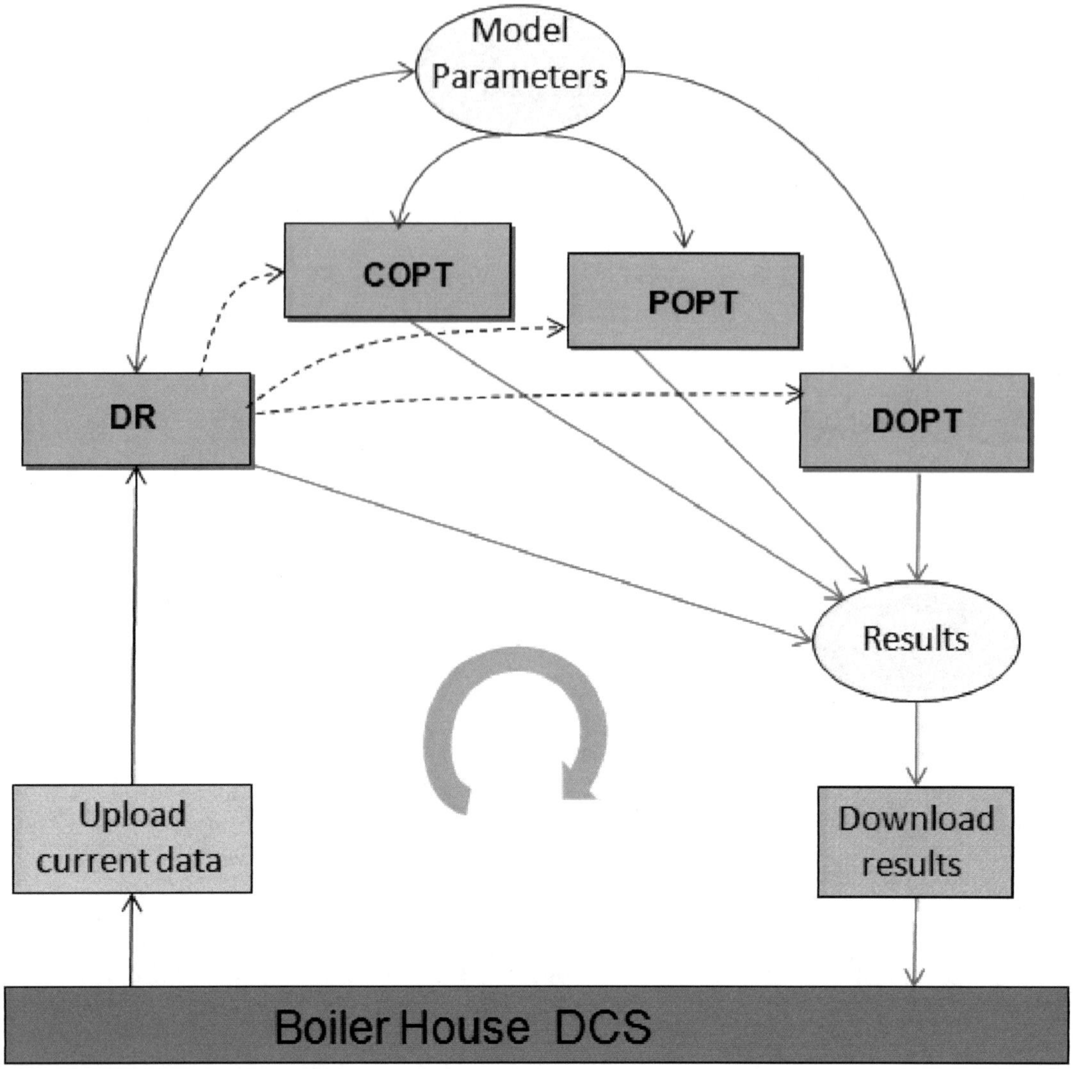

Figure 10: Case Study: IPP Boiler House – ELA Optimizer Execution Cycle

User Interface

All operator interactions with the Optimizer were via an Operator Station on the DCS. Custom graphics per client specification were configured that displayed both the inputs and the outputs. Economic parameters were also displayed on the input graphics that could be modified by the operator. The results of the three optimization runs (COPT, POPT and DOPT) were shown side by side along with the current operations so the operator could clearly see the optimization recommendations and make the necessary moves.

Summary and engineering reports are also produced on the engineering station that are accessible to the engineers from anywhere in the plant.

Optimizer Performance

Benefits of Optimization have been two folds;

- *One time Benefits*: During the modeling stage the current operating practices were thoroughly reviewed and examined. This led to a better understanding of the current operations and some modifications in operating philosophy were made which resulted in quick improvement in certain operations.

- *Continuous Benefits*: The optimizer runs every two minutes so it responds quickly to changes in operating conditions and identifies opportunities as they arise. It clearly summarizes the expected benefits and the way to realize them. The optimizer has been well received by the operators because of the ease of it's of use. It has been integrated in the daily work flow and has resulted in a more streamlined operation of the refinery boiler house. The economic benefits have been estimated to be about 1% reduction in the Industrial Power Plant operating cost.

Future Plans

The boiler house is in the midst of an upgrade; a new boiler is being commissioned and two of the older boilers are to be decommissioned. The optimizer model has been designed with these changes in mind and should continue to operate in the new configuration with minimal changes, if any. It is expected that the new configuration will have more optimization opportunities and then closing the loop with COPT results may be worthwhile.

VII. CONCLUSIONS

Operations optimization of Industrial Power Plants is often overlooked for a variety of reasons. However, for medium to large industrial production facilities, energy usage is the most significant cost after raw material cost, so operations optimization of the industrial power plants is well worth the effort. Non-traditional approach, such as MILP, is well suited for operations optimization of industrial power plants as has been demonstrated in a recently completed Refinery IPP Optimization project.

ACKNOWLEDGEMENTS

We would like to thank J. Schilling, J. Charr, N. Badgujar, D. Boudreax and P. Lynch for their support of this project. We would also like to thank the management of Honeywell for permission to publish this work.

ABBREVIATIONS

A	Constant coefficient
BD	Boiler Blow Down
BFW	Boiler Feed Water
COND	Condensate
COPT	Continuous Optimization
CTG	Condensing Turbo-Generator
DCS	Distributed Control System
DMW	De-Mineralized Water
DOPT	Discrete Optimization
DR	Data Reconciliation and parameter estimation
ELA	Economic Load Allocation

EU	Engineering Unit of measure
F	Objective Function
FG	Fuel Gas
G	Constraint equation
GT	Gs Turbine
HP	High Pressure
HRSG	Heat Recovery Steam Generator
IPP	Industrial Power Plant
LP	Low Pressure or Linear Programming
MILP	Mixed Integer Linear Programming
MINLP	Mixed Integer Non Linear Programming
MMBtu	Million (10^6) BTU
MPPH	Thousand (10^3) Pound Per Hour
MW	Mega (10^6) Watts
MP	Medium Pressure
NLP	Non Linear Programming
POPT	Pump Optimization
PR	Pressure Reducing station
RHS	Right Hand Side (of an equation)
STG	Steam Turbo-Generator
X	Continuous variable
Y	Binary variable

REFERENCES

1. http://en.wikipedia.org/wiki/Energy_in_the_United_States
2. Luenberger, D. G., "*Linear and Nonlinear Programming,*" Addison-Wesley (1984).
3. Anonymous, "*LINGO User's Guide*", Lindo Systems Inc, (2011).
4. Salbidegoitia, I. et. al, "*Operations Optimization of Concentrating Solar Power Plants*", paper presented at the AIChE Spring Meeting, Houston, TX (2013).

(26a) Sulphur Technology Tutorial – Conventional Wisdom and Innovative Frontiers

David Stevens, KPS Technology & Engineering, Overland Park, KS and **Angela Slavens**, WorleyParsons, Monrovia, CA

This tutorial will provide a brief overview of conventional sulphur recovery, tail gas treating and sulphur degassing technologies which are routinely employed to achieve environmental and operational requirements. The intent is to provide participants with a broad understanding of the types of processes which are available, and how they are typically applied, as a foundation for exploring possibilities for improvements upon these traditional practices.

Following the review of conventional technologies, an exploration of some recently-developed, innovative technologies will be undertaken. The technical and commercial merits of each new technology will be compared to conventional technologies, and the advantages / disadvantages will be explored. Examples of some of the new technologies to be presented include:

- SMARTSULF – this technology applies the sub-dewpoint principle, which is widely-known to increase sulphur recovery efficiencies in the SRU due to operation of the Claus reactors at lower temperatures, which increases the extent of the exothermic Claus process. The SMARTSULF technology further improves sulphur recovery through the use of internally-cooled reactors, which further reduce catalyst bed temperatures and therefore increase sulphur recovery efficiency.
- Low Temperature Tail Gas Oxidation Process – with this technology, the conventional fuel gas fired incineration process is omitted and indigenous hydrogen and carbon monoxide in SRU tail gas are consumed in a catalytic oxidation process, avoiding additional fuel consumption.
- In-Situ Sulfur Degassing Process – this technology removes the dissolved H2S present in Claus-produced sulfur within the sulfur condenser itself, as opposed to external, conventional degassing methods, which may require additional plot space, utility consumption and/or which may pose HSE concerns.
- RSR Production of Elemental Sulphur from SO2 – this technology converts post-combustion SO2 into an elemental sulphur product, rather requiring H2S as the feed to the sulphur production process. The technology combines two well-established processes: 1) reaction of CH4 and sulphur vapor to produce CS2, followed by 2) catalytic hydrolysis of CS2 to H2S, along with Claus reaction of H2S and SO2 to form elemental sulphur.

Successful Sulfur Unit Startups

Doug Cicerone
Cicerone & Associates, LLC
1302 Waugh Drive, Suite 248
Houston, TX 77019 USA
doug.cicerone@gmail.com

Prepared for Presentation at
American Institute of Chemical Engineers
2013 Spring Meeting
San Antonio, Texas
April 28-May 2, 2013

1. Introduction

The startup of any facility can have a big effect on the project or turnaround schedule and budget. As facilities become more integrated, the smooth startup of each unit within the facility becomes more important as it directly affects other units. Sulfur Recovery Units (SRUs) are often critical for the overall facility startup as their operation is a requirement from an environmental permitting standpoint.

This paper gives general guidelines and information for the safe and successful startup of an SRU. These guidelines do not supersede any site or vendor procedures or guidelines. Specific startup procedures must be developed for each facility prior to startup.

2. Sulfur Recovery

Although many sulfur recovery and tail gas cleaning technologies exist, only a standard straight-through Claus process will be discussed. (See reference [1] for more details on the sulfur recovery process.)

2.1 Process Chemistry

The modern Modified Claus process for the conversion of hydrogen sulfide (H_2S) to elemental sulfur (S) was first used in 1936. The process is divided into a high temperature thermal conversion section and a lower temperature catalytic conversion section. Although many reactions take place in the thermal section, the majority of the thermal conversion occurs according to the following two reactions:

$$\mathbf{H_2S + 3/2\ O_2 \rightarrow SO_2 + H_2O} \text{ [Eq. 1]}$$

$$\mathbf{2H_2S + SO_2 \rightarrow 3/x\ S_x + 2H_2O} \text{ [Eq. 2]}$$

In the catalytic section, the conversion of hydrogen sulfide occurs according to Equation 2 (also known as the Claus reaction) over a catalyst that is typically alumina or titania based.

2.2 Process Description

Figure 1 below shows a simplified process flow diagram for a typical sulfur recovery unit with two catalyst beds. The H_2S bearing feed gas (acid gas) flows to a Reaction Furnace where it is combusted with air supplied by a Combustion Air Blower. The air is metered such that one-third of the H_2S is burned to sulfur dioxide (SO_2). The unburned H_2S reacts with the SO_2 according to Equation 2 and forms elemental sulfur. Temperatures in the Reaction Furnace range from 1,500°F to 2,500°F depending on acid gas composition. Heat is recovered in the Waste Heat Boiler, typically by generating high pressure steam.

Figure 1: Sulfur Recovery Unit Simplified Process Flow Diagram.

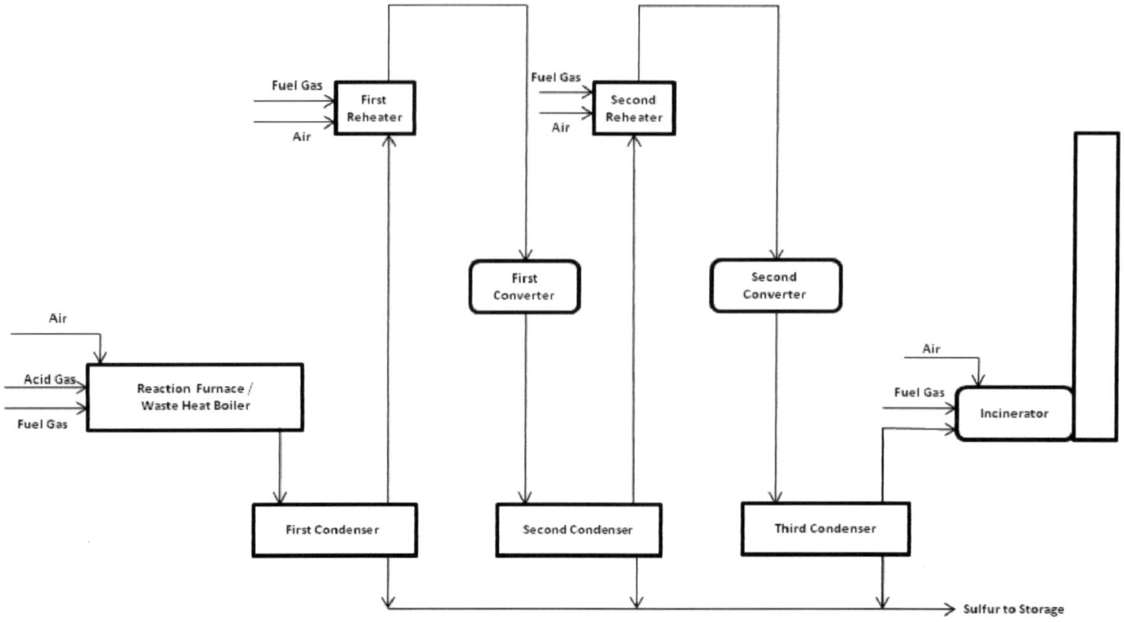

The process gas flows to a Condenser where the sulfur vapor is condensed to liquid sulfur, which is sent to storage. Heat is recovered in the Condenser, typically by generating low pressure steam. The process gas exits the condenser at its sulfur dewpoint and passes through a Reheater. This Reheater raises the process gas temperature above the dewpoint to prevent sulfur from condensing on the downstream catalyst. Sulfur condensation causes a reduction in conversion by blocking active catalyst sites. The heat for the Reheater is typically supplied by high pressure steam or by the direct firing of fuel gas or acid gas.

The heated process gas flows to the Converter where Equation 2 takes place across a catalyst bed. Temperatures in the Converter are typically in the range of 525°F to 625°F. The sulfur vapor is condensed in the downstream Condenser, and the Reheat, Conversion and Condensing steps are repeated.

The process gas from the Final Condenser (tail gas) flows to the Incinerator where any remaining H_2S (along with any other reduced sulfur compounds) is oxidized to SO_2 before exiting to the atmosphere through a stack.

2.3 Process Safety

Note that this section is not in any way meant to be a discussion of all SRU safety concerns, and is only meant to give the background necessary for the remainder of the paper.

The principle safety concern in a sulfur recovery unit is personnel exposure to and environmental release of H_2S. The Threshold Limit Value (TLV) for H_2S is 10 ppm, and it is lethal in concentrations of 1,000 ppm and above. Although sulfur recovery units operate at low pressures (5 to 15 psig), H_2S concentrations are very high. Typical concentrations in the front end of an SRU are in the range of 50% to 90%, and in the back end of the unit they range from 1% to 2%. Consequently, any loss of containment of the process gas has the potential to cause injury or death.

Wind socks should be placed at high points throughout the unit to indicate wind direction. All personnel in an operating unit must be aware of the wind direction. If an H_2S release occurs, the plant evacuation should proceed upwind or crosswind. Stationary and personal H_2S monitors should also be used.

Burns are another safety concern as liquid sulfur is in the temperature range of 300°F to 350°F. Special care must be taken to prevent burns when checking the sulfur lookboxes.

3. Startup Planning

As early as possible during the detail design phase of a project, a Startup Plan should be developed. This plan should be refined as the project progresses through construction and approaches startup. The Startup Plan becomes the road map for turning the plant over from Construction to Operations. It should address issues such as:

- Organizational Chart
- Matrix of Responsibilities
- Overall Schedule
- Budget
- Spare Parts
- Acceptance criteria for mechanical and electrical equipment
- Order in which utility and process systems will be completed mechanically
- Specific procedures for line flushing, steam bows, equipment run-ins, refractory dryout, etc.

This plan must be agreed upon by all parties involved prior to Startup activities. See reference [2] for detailed information regarding startup planning.

4. Precommissioning

Precommissioning is the project phase that occurs between construction and startup. During this period the equipment and piping are cleaned of the greases, dirt and debris that accumulate during manufacturing and construction.

The overall smoothness and success of a startup is almost always a direct result of the care taken by the commissioning team during precommissioning. While it can be very tempting to relax standards and cut corners in order to compress the schedule in the short term, these compromises will often have the opposite effect on the schedule in the long term.

For example, a proper instrument loop check includes the testing of all aspects of the loop, including controller action, valve stroke and DCS display. It is easier and takes less time to check only the point-to-point continuity of the control loop, which only ensures that the right instruments are wired correctly in the loop. This leaves the bulk of the instrument troubleshooting to occur during startup which can cause delays.

The following sections list items to consider with regards to the precommissioning of equipment, piping and instrumentation. Typically experienced operations, maintenance and engineering personnel are required to sign off on all items prior to the plant being deemed "ready for startup."

4.1 Piping & Equipment

General Considerations:

- Insides of all vessels cleaned of rust, dirt and debris.
- All vessel internals installed.
- Proper gaskets installed at flanges (and spare gaskets in warehouse).
- All fired equipment refractory installed and dust removed.
- All vessel slide plates free of debris and anchor bolts are loose or tight as per mechanical drawings.
- All motors greased.
- All motors checked for proper rotation.
- All seals installed with proper seal fluid.
- All lubricating systems full of oil.
- All temporary and/or permanent strainers installed.
- All compressor and pump alignments checked.
- Filter elements installed (and spares in warehouse).

Special Considerations:

- Ensure the manufacturer's recommended spare parts for all rotating and fired equipment are in good condition and available in the warehouse. This is particularly important if the site is in a remote location.
- When flushing or blowing pipe and equipment, ensure control valves and regulators are protected by removing or isolating. Orifice plates should also be removed when flushing.
- All utility and process piping should be blown with air or nitrogen to remove dirt and debris. Blinds should be used to prevent the transfer of dirt from one section to another or to equipment.
- For the boiler feedwater and cooling water systems, water can be used to flush the piping.
- For the condensate systems, condensate should be back-flushed to the downstream side of the steam traps.
- For the steam systems, steam should be used to blow the lines (and special steam blow procedures may be necessary if the system includes steam turbines).
- The catalyst support grids and screens in the Converter vessels must be installed properly such that catalyst cannot slip between the grid/screen and vessel wall.
- The sulfur pit should be cleaned of all debris. Any standing water should be pumped out.

4.2 Electrical & Instrumentation

General Considerations:

- All switchgears and breakers tested.
- Heater settings verified for motors.
- Acceleration times set for motors.
- All transmitters calibrated.
- All control valves stroked.
- All orifice plates installed in proper orientation.
- All PSVs tested and installed with block valves car sealed or locked as necessary.
- All instrument loops properly checked.

Special Considerations:

- Machinery monitoring and anti-surge systems for air blowers and compressors must be set up, typically by the vendor.
- The Burner Management System for each piece of fired equipment must be functionally checked prior to the introduction of fuel gas. This means verifying each line of the Cause & Effect Diagram with all automated valves moving to their proper positions. Additionally, any communication between PLC systems and DCS systems must be fully checked.
- The flame scanners on the Reaction Furnace, Incinerator and any other fired equipment may require calibration as the burners are fired on fuel gas. They may need additional calibration when acid gas is introduced. Ensure the calibration procedures are understood and/or vendor support is available.

4.3 Refractory Dryout & Boilout

The Reaction Furnace, Incinerator and any fired heaters will contain refractory to protect the vessel shell from the heat of the flame. For small units, the refractory may be installed and dried in the shop. For larger units, the refractory is installed and dried in the field.

The dryout can be done with the Reaction Furnace burner or by a third party dryout company. The third parties use their own burners and normally exhaust out of a manway on the furnace or boiler. The advantage to using the Reaction Furnace burner is that it will allow for run time on the burner and Burner Management System before startup. The advantage to using the third party is that they can normally provide tighter temperature control.

The dryout procedure includes periods of heating up slowly, followed by periods of holding temperatures. These heat-up and hold periods allow water contained within the refractory to boil off without damaging the refractory. Heating up too quickly, or shortening the hold times, can cause refractory damage due to the rapid escape of the water.

The refractory vendor will supply the specific procedure for drying out their refractory. A typical procedure increases the temperature at 50°F per hour with hold points at 250°F, 500°F and 1,000°F, holding each temperature for one hour per inch of refractory. Cool down rates are typically 50°F to 100°F per hour. For example, a Reaction Furnace with 14 inches of refractory, the dryout will take about three days for the heat-up and hold points, and another three or four days to cool down to ambient. The cool down proceeds at the prescribed rate to the discharge temperature of the Combustion Air Blower, after which it cools down to ambient much more slowly. It is advisable to enter the vessels and inspect all of the refractory after dryout. This will allow for repairs if necessary.

If the Reaction Furnace burner is used for the refractory dryout, it is common to perform a boilout of the Waste Heat Boiler and Condensers at the same time. Otherwise the boilout is done just before startup. The boiler vendor will provide a detailed boilout procedure. It typically calls for a boilout solution of detergent and alkaline cleaning agents to be added to the boiler feedwater in the boilers. This boilout solution helps to clean any greases or oils that have accumulated during manufacturing and construction, and ensures the heat transfer surfaces are clean. This solution is flushed out with boiler feedwater during the boilout. It typically takes three to five volume change-outs to flush out the solution.

The refractory dryout of the Incinerator and fired Reheaters will follow the same basic procedure, although the hold and end temperatures will vary based on the normal operating temperature. Often the Converters have several inches of refractory for fire protection. This can be dried out as the Reaction Furnace is dried out, or as part of the startup.

4.4 Catalyst Loading

Catalyst is typically loaded as the last step of precommissioning in order to protect it from becoming fouled with dirt and debris from line blowing, and soot from the refractory dryout. There are usually at least two layers of catalyst, with a larger diameter support layer on the bottom, and smaller diameter catalyst balls on top of the support. Some catalyst systems use two types of catalyst, and some include a larger diameter hold-down layer on top. Prior to loading, markings should be made on the Converter vessel walls to show the heights of the layers. These markings are used when loading.

The catalyst vendor will supply a loading procedure. The procedures typically call for the use of a hopper and sock. The catalyst is dumped into the hopper, and flows through the sock into the Converter vessel. The goal is to limit the free-fall of the catalyst balls to prevent them from breaking apart.

Periodically during the loading process, the catalyst should be leveled inside the converter. Catalyst loading can create a lot of dust, so the vessel needs to be ventilated and workers in the vessel must wear proper respirators. Plywood can be used to facilitate walking on the catalyst. When the loading is complete, the catalyst bed should be a uniform level across the length and width of the vessel.

When the catalyst loading is complete, the beds will still contain dust generated during the loading process. This dust should be blown out of the unit prior to startup. This is normally done using the Reaction Furnace Air Blower to blow air through the unit. It is preferable to expel the dust from each converter upstream of its condenser. If this is not possible, then the dust can be blown through each condenser and exhausted through the rundown line. In this case, the sulfur seals must be blinded to ensure the dust does not enter them.

5. Startup

Once the piping and equipment cleaning, instrument checks, refractory dryout and catalyst loading have been completed, the unit is "ready for startup." At this point there are several final preparations that are necessary before the SRU can be heated and acid gas admitted.

5.1 Final Preparation

The following are items to consider while preparing the unit for Acid Gas.

- Ensure all firefighting and safety equipment (including breathing air and/or rescue packs) is in place and in good working condition.
- Heat up all steam lines draining condensate from boots and low points. Do not line up steam traps until the condensate drains clear.
- Warm up the Sulfur Pit steam coils 24 to 48 hours before startup in order to boil off any water that has collected in the pit.
- Warm up all steam jackets and steam tracing at least 24 hours before startup, and check all steam traps for proper operation.
- The fuel gas and acid gas systems should be purged up to the final block valves and have less than 0.5% oxygen (typically nitrogen is used to purge).
- The acid gas system from the battery limit to the Reaction Furnace should be leak checked at normal process pressure (normally around 15 psig). All flanges and instrument connections should be taped and checked with soapy water solution. The same leak check should be performed on the process gas pipework and equipment from the Reaction Furnace to the Incinerator.
- If seal pots are used, it is recommended that they be filled with sulfur prior to startup. Solid sulfur pellets can be used, which then melt when the seal pot jacket steam is turned on.
- Ensure all boilers are full of boiler feedwater and boiler feedwater treatment systems are ready for operation.
- Ensure the Burner Management System and all other Safety Systems are fully functional and all bypasses have been removed.
- If possible, analyze samples of the fuel gas and acid gas to determine the stoichiometric air ratios for each. For fuel gas the target is 95% to 98% of stoichiometric. For acid gas, the target is to burn 1/3 of the H_2S and all of the hydrocarbons (if any).
- Follow the vendor instructions for the startup of the Air Demand Analyzer (also known as the Tail Gas Analyzer). The analyzer sample systems are typically isolated until the unit operation has stabilized on acid gas.

5.2 Unit Heat-Up

Most SRU licensor and operating companies heat their SRUs above the acid dewpoint and sulfur melting temperature before admitting acid gas to the unit (all temperatures above 250°F is the normal target). The Incinerator is lit and heated according to the refractory vendor's schedule. Then the Reaction Furnace is fired with fuel gas to heat the furnace and to provide flow through the unit. The refractory vendor should supply a heat-up curve for normal startups after the refractory has been dried, and this curve should be followed. Heating the refractory too quickly can cause the inner layer(s) to expand more quickly than the outer layer(s) which can cause refractory damage.

Note that excess air can be used in the Reaction Furnace and fired Reheaters to control temperature ONLY during the initial startup or after a shutdown when the unit has been freed of sulfur and the equipment open to atmosphere. Once acid gas has been admitted to the unit, a layer of iron sulfide forms on the piping and equipment. Iron sulfide is pyrophoric and will easily catch fire if excess oxygen is present downstream of the Reaction Furnace.

As the Reaction Furnace heats up, the Reheaters are used to heat the catalyst beds. A heat-up rate of 50°F per hour is typical. All process temperatures should be brought to their normal operating temperature based on the Heat and Material Balance.

The site ports of all the fired equipment should be checked regularly during startup. The flame should be checked for stability and soot (which normally appears as white-hot specs in the flame), and the refractory within the viewing areas should be checked for cracks.

Most Reaction Furnaces have a tempering steam (or sometimes nitrogen) connection to help control the furnace temperature. Once all downstream process temperatures are above 250°F, the tempering steam should be started. Admitting steam when the rest of the piping and equipment is cold can cause corrosion from the condensation of the steam.

As the unit heats up, the level controls on the boilers can be put in automatic, along with the temperature controls of the Reheaters. The Reaction Furnace and Incinerator burner controls should also be put in automatic.

The final step before admitting acid gas is to perform a stoichiometric fire test in the Reaction Furnace. During this test, the air flow is adjusted until the process gas at the First Condenser outlet contains no oxygen. The resulting air to fuel gas ratio is used on subsequent startups to ensure oxygen does not escape the Reaction Furnace to cause fires in the downstream equipment. A portable oxygen analyzer can be used at the Condenser outlet to check the oxygen concentrations. Note that as the furnace approaches stoichiometric conditions, the furnace temperature will increase and tempering steam will be necessary to control the temperature to the normal operating range. Be sure the temperatures do not increase above the design temperature of the refractory.

Once all of the process temperatures are within their normal operating range, the Reaction Furnace is firing at a stoichiometric ratio and there is no oxygen measured downstream, the unit is "ready for acid gas."

5.3 Admitting Acid Gas

As Acid Gas is admitted to the plant, the air requirements in the furnace will change. If an acid gas sample has been analyzed, use the calculated air ratio. If not, use the Heat and Material Balance ratio. It can be helpful to create a chart with the fuel gas flow on axis, the acid gas flow on the other axis, and the calculated air flow required for the combined acid gas and fuel gas flows in the body of the chart.

Slowly admit acid gas to the Reaction Furnace, and increase the air flow by a corresponding amount, based on the air-to-acid gas ratio. Once the acid gas is above its minimum flow (above 50% of design flow is preferred), the fuel gas flow can be decreased and the air flow decreased correspondingly based on the air-to-fuel gas ratio. When the fuel gas is below its minimum flow, it can be stopped and the fuel gas system isolated. The tempering steam flow should be stopped at the same time.

Monitor all furnace and catalyst temperatures as the operation is switched from fuel gas to acid gas. Rapid increases in catalyst temperatures can indicate oxygen breakthrough from the furnace. In this case, reduce the air flow.

The Reaction Furnace flame will switch from blue or clear to a blue-green color when acid gas is admitted. The flame should be checked regularly to monitor its stability and to check for soot.

5.4 Initial Operations

Once the Reaction Furnace is operating on acid gas only and operation has stabilized, the process temperatures should be closely watched. The temperature differences in the catalyst beds should approach their normal operating ranges. The temperature rise in the catalyst beds is a direct indication of the amount of reaction that is taking place.

When the unit operation has stabilized, place the Air Demand Analyzer on-line and the remainder of the Reaction Furnace controls in automatic or cascade. Most SRUs run with a tail gas H_2S:SO_2 ratio of 2:1 or 4:1. A ratio of 2:1 maximizes the efficiency of the SRU, while 4:1 provides operating stability for a downstream Tail Gas Unit.

Watch the process pressure measured upstream of the Reaction Furnace, typically on the air line just upstream of the burner. If this backpressure starts to rise while the feed flows are constant, it is an indication that sulfur is collecting in at least one of the condensers. Check all of the steam traps associated with the liquid sulfur lines to ensure they are hot. Often the plug will happen at the last Condenser where sulfur flows are the lowest.

Other items to consider during initial operations:

- As the catalyst temperatures increase, sulfur should start flowing in the look boxes (or site glasses). These flows should be checked regularly.
- Some sulfur pit pumps have a minimum sulfur level requirement for operation. Be sure the sulfur pit is at or above this level before starting the pumps.

- Check all steam traps regularly, including any steam traps associated with Reheaters.
- Check all fired equipment site ports regularly.
- Blow down the boilers regularly.
- If catalyst temperatures rise quickly, indicating a sulfur fire, do not shut down the Reaction Furnace. Reduce the Reaction Furnace air flow until the temperatures start to decrease.

5.5 Restarts

There are two types of SRU restarts: hot and cold. A hot restart typically occurs soon after a shutdown, when Reaction Furnace temperatures are above 1,500°F and all downstream process temperatures are above 250°F. In this case, the Reaction Furnace is typically lit with fuel gas and acid gas is reintroduced as soon as possible.

Cold restarts happen after the unit has been shut down for at least several days, and temperatures have cooled to below the hot restart criteria. In this case the Reaction Furnace should be heated using the same schedule followed in section 5.2. Unlike the initial startup, however, excess air cannot be used during a cold restart because the piping and equipment will contain sulfur and iron sulfide. The stoichiometric air-to-fuel gas ratio determined during the initial startup must be used while firing fuel gas during a cold restart. Tempering steam is used to control furnace temperatures.

5.6 Hot Standby

Hot Standby is the practice of firing fuel gas in the Reaction Furnace to keep the furnace and downstream equipment hot and ready for acid gas. This can be done relatively safely if the fuel gas composition is stable. If the fuel gas composition changes frequently, it can be more challenging to maintain a stoichiometric air-to-fuel gas ratio.

In either case it is not recommended to keep the unit in Hot Standby for long periods of time because of the potential for the following to happen:

- Temperature spikes in the Reaction Furnace due to changes in fuel gas composition, air-to-fuel gas ratio or tempering steam flow.
- Fires in the catalyst bed due to oxygen breakthrough from the Reaction Furnace.
- Soot deposition on the catalyst from firing the Reaction Furnace substoichiometric, leading to catalyst deactivation.
- Corrosion in the Final Condenser from the condensation of the water from combustion in the Reaction Furnace. The Final Condenser is the coldest part of the plant, and the potential for condensation and corrosion is the greatest when the mass flow through the unit is low, such as during Hot Standby operation.

6. Recommendations

The following are some final recommendations for a smooth and successful startup:

- Take care during precommissioning to ensure all pipework and equipment is clean.
- Ensure all spare parts are available and in good working condition.
- Follow the refractory vendor's guidelines for refractory dryout and do not try to shorten the schedule.
- Use excess air in the Reaction Furnace only during the initial startup or after the unit has been swept of sulfur. Use the stoichiometric air-to-fuel gas ratio any other time fuel gas is fired in the Reaction Furnace with tempering steam to control the temperature.
- Monitor catalyst temperatures closely during the startup and initial operations. Reduce the air flow to the Reaction Furnace if the temperatures rise quickly.
- Check all steam traps regularly, particularly those associated with liquid sulfur lines.
- Minimize periods of Hot Standby operation in the Reaction Furnace.

7. References

[1] Engineering Data Book, 10th Edition, Gas Processors Suppliers Association, Tulsa, OK, 1994.

[2] Planning for Startup, Implementation Resource 121-2, Construction Industry Institute, Austin, TX, 1998.

Sulfur Pit Explosion Protection by Deflagration Venting

WorleyParsons

Bill DeWees, Angie Slavens, Shadi Al Adel, Vincent Yih, Dan Samani

Disclaimer

The authors have prepared this paper for the purpose of presenting available information on the subject matter and discussing and evaluating the advantages and possible disadvantages of utilizing deflagration vents on sulfur pits. Neither the authors nor WorleyParsons make any recommendations or representations as to the merits of deflagration vents, and hereby disclaim any responsibility or liability with respect to decisions with respect to use or non-use of such vents or any other use of the information contained in this paper.

Contents

Introduction

Liquid sulfur produced by the Claus process contains dissolved H_2S, which when liberated into the vapor space of its container can form a flammable mixture with air. Concrete, underground storage pits have been the industry standard for the collection, degassing and storage of elemental sulfur, produced in the Claus process, for many years. Most sulfur pits include features designed to prevent deflagrations using the principle of combustible concentration reduction, whereby the concentration of H_2S is maintained below the lower flammability limit (LFL) by dilution with sweep air drawn through the vapor space of the pit using a steam ejector. Backup ejectors and natural draft "chimneys" (steam jacketed vents to atmosphere), together with safety interlocks monitoring H_2S concentration, sweep air and/or motive steam flow, are commonly used features designed to ensure that sweep air keeps flowing at an adequate rate, and to alert operators if there is a deficiency.

Should all of these protective features fail, many sulfur pits also include deflagration vents (or "explosion hatches") designed to relieve pressure resulting from a deflagration, thereby preventing a full-blown explosion event that could compromise the structural integrity of the pit and injure personnel. NFPA 68 – Standard on Explosion Protection by Deflagration Venting, as well as some European standards, contain guidelines for designing deflagration vents. However, these standards can be challenging to interpret and apply. Something as seemingly straightforward as determining the strength of a concrete enclosure can be difficult. There are also different types of venting devices to choose from, ranging from explosion doors or hatches to rupture diaphragms, each with different efficiency, costs, operability and maintenance concerns.

Complicating matters even further is the fact that owners, operators and engineers don't always agree on the need for deflagration vents and often ask, "are these really required?" This straightforward question is not always simple to answer, as some consider the sweep air and monitoring features to be sufficient, while others consider deflagration vents to be essential, similar to a relief valve on a pressure vessel.

This paper discusses some of the arguments for and against deflagration vents. It also provides interpretation of the guidelines and design methods included in the applicable NFPA standards and reviews the different types of vent designs commonly used. Alternative sulfur storage designs that can eliminate the need for deflagration vents altogether are also explored.

Sulfur industry literature is noticeably silent on this subject. This paper seeks to raise the profile of this important issue within the sulfur industry.

Typical Sulfur Storage Pit Design and Safety Features

Liquid sulfur produced by the Claus process typically contains about 300 ppmw of dissolved hydrogen sulfide (H_2S) and hydrogen polysulfides (H_2S_x) [1]. Over time, polysulfides decompose to form H_2S and sulfur by Reaction 1 below. As the sulfur cools, either from heat lost to the surroundings or with the aid of a sulfur cooler, the equilibrium of Reaction 1 shifts to the right. H_2S is physically desorbed into the gas phase by Reaction 2, aided by any agitation that may take place in the rundown container.

$$H_2S_x \leftrightarrow H_2S(l) + S_{x-1} \tag{1}$$

$$H_2S(l) \leftrightarrow H_2S(g) \tag{2}$$

Degassing technologies of various types may be used to speed up the process by increasing the level of agitation, improving the efficiency of vapor/liquid contact and/or via the addition of a catalyst to promote Reaction 1. However, even without a special degassing process, H_2S will be released into the vapor space of the rundown container and, if not removed can accumulate and potentially form a flammable mixture with air. The lower and upper flammability limits (LFL and UFL) of H_2S in air at 135 °C (275 °F) are 3.3 mol% and 50.4 mol%, respectively [11, 12]. Therefore, the rundown storage system must be designed to avoid this potentially flammable mixture, as well as possible ignition sources, during normal operation and maintenance activities.

NFPA 69 – Standard on Explosion Prevention Systems, describes systems for the prevention of explosions in enclosures that contain flammable concentrations of gases, vapors, mists, dusts or hybrid mixtures. These systems fall into six general prevention or control methods:

1. **Deflagration prevention by oxidant concentration reduction**
2. **Deflagration prevention by combustible concentration reduction**
3. Deflagration prevention by hot particle detection and intervention systems
4. Deflagration control by suppression
5. Deflagration control by isolation
6. **Deflagration control by pressure containment**

As explained above, it is possible for the vapor space of a molten sulfur enclosure to contain an explosive mixture of H_2S and air; therefore, the prevention methods described in NFPA 69 should be applied. While methods 1, 2 and 6 are commonly employed in this service, the authors are not aware of any sulfur storage systems using methods 3, 4 or 5.

Method 1, deflagration prevention by oxidant concentration reduction, involves maintaining the oxidant concentration in the system low enough to prevent deflagration [NFPA 69 5.1.1]. This method has been employed in some sulfur plants using fuel gas, nitrogen or other inert gas as the purge gas [1] to maintain the oxidant concentration below the *Limiting Oxidant Concentration (LOC)*, below which a deflagration cannot occur [NFPA 69 3.3.23]. The primary drawback to this method is that iron sulfide is allowed to form and accumulate on steel surfaces when air is not continually present to oxidize it. Upon eventual contact with air (during maintenance, for example) the built-up, pyrophoric iron sulfide can spontaneously ignite, causing serious safety concerns [1].

Most sulfur pits utilize **method 2**, combustible concentration reduction, for the prevention of explosive mixtures within the vapor space. In this method, the concentration of H_2S is maintained below the LFL by dilution with ambient air. By maintaining an oxidizing atmosphere, the buildup of iron sulfide is prevented.

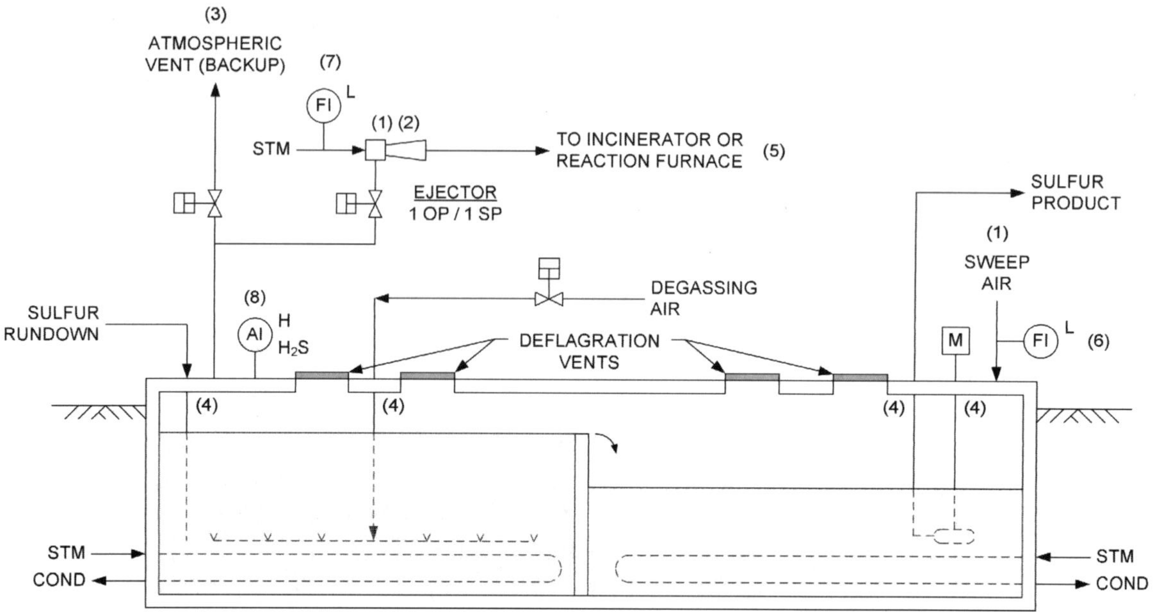

Figure 1 – Typical Sulfur Pit with Sweep Air

Figure 1 illustrates a typical scheme of this type, including the following features:

(1) A steam ejector located at one end of the pit is used to draw atmospheric air through air intakes at opposite corners of the pit, sweeping the entire vapor space and ensuring that no stagnant zones exist.

(2) A backup steam ejector is often installed which can either be started manually or automatically, in the event that the normally-operating ejector fails.

(3) A backup natural draft vent to atmosphere that can either be opened manually or automatically.

(4) Well sealed pit roof penetrations to prevent air ingress through these connections rather than through the dedicated air-intakes .

(5) In most systems, the sulfur pit ejector discharges the vent gas directly to the incinerator; however, tightening environmental regulations may force some systems to discharge the pit vent to the SRU Reaction Furnace for recovery of the H_2S and sulfur vapor in the stream. Implementation of a pit vent recycle scheme to the SRU Reaction Furnace requires additional controls and safeguards to prevent the backflow of hot furnace gas. Sulfur pit deflagration vents would not typically be designed to protect against this scenario because the carbon steel piping between the pit ejector and the reaction furnace would most likely fail before the furnace gases ever made it to the sulfur pit.

(6) The flow rate of sweep air is monitored, typically at the air inlet(s) rather than at the ejector suction or discharge. Sweep air flow measurement at the air inlet

location is selected for two reasons: 1) to ensure that sufficient air is entering through the air intake line rather than through pit penetration leaks, creating the opportunity for stagnant zones and inadequate sweeping despite achieving what appears to be an adequate pit sweep flow rate through the ejector, and 2) to avoid plugging the instrument taps with sulfur. A low air intake flow alarm alerts operators or trigger an automated response to start the backup ejector or open the backup atmospheric vent.

(7) The flow rate of ejector steam is monitored and a low flow alarm alerts operators or triggers an automated response similar to the low sweep air flow response.

(8) An H_2S analyzer monitors the vapor space and a high alarm alerts operators or triggers an automated response to shut-off the degassing air, if present, start the backup ejector or open the backup atmospheric vent. NFPA provides guidelines on acceptable H_2S concentration limits. The combustible gas concentration shall be maintained at or below 25% of the LFL (up to 60% of LFL with automatic safety interlocks) [NFPA 69 6.3.1]. Operations shall be discontinued whenever instruments show a combustible gas concentration of 35% or more of the LFL and should not be resumed until the instruments indicate a concentration of 15% or less of the LFL [NFPA 655 5.2.3 and 5.2.4].

The above safety features are aimed at preventing the possibility of forming an explosive mixture. For additional precaution, the following safety features are designed to prevent a source of ignition:

- The liquid sulfur inlets use a dip tube to deposit the sulfur below the liquid level and adequate grounding is provided to avoid static charge.
- The oxygen-rich environment prevents the buildup of pyrophoric iron sulfide.
- All instruments use explosion-proof housings in accordance with the electrical area classification.

Other typical safety features, such as monitoring the vapor space temperature to detect a fire, or providing snuffing steam to extinguish a fire, while important, do not specifically prevent deflagrations.

Protection against overpressure due to phenomena other than internal deflagrations (e.g. water leaks from heating coils or a sulfur cooler which may result in overpressure due to the rapid vaporization of water) are not addressed by NFPA 68.

Are Deflagration Vents Required?

If all of the protective features designed to prevent a deflagration from occurring should fail, many sulfur pits also include deflagration vents (or "explosion hatches") designed to relieve pressure resulting from a deflagration, thereby preventing a full-blown explosion event that could compromise the structural integrity of the pit and injure personnel.

Owners, operators and engineers don't always agree on the need for deflagration vents and often ask, "are these really required?" This straightforward question is not always simple to answer, as some consider the sweep air and monitoring features to be sufficient, while others consider deflagration vents to be essential, similar to a relief valve on a pressure vessel.

Sulfur Industry Literature Sources

While a great deal of information related to deflagration venting is available in general literature, it is surprising that sulfur industry literature is noticeably silent on this subject. Many papers have been written on safe sulfur storage practices for sulfur pits [1, 2, 3], but few mention the role of explosion hatches in the overall safety scheme of the sulfur pit. Those that do, mention them only briefly and don't discuss the specific details and challenges associated with their implementation in this service.

A literature search including Brimstone and Laurence Reid conference archives, ASRL Quarterly Bulletins, and other sulfur industry resources revealed only two references to explosion hatches specifically for sulfur storage. One was a slide presentation about safe sulfur storage in which a single bullet on a single slide said, "Include explosion hatches in design" [2]. The second was a Q & A style article about safe sulfur storage in which a question was asked, "Does anyone use explosion doors to minimize damage?" The answer was simply, "We have never used blow-off doors on pits" [3].

NFPA Standards

The National Fire Protection Association (NFPA) develops codes and standards intended to minimize the possibility and effects of fire and electrical risks in buildings, processing facilities and other industrial applications in the United States, as well as many other countries. These codes & standards may be adopted into law or adopted as design standards by private companies. NFPA is the recognized authority in the US for fire and explosion prevention.

The main NFPA standards applicable to deflagration prevention and control within sulfur storage enclosures are described below.

- NFPA 655 – *Standard for Prevention of Sulfur Fires and Explosions*, 2012 Edition
- NFPA 69 – *Standard on Explosion Prevention Systems*, 2002 Edition
- NFPA 68 – *Standard on Explosion Protection by Deflagration Venting*, 2007 Edition

NFPA 655

NFPA 655 applies to the handling of sulfur in any form, liquid or solid [NFPA 655 1.1.1]. Chapter 5, *Handling of Liquid Sulfur at Normal Handling Temperatures (246 °F to 309 °F)*, covers the "typical" deflagration prevention features of a sulfur pit, such as the need for an H_2S vent system, H_2S monitoring requirements, safe concentration limits of H_2S in the vapor space, avoiding static buildup/discharge, avoiding exposure of heating coils to air, plus other safety features unrelated to deflagration prevention. However, Chapter 5 does not require, or even mention, deflagration vents.

Chapter 6, *Handling of Liquid Sulfur and Sulfur Vapor at Temperatures above 309 °F*, explicitly requires that process equipment *shall be* provided with heated deflagration venting [NFPA 655 6.2.2.1]. This is perhaps because at temperatures above 309 °F, sulfur is nearer the *flash point* - the minimum temperature at which a liquid or solid emits vapor sufficient to form an ignitable mixture with air near the surface [NFPA 68 3.3.14]. Reported flash points of sulfur vary but are consistently greater than 309 °F. However, since most sulfur pits operate well below 309 °F, NFPA 655 does not provide a clear mandate for deflagration vents in normal service.

NFPA 69

NFPA 69 covers the design of the H_2S vent systems, which are required by NFPA 655 Chapter 5. As discussed previously, the most commonly used deflagration prevention method for sulfur pits is combustible concentration reduction with sweep air.

NFPA 69 also covers deflagration control by **method 6**, pressure containment, which may be an alternative to deflagration venting. It is only practical where a pressure vessel is used instead of a concrete pit. However, there is no mandate in NFPA 69 that deflagration vents or containment be provided in addition to a sweep air system.

NFPA 68

NFPA 68 applies to the design, location, installation, maintenance and use of devices and systems that vent the combustion gases and pressures resulting from a deflagration within an enclosure so that structural and mechanical damage is minimized [NFPA 68 1.1].

NFPA 68 only applies where the need for deflagration venting has already been established [NFPA 68 1.3]. It does not itself establish the need for deflagration vents. That responsibility is left to the authority having jurisdiction (AHJ), which is a phrase used throughout NFPA documents to refer to the organization, office, or individual responsible for approving equipment, materials, an installation, or a procedure. The AHJ could be a government agency, insurance agency or the property owner [NFPA 68 A.3.2.2].

European Standards

NFPA is not the only organization which addresses electrical and fire safety. The European Committee for Standardization (CEN) is a significant contributor to the development and implementation of European standards and technical specifications. The CEN's 33 national members work together to develop voluntary European Standards (ENs).

The main European standards that can be interpreted as being applicable to deflagration prevention and control within sulfur storage enclosures are:

- EN 1127-1:2007 – Explosive atmospheres - Explosion prevention and protection
- EN 14994:2007 – Gas Explosion Venting Protective Systems
- EN 14460:2006 – Explosion Resistant Equipment
- EN 14797:2006 – Explosion Venting Devices

The concepts in the European standards are similar to those in the NFPA standards but the specific requirements for their implementation and the formulas used for sizing, location, etc. are sometimes different. Nevertheless, one key consistency between the NFPA and CEN standards is that mandates for the installation of deflagration vents on sulfur pits are noticeably absent in each, despite that guidelines for their sizing and placement are provided.

In summary, there is no industry consensus with respect to the need for deflagration vents on sulfur storage enclosures, nor does there appear to be any definitive obligation to provide them in published safety standards. Therefore, the responsibility for deciding whether they are required belongs to the authority having jurisdiction, which is likely the facility owner in this case.

Making an Informed Decision

In order for the owner to make an informed decision about the need for deflagration vents, the topic should be investigated in-depth during the design phase of a project as part of the HAZOP review. However, it is possible that deflagration may not be properly considered during these safety reviews, as the vent devices may not always be shown prominently on P&IDs, which provide the primary basis for the HAZOP. They may be shown as combination maintenance/explosion hatch devices, which masks their importance as safety features. Furthermore, deflagration vent devices are rarely assigned tag numbers, possibly because they are considered an integral part of the concrete pit and so are designed by civil/structural engineers, who do not typically participate in HAZOP meetings.

The front-end designer typically does not have access to the details that will be developed in later phases of a project, such as sulfur pit detailed design calculations and construction techniques. Additionally, as mentioned previously, the owner's operating, maintenance and safety philosophies/procedures are also very important in determining whether the installation of deflagration vents should be considered, yet the front-end designer may not have full access to the specific owner's vigilance with respect to implementing and maintaining such practices. In the absence of these important details, in order to ensure a conservatively safe design, during the Basic Design stage of a project, deflagration vents may be suggested with a brief note on the sulfur pit data sheet; something like "Provide maintenance/explosion hatch. Size and number to be determined during detail engineering."

Some of the detailed design and/or operational factors that may make it necessary to consider sulfur pit deflagration vents include:

- Backup protection against deflagrations is needed due to inadequate safeguards to ensure the reliability of the sweep air system.
- Lower than normal sulfur pit strength due to design, construction materials, field installation procedures and/or soil conditions.

Some of the reasons for attempting to avoid installing sulfur pit deflagration vents include:

- In the event of a deflagration relief, toxic gas must be vented into the atmosphere, possibly putting personnel at risk of exposure to flame, heat and/or SO2.

- The strength of the concrete pit may weaken over time to lower than the relieving pressure that the vents were originally sized for.
- Heated vents may be required to prevent plugging and/or maloperation.
- Onerous inspection and maintenance, of both the pit and the relieving devices, is required to ensure adequate protection as the facility ages.
- Depending on the strength of the pit, there may be insufficient area available on the pit roof for adequate venting.
- A sulfur pit equipped with deflagration vents may be more expensive, in terms of engineering and construction, than a sulfur pit without them.

Designing Deflagration Vents

If, in fact, it is concluded that deflagration vents are required, NFPA 68 provides guidelines for their design and implementation. The following sections discuss the basics of deflagration vent design, as described in NFPA 68. Many specific NFPA citations are provided, with accompanying interpretation which is specific to sulfur pit overpressure protection applications. The discussion and examples are intended to provide a general overview and point out some of the nuances that are specific to sulfur pits, but should not be considered a substitute for NFPA 68.

Fundamental Properties of Combustible Gases

The equations provided in NFPA 68 for sizing deflagration vents rely on two fundamental measured properties of the combustible gas: The *Deflagration Index for Gases*, K_G, and the *Fundamental Burning Velocity*, S_u.

The *Deflagration Index for Gases*, K_G, is determined by the maximum rate of pressure rise attained by combustion in a closed test vessel with volume, V, and is defined by the following equation:

$$K_G = \left(\frac{dP}{dt}\right)_{\max} \cdot V^{1/3}$$ **NFPA 68 Equation 6.1.1**

The value of K_G for H_2S is provided in NFPA 68, Annex E, Table E.1. The higher the value of K_G, the higher the rate of pressure rise upon deflagration of the particular gas. The value of K_G for H_2S in Table E.1 is 45 bar-m/sec. The gas with the next closest value of K_G in Table E.1 is CH_4 at 55 bar-m/sec.

The *Fundamental Burning Velocity*, S_u, is the rate of flame propagation, relative to the velocity of the unburned gas ahead of it, of a laminar flame under stated conditions of composition, temperature, and pressure of the unburned gas. The values of S_u for many gases have been measured and published in NFPA 68, Annex D, Table D.1(a). The value of S_u for H_2S is not provided in this table; however, the value of S_u for one flammable gas can be approximated from that of another flammable gas by the following equation:

$$(S_u)_2 = (S_u)_1 \cdot \frac{(K_G)_2}{(K_G)_1}$$ **NFPA 68 Equation E.1**

Equation E.1 produces the most accurate values where the two flammable gases have similar values of K_G [NFPA 68 E.1]; thus, CH_4 properties can be utilized to derive the value of S_u for H_2S. The value of S_u for CH_4 in Table D.1(a) is 40 cm/sec. Therefore, the value of S_u for H_2S calculated by equation E.1 is 33 cm/sec.

Gas Mixtures and Sulfur Mist

For the purpose of this paper, the flammable/explosive mixture is considered to be H_2S in air. The concentration of sulfur vapor in the vapor space is assumed to be negligible compared to the concentration of H_2S. Sulfur mist in the vapor space is also considered negligible, as would be the case without sulfur degassing inside the sulfur pit and/or when in-pit degassing is limited to air sparging from the bottom of the pit, below the liquid sulfur level. However, where a significant quantity of sulfur mist is present, which may be the case for some degassing processes where liquid sulfur is sprayed into the vapor space through nozzles, a different approach may be required. For combustible mists, the K_G for propane of 100 bar-m/sec or the equivalent S_u for propane of 46 cm/sec shall be used unless specific test data are available [NFPA 68 7.1.3]. It is observed that these values are substantially higher than those for H_2S, resulting in greater consequences, should a deflagration occur.

It may also be necessary, in some circumstances, to consider the case where a sulfur seal fails and process gas from the sulfur condenser flows into the pit. In this case, the combustible gas might be a mixture of H_2S, SO_2, CO_2, CO, COS, CS_2, H_2, N_2, H_2O, sulfur vapor and air. The mixture may or may not be flammable depending on the dilution effect of inert species and quantity of oxygen available.

Another scenario that may be considered is the presence of contaminants, such as unburned hydrocarbons, dissolved in the liquid sulfur, which may be subsequently released into the pit vapor space. Each credible scenario should be evaluated and the controlling case chosen for the design of the deflagration vents.

Where the hazard consists of a flammable gas mixture, the vent size shall be based on the K_G or S_u of the mixture [NFPA 68 6.2.1.1]. Where the gas mixture composition is not certain, the vent size shall be based on the component having the highest K_G or fundamental burning velocity [NFPA 6.2.1.2]. Hydrogen may present a serious concern in this case because it is a fast-burning gas with K_G of 550 bar-m/sec and S_u of 312 cm/sec. Fast-burning gas deflagrations can readily undergo transition to detonation, in which case NFPA recommends using alternative measures discussed in NFPA 69 [NPFA 68 A.7.1.1].

Effect of Turbulence

If the gas system is initially turbulent, or if internal structures promote turbulence, the rate of deflagration increases [NFPA A.7.5]. In most instances, the vapor space of the sulfur pit can reasonably be considered quiescent, or non-turbulent. However, for some upset scenarios that may be considered, such as failure of sulfur seals, turbulence may exist, hence requiring consideration in the deflagration calculations. To account for the effects of initial turbulence and internal structures for enclosures with initial pressures near atmospheric, the K_G for hydrogen (550 bar-m/sec) shall be used for gases that have values, in the quiescent state, that are close to or less than that of propane (100 bar-m/sec) [NFPA 7.5]. Thus, similar to the scenario where the presence of sulfur mist is considered, greater deflagration consequences would also be expected if turbulence was present in the sulfur pit vapor space.

Enclosure Design Pressure

NFPA 68 defines several terms related to the pressure inside on an enclosure, as follows:

- *Maximum Pressure* (P_{max}) is the maximum pressure developed in a contained deflagration of an optimum mixture.
- *Reduced Pressure* (P_{red}) is the maximum pressure developed in a vented enclosure during a vented deflagration.
- *Static Activation Pressure* (P_{stat}) is the pressure that activates a vent closure [NFPA 68 3.3.27].
- *Enclosure Strength* (P_{es}) is the static pressure that results in either the deformation (yield strength) or failure (ultimate strength) of the weakest structural component of an enclosure [NFPA 68 6.3.2].

The amount by which P_{red} exceeds P_{stat} is a complicated function of the rate of pressure development within the enclosure, vent size, and mass of the vent closure device. These relationships are illustrated in Figure 2. Where deflagration vent area to enclosure volume ratio is large, P_{red} approaches P_{stat} (C). As the vent area is reduced, P_{red} increases (B) and approaches P_{max} as the vent area goes to zero (A) [NFPA 68 A.6.3.1.1].

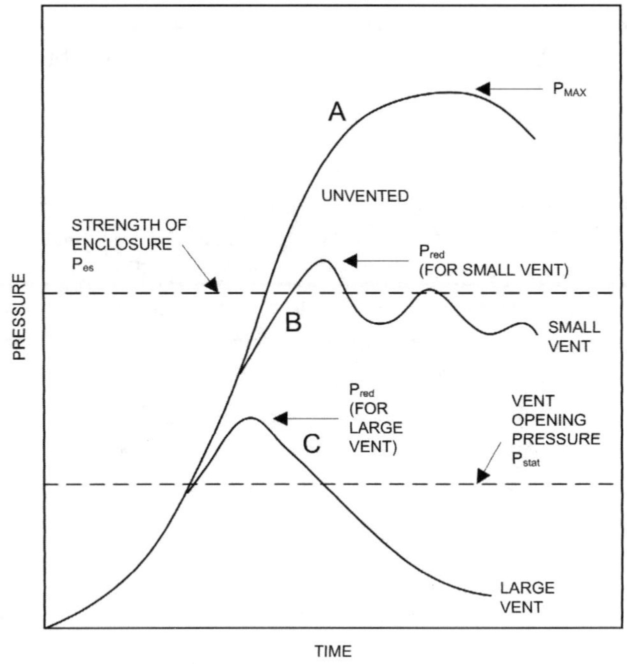

Figure 2 – Pressure-time characteristics of vented and unvented deflagrations from initially closed vessels illustrating the relationship of P_{max}, P_{red}, P_{stat}, and P_{es}.
(Adapted from [13])

As illustrated in Figure 2, the pressure from a vented deflagration rises to a maximum then decreases over a short time period after the vent opens. The load on the enclosure is therefore dynamic rather than static, causing the peak load to be amplified. The degree of amplification is called the *Dynamic Load Factor (DLF)* and depends on factors such as the duration of the load and the natural frequency of the system. In the absence of detailed structural response analysis, it shall be permitted to assume a worst-case value of $DLF = 1.5$ [NFPA 68 6.3.2.1]. Therefore, P_{red} shall not exceed two-thirds of the ultimate strength of the vented enclosure where deformation of the equipment can be tolerated, or yield strength where deformation cannot be tolerated [NFPA 68 6.3.1]. For a concrete sulfur pit, deformation cannot be tolerated, and therefore yield strength would apply.

$$P_{red} \leq \frac{P_{es}}{DLF} \qquad \textbf{NFPA 68 6.3.2}$$

The structural design of the concrete pit and the design of deflagration vents are inseparable because the strength and geometry of the pit impacts the size and location of the vents. The following sections provide the methodology for determining the strength of the pit and corresponding venting requirements.

NFPA 68 provides different methods for sizing deflagration vents depending on the reduced pressure, P_{red}. Low-Strength Enclosures are defined as those that are capable of withstanding reduced pressures of not more than 0.1 barg (1.5 psig). High-Strength Enclosures are defined as those that are capable of withstanding reduced pressures from 0.1 barg (1.5 psig) up to 2 barg (29 psig).

In general, the higher the reduced pressure, the smaller the vent area required for both high-strength and low-strength enclosures. Therefore, there is incentive to design the pit for a high strength. However, at some point, the cost of the extra concrete and reinforcing steel that is required to reinforce the pit will eventually outweigh the benefits of smaller vent area.

Figure 3 plots reduced pressure, P_{red} on the x-axis against required deflagration vent area, A_v, on the y-axis for two sections of a hypothetical sulfur pit: a 30 ft wide x 30 ft long x 10 ft deep degassing section followed by a 30 ft wide x 80 ft long x 10 ft deep storage section. The plotted values were determined by calculation methodology provided in NFPA 68, which will be discussed in detail later in the paper. Figure 3 illustrates that for low strength enclosures, required vent area increases exponentially, as reduced pressure decreases. At the upper pressure limit of 1.5 psig for a low-strength enclosure, the deflagration vent area extends across 30-35% of the roof area. Increasing the reduced pressure to 3 psig (perhaps a practical maximum for a concrete sulfur pit), upgrades the pit to a high-strength enclosure, and allows the vent area to be reduced to about 12-17%.

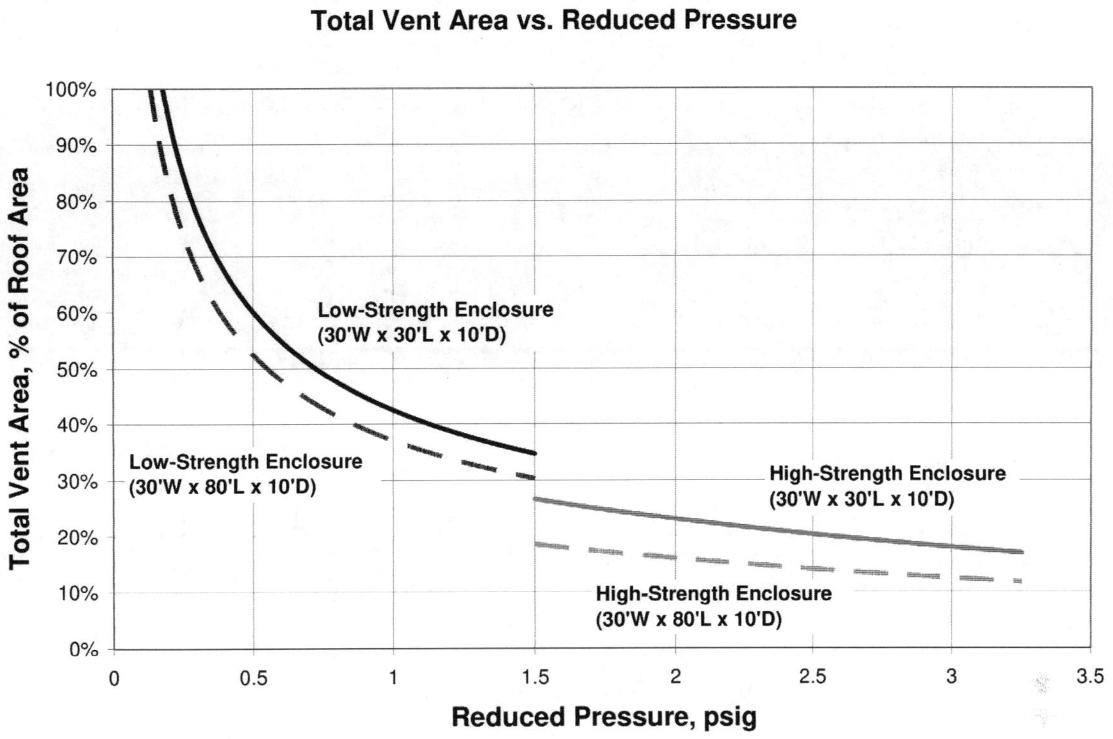

Figure 3 – Vent Area vs. Reduced Pressure

Figure 3 begs the question, "can the pit be designed to withstand a higher pressure without increasing the cost significantly?" The answer to this question is not simple because the strength depends on many factors. Pit geometry, design of the roof, number and size of roof penetrations, the type of roof-to-wall joint, air and soil temperatures and the internal reduced pressure, all contribute to the ultimate pit design strength.

In most cases, if a low-strength design is chosen, based on the geometry and configuration of the enclosure, the sizing of the walls and slabs are minimally impacted by the internal design pressure. This is due to the fact that the thermal stress caused by the temperature gradient between the hot molten sulfur side of the concrete and the colder soil side will govern the design. Referring back to Figure 3, recall that this option will require at least 30-35% of the roof area to be apportioned for deflagration venting devices. With so much open area, along with other roof penetration requirements, it is challenging to incorporate adequate support which will ensure high structural integrity. On the other hand, if a high-strength design is selected, the higher internal design pressure (in excess of 1.5 psig) will reduce the required venting area, but may have a significant impact on the sizing of the walls and slabs, and may result in additional concrete and reinforcing steel.

Unlike a vessel that is fabricated and tested in a shop, field fabrication of sulfur pits results in many uncertainties, such as quality, storage and preparation of materials, installation skill level of laborers, and weather conditions, to name a few, that can detrimentally impact the actual strength of the pit, regardless of its design strength. With so many variables in play, it is impossible to determine the optimum enclosure strength that works for every situation. Thus, specific calculations should be conducted for each project during detailed design. And, if necessary, calculations should be revised following construction, if it is suspected that field fabrication conditions may have contributed to a reduction in design pressure below the intended design value.

The calculation methodology which was used to derive the data included in Figure 3 is described in detail in the sections that follow.

Enclosure Length-to-Diameter Ratio (L/D)

The *L/D* ratio of the enclosure is an important consideration, especially for high-strength enclosures where it can have an impact on the vent area sizing calculation.

$$L/D = \frac{H}{D_{he}}$$ **NFPA 68 6.4.3.7**

where:

H = Maximum flame length (in meters) along which a flame can travel based on the maximum distance, taken along the central axis, from the farthest end of the enclosure to the opposite end of the vent [NFPA 68 6.4.3.2]. The effective volume, V_{eff}, is the volume of that part of the enclosure defined by H.

D_{he} = Effective hydraulic diameter (in meters) and is calculated by following equation:

$$D_{he} = 4 \cdot \frac{A_{eff}}{p}$$ **NFPA 68 6.4.3.6**

Hydraulic diameter is used for noncircular cross-sections, where A_{eff} is the cross-sectional area normal to the longitudinal axis of the space and p is the perimeter of the cross-section [NFPA 68 3.3.18]. See Figures 4, 5 & 6.

The most conservative method for calculating *L/D* is to determine H and V_{eff} based on the total enclosure, irrespective of vent location [NFPA 68 6.4.3.4], as illustrated in Figure 4. However, this method may result in a larger vent area than is actually required if *L/D* greater than 2.

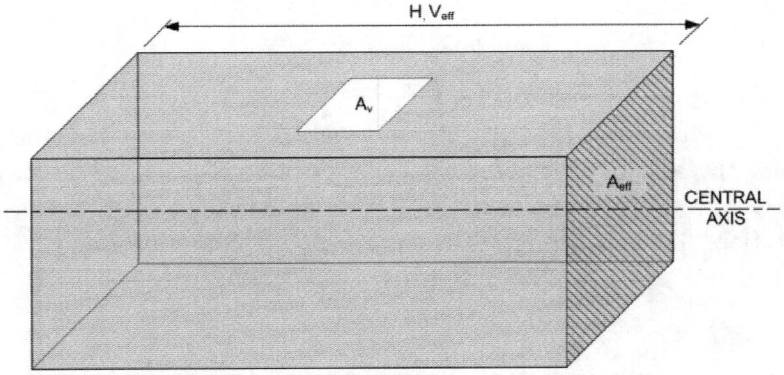

Figure 4 – *L/D* Method 1

When multiple vents are provided, a single value of *H*, and *L/D*, shall be permitted to be determined for the enclosure based on the farthest vent [NFPA 68 6.4.3.2.1], as shown in Figure 5.

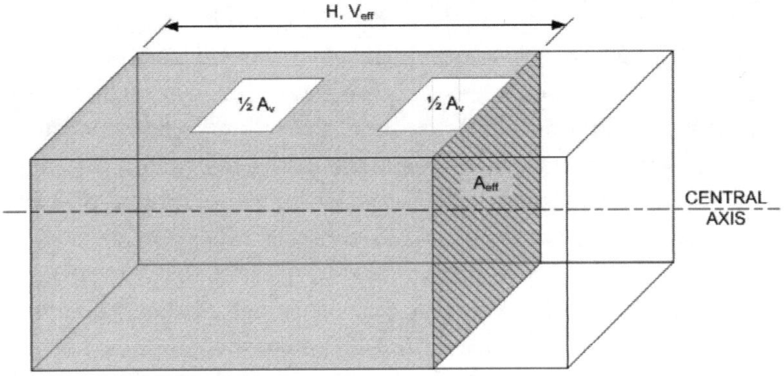

Figure 5 – *L/D* Method 2

The value of *H*, and *L/D*, can also be determined for each section using the maximum distance from the closest end of one vent to the opposite end of the next vent [NFPA 68 6.4.3.2.2], as illustrated in Figure 6.

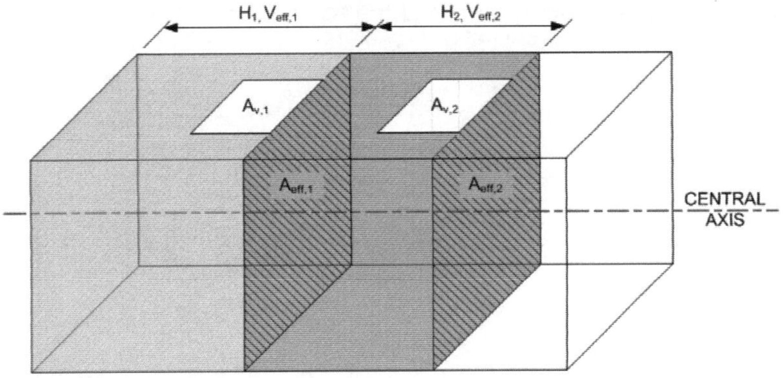

Figure 6 – *L/D* Method 3

Enclosure Internal Surface Area (A_s)

Internal surface area, A_s, should include the areas of the roof, floor and walls [NFPA 68 7.2.4.1]. Since the liquid sulfur in the bottom of the pit is incompressible, it acts like a "floor;" thus, the level will be a factor in determining the effective wall area. The choice of liquid level to use for the worst case scenario depends on the circumstances and should consider startup and shutdown as well as normal operation. The most conservative approach is to base the wall area on the empty pit dimensions [10].

Enclosure Compartments

If the pit is subdivided into separate compartments, each compartment shall be provided with its own vent(s) [NFPA 68 7.2.6.2]. This may be the case where separate degassing and storage compartments are provided, as shown in Figure 1. The wall between the two compartments extends from the pit floor to within 1-2 feet of the roof. While the vapor space of the two compartments is contiguous, the relatively small opening acts as a restriction to movement of expanding gases during a deflagration. NFPA 68 is not entirely clear if the two compartments must be treated separately in this specific case, but it seems prudent to do so. One reason is that this geometry may lead to a phenomenon called *pressure piling*, which can occur when a deflagration takes place in a vessel that is interconnected, through a pipe or duct, to another vessel, as illustrated in Figure 7. The pressure front accelerates through the restriction, causing it to move faster than the reaction front, which increases the initial pressure, P_o, ahead of it by about 8 x P_o. When the reaction front arrives, the final pressure becomes 8 x 8 x P_o = 64 P_o. The restriction also increases the combustion-generated turbulence, which amplifies the combustion properties of the gas. These phenomena may lead to a *deflagration-to-detonation transition* (DDT) if enough energy accumulates in the pressure wave so that the adiabatic compression of the gas leads to autoignition and initiation of a detonation [12]. Given these possibilities, it is recommended that each compartment be treated as a separate enclosure for the purpose of sizing the deflagration vents in that particular compartment.

When carrying out vent area calculations for an individual compartment, the designer must keep in mind that some of the calculations depend upon the longest dimension of the enclosure or the L/D of the entire pit. Thus, while the overall pit may be rectangular with the longest dimension in the direction of the x-axis, for example, an individual compartment may have its longest dimension in the direction of the y-axis.

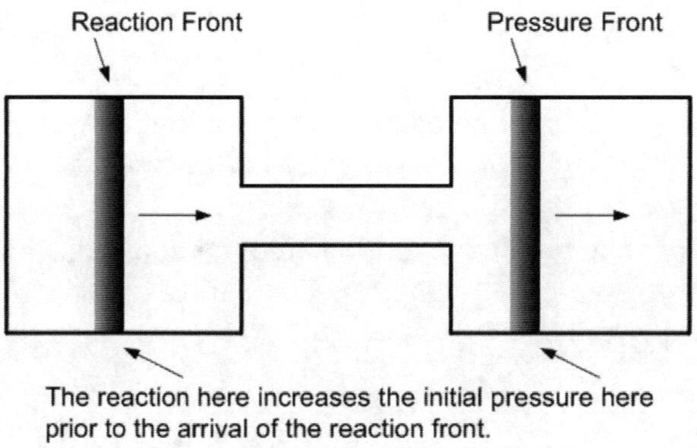

Figure 7 – Pressure Piling in Interconnected Vessels
(Adapted from [12])

Venting Area Requirement for Low-Strength Enclosures

The following equation determines the minimum venting area required for a low-strength enclosure [NFPA 68]:

$$A_v = \frac{C \cdot A_s}{P_{red}^{1/2}}$$ **NFPA 68 Equation 7.2.2**

where:

A_v = vent area (m^2)
C = venting parameter ($barg^{1/2}$)
A_S = internal surface area of enclosure (m^2)
P_{red} = maximum venting pressure (barg)

The venting parameter, *C,* is determined by the following equation for a fundamental burning velocity, S_u, less than 60 cm/s [NFPA 68]:

$$C = 1.57x10^{-5} \cdot (S_u)^2 + 1.57x10^{-4} \cdot (S_u) + 0.0109$$ **NFPA 68 Equation 7.2.2.1a**

Where:

S_u = fundamental burning velocity (cm/sec)

Note that for low-strength enclosures there are no dimensional restrictions on the enclosure or the placement of the vents as long as the vents are distributed as evenly as possible with respect to the longest dimension. If the vents are restricted to one end of the pit, the *L/D* ratio should not exceed 3 [NFPA 68 7.2.3.1].

It is observed that Equation 7.2.2 in NFPA 68 for venting gas or mist deflagration in low-strength enclosures is the same one used in EN 14491:2006 for venting dust deflagrations in buildings.

Venting Area Requirement for High-Strength Enclosures

For high-strength enclosures, the *L/D* ratio of the enclosure, and of the sections between vents, can have an impact on the vent sizing.

If *L/D* is 2 or less, the following equation determines the minimum venting area, A_v required:

$$A_v = \{(0.127\log_{10}(K_G) - 0.0567) \cdot P_{red}^{-0.582} + 0.175P_{red}^{-0.572}(P_{stat} - 0.1)\} \cdot V^{2/3}$$

NFPA 68 Equation 7.3.3.2

where:

K_G ≤ 550 bar-m/sec
P_{red} ≤ 2 barg and at least 0.05 barg > P_{stat}
P_{stat} ≤ 0.5 barg
V ≤ 1000 m^3
Initial pressure before ignition ≤ 0.2 barg

If *L/D* is 2-5, A_v shall be increased by adding more venting area, *ΔA*, which is calculated using the following equation:

$$\Delta A = \frac{A_v \cdot K_G \cdot (L/D - 2)^2}{750}$$

NFPA 68 Equation 7.3.3.3.1

Notice that the larger *L/D* becomes, the larger *ΔA* becomes, up to a maximum of 54% additional vent area for *L/D* = 5, in the case of H_2S. Therefore, when locating the vents along the major axis of the pit, as shown in Figure 6, it is best to limit the spacing so that the *L/D* ratio of each section is 2 or less, if possible, to avoid the need for extra vent area.

Effects of Vent Closure Panel Inertia

The free area of a vent does not become fully effective in relieving pressure until the vent closure moves completely out of the way of the vent opening. Until this occurs, the closure obstructs the combustion gases that are issuing from the vent [NFPA 68 A.6.7.4]. For a given vent area, the reduced pressure increases with the specific mass (kg/m^2) of the vent closure panel because heavy panels take longer to clear the vent opening than light panels do [NFPA A.6.7.2].

The specific mass of a vent panel, *M*, is the total mass of the closure (including counterweights or insulation) divided by the area of the vent opening [NFPA 68 6.7.1, 6.7.3 and EN 14797:2006 3.10]. The threshold mass, M_T, is the specific mass above which a correction to the vent opening area is required to account for inertia effects. Where *M* is less than M_T, no correction is required to the vent sizing equations given in NFPA 68 [NFPA 68 6.7.3]. Where M is greater than M_T, but less than 40 kg/m^2, the venting area must be increased by *ΔA* as shown below.

$$M_T = \left[6.67 \cdot \left(P_{red}^{0.2} \right) \cdot \left(n^{0.3} \right) \cdot \left(\frac{V}{K_G^{0.5}} \right) \right]^{1.67}$$ **NFPA 68 Equation 7.2.2.5.2**

where:

M_T = threshold mass (kg/m^2)
P_{red} is in barg
n = number of panels
V > 1 m^3

$$\Delta A = A_v \cdot (0.0075) \cdot M^{0.6} \cdot \frac{K_G^{\ 0.5}}{n^{0.3} \cdot V \cdot P_{red}^{0.2}}$$ **NFPA 68 Equation 7.2.2.6**

where:

ΔA = additional vent area required (m^2)
M = mass of vent panel (kg/m^2)
A_v = vent area (m^2) calculated by Equation 7.2.2 or 7.3.3.2

The value of K_G used in equation 7.2.2.6 shall be 75 bar-m/sec, if K_G for a certain component is less than 75 bar-m/sec.

Where M is greater than 40 kg/m^2, it is necessary to perform testing to determine the efficiency of the vent closure or apply alternate explosion protection methods per NFPA 69, Standard on Explosion Prevention Systems [NFPA 68 A.7.2.2.5.2].

European Standards differ from NFPA 68 in the limits imposed on the value of M before it becomes necessary to perform testing.

$$\Delta A = A_v \cdot \left(\frac{1}{E_f} - 1\right)$$ EN 14994:2007 5.2 (2)

$$E_f = \frac{A_E}{A_d}$$ EN 14797:2006 7.3.4

where:

E_f is the dimensionless venting efficiency of the venting device
A_E is the effective vent area of the device, m^2
A_d is the actual vent area of the device, m^2

- E_f = 1 for explosion venting devices with M less than 0.5 kg/m^2
- E_f = 1 for explosion venting devices with M greater than 0.5 kg/m^2 and smaller or equal to 10 kg/m^2 provided $A_v/V^{0.753}$ < 0.07, where A_V is the vent area and V the vessel volume. This is valid for $P_{stat} \leq 0.1$ barg and 0.1 barg < P_{red} < 2 barg

- for all other conditions and for explosion venting devices with an M greater than 10 kg/m^2 the efficiency E_f has to be determined by tests.

In summary, NFPA says that vent closures with $M > 40$ kg/m^2 require testing to determine their efficiency while European Standards set the limit much lower at $M > 10$ kg/m^2.

Hazard Zone

NFPA 68 discusses the hazard zone around the discharge of deflagration vents in the context of protecting personnel from the hot gases and material that may be ejected. It does not address the potentially toxic cloud that may result from the combustion of H_2S and/or other sulfur species, in the case of a sulfur pit deflagration. The determination of the hazard zone area is important because it may have some bearing on the location of deflagration vents on the sulfur pit roof and/or may present a requirement for cordoning off certain areas around the pit to prevent regular personnel access.

The fireball from a vented gas or dust deflagration presents a hazard to personnel who may be in the vicinity. People caught in the flame itself will be at obvious risk from burns, but those who are outside the flame area can be at risk from thermal radiation effects. The heat flux produced by the fireball, the exposure time, and the distance from the fireball are important variables to determine the hazard [NFPA 68 A.7.6.4].

The hazard zone from a vented gas deflagration is calculated by the following equation:

$$D = 3.1 \cdot \left(\frac{V}{n}\right)^{0.402}$$ **NFPA 68 Equation 7.6.4.1**

where:

D = axial distance (front-centerline) from vent (m)
V = volume of vented enclosure (m^3)
n = number of evenly distributed vents

The hazard zone measured radially (to the sides, measured from the centerline of the vent) shall be calculated as one-half D [NFPA 68 7.6.4.2]. Ducts should be carefully designed because they can impact the vent area sizing [NFPA 68 7.4].

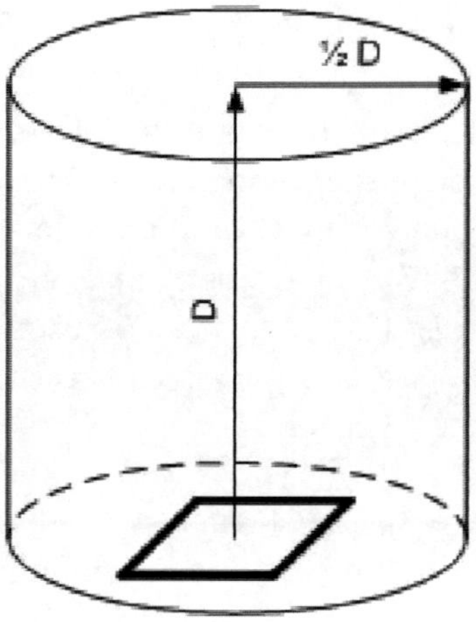

Figure 8 – Hazard Zone

Vent Closure Static Activation Pressure (P_{stat})

Similar to API limitations for relief valve overpressure, NFPA specifies permissible limits for deflagration vent activation pressure. Low strength and high strength enclosures each have their own specific requirements.

- For low strength enclosures, P_{red} shall exceed P_{stat} by at least 0.024 barg (0.35 psig) [NFPA 68 7.2.6.1].
- For high strength enclosures, P_{red} shall exceed P_{stat} by at least 0.05 barg (0.73 psig) and P_{stat} shall be less than 0.5 barg (7.3 psig) [NFPA 68 7.3.3.2].

Deflagration vent closures should release at a P_{stat} value that is as low as practical, yet should remain in place when subjected to external wind forces that produce negative pressures, to prevent vents from being pulled off. In most cases, a P_{stat} value of 0.01 barg (0.14 psig) is acceptable. In areas subject to severe windstorms, release pressures up to 0.015 barg (0.21 psig) are used [NFPA 68 A.7.6.2.1].

Vent Closure Types

Similar to a relief valve, a vent closure is a device that is installed on a sulfur pit vent opening that is designed to open at a pre-determined pressure, P_{stat}, to prevent the pressure in the enclosure from exceeding P_{red}. Vent closures should withstand pressure fluctuations that are below P_{stat} [NFPA 68 6.5.5]. For sulfur pits, this would include any slight vacuum potentially created by the sweep ejector.

Vent closures may have non-reusable elements or reusable elements. Those with reusable elements may automatically reclose following a deflagration or may require manual repositioning. Some common types of vent closures are described in the following sections.

Rupture Diaphragms

Rupture diaphragms (also called rupture or bursting panels, disks or membranes) are non-resealing, one-time use devices that rupture at a pre-determined differential pressure (P_{stat}). They can be round, square or rectangular and are typically mounted in a support frame which is attached to the enclosure (Figures 9-13). Individual panels are available in a range of sizes up to a few square meters in area. Flat panels can be used for vacuum service if backed by a support grid (Figure 10). Domed panels can resist vacuum conditions on their own without a support grid (Figure 11).

Thermally insulated vent panels are available (Figure 13) to reduce heat loss to the atmosphere, which can lead to sulfur condensation and/or solidification on and around the device, potentially hindering the relief device's performance. Heat loss and sulfur condensation/solidification concerns associated with sulfur pit deflagration vents are specifically discussed in a subsequent section of this paper.

Rupture diaphragms are almost inertia-free, with a specific mass less than 0.5 kg/m^2, and do not impede the venting process. Therefore, they have a venting efficiency, $E_f = 1$. [EN 14797:2006 7.3.3.1].

Some benefits of rupture diaphragms include:

- They are cost effective when infrequent activations are expected. For example, indicative cost of materials for a total vent area of 70 m^2 is about $58,000 for rupture diaphragms, compared to $525,000 for explosion doors [15].
- They have low inertia and therefore no penalty on vent area.

Some drawbacks of rupture diaphragms include:

- Their service life may be negatively impacted by pulsation or vibration.
- They must be replaced after each incident.

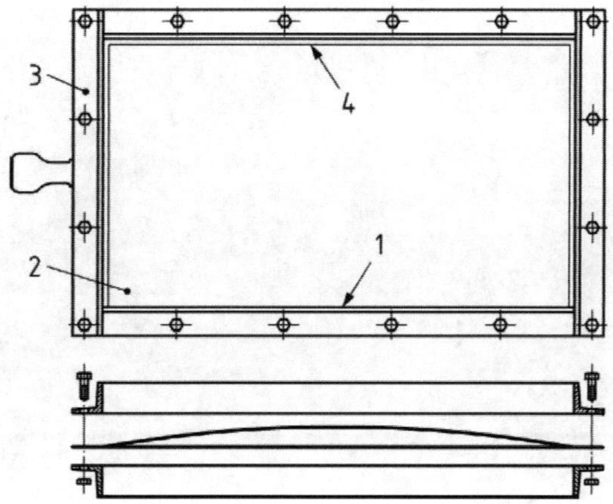

Key

1 fixed side

2 area of the explosion venting device

3 supporting frame

4 sides yielding during venting process

Figure 9 – Rectangular Rupture Diagram

(Source: EN 14797:2006)

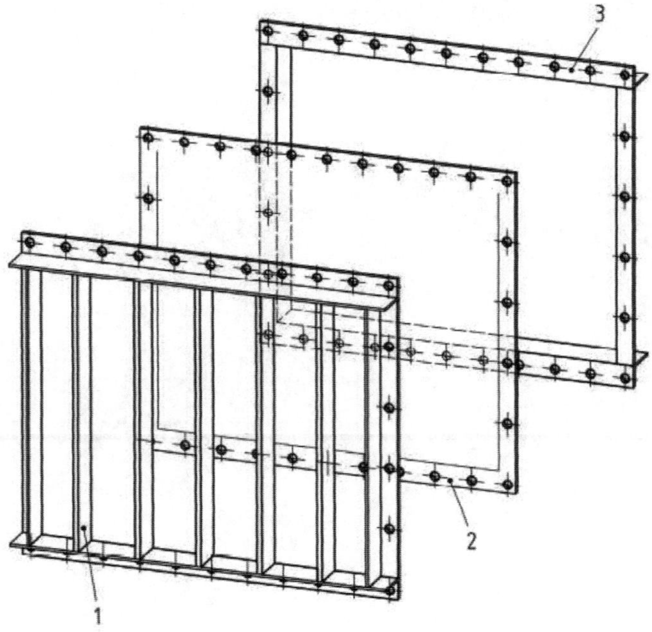

Key

1 back pressure support

2 vent panel

3 supporting frame

Figure 10 – Back Pressure Support

(Source: EN 14797:2006)

Figure 11 – Typical Domed Rupture Diaphragm (Round)
(Photo provided by BS&B Safety Systems)

Figure 12 – Typical Rupture Diaphragm (Square)
(Photo provided by BS&B Safety Systems)

Figure 13 – Thermally Insulated Rupture Diaphragm
(Photo provided by BS&B Safety Systems)

Pop-Out Panels

Pop-out panels come in many forms; a few examples are shown in Figures 14 and 15. They may be semi-rigid panels, secured at the edges to a frame, designed to bend and become detached at the desired P_{stat}. They may also use rigid panels secured to the frame at the edges with clamping devices, designed to break or release at the desired P_{stat}. Pop-out panels may be of single-wall or double-wall construction with insulating material between, so called "sandwich panels".

Pop-out panels may be installed in the vertical or horizontal orientation. If installed in the horizontal position, provisions to prevent the accumulation of snow, ice, rainwater or debris must be included in the design.

Pop-out panels have restraining devices (chain, wire, etc.) to prevent them from becoming projectiles. Restraining devices shall not impede the operation of the vent of vent closure device [NFPA 68 6.5.3].

The primary benefit of pop-out panels is that they may be reusable following an incident. The primary drawback is that broken clamps must be replaced after each incident.

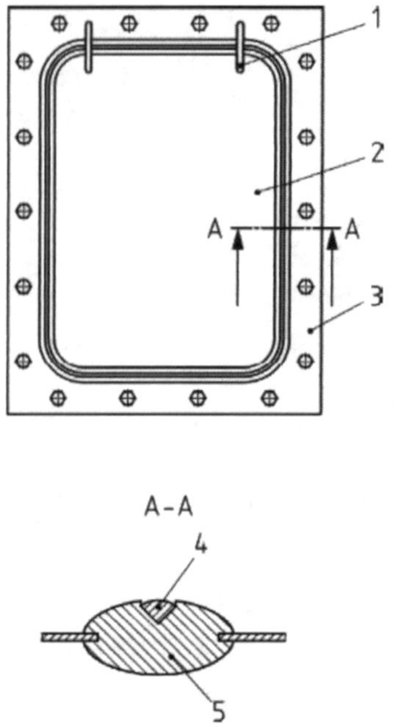

Key

1 restraining element
2 venting element
3 supporting frame
4 clamped device
5 plastic or rubber seal

Figure 14 – Restrained Pop-out Panel
(Source: EN 14797:2006)

Figure 15 – ATEX Reflex Pop-out Panel
(Photo provided by ATEX Explosion Protection L.P.)

Explosion Doors

Explosion doors are hinged devices, typically held shut with spring-loaded, magnetic or friction latches; some examples are shown in Figures 16 and 17. Hinged closures shall be permitted to be used, provided the following conditions are met [NFPA 68 6.7.4]:

(1) There are no obstructions in the path of the closure that prevent it from opening.
(2) Operation of the closure is not restrained by corrosion, sticky process materials, or paint.

Some benefits of explosion doors include:

- They are cost effective when multiple activations are expected or when the replacement of vents after opening is very difficult and/or time consuming.
- They can be made self-closing, which can limit oxygen supply and minimize the damage from fire.
- They can be field-calibrated for easy setup and maintenance testing.

Some drawbacks of explosion doors include:

- Explosion doors have greater inertia than rupture diaphragms and so may require additional vent area.
- Self-closing doors may create a vacuum in the pit resulting from cooling of the hot combustion gases following a deflagration.
- Explosion doors may be more challenging to heat, if required (this is discussed in more detail in a subsequent section of the paper).
- Maintenance of the hinge and latch mechanism is required.

Figure 16 – Explosion Doors
(Photo provided by ATEX Explosion Protection L.P.)

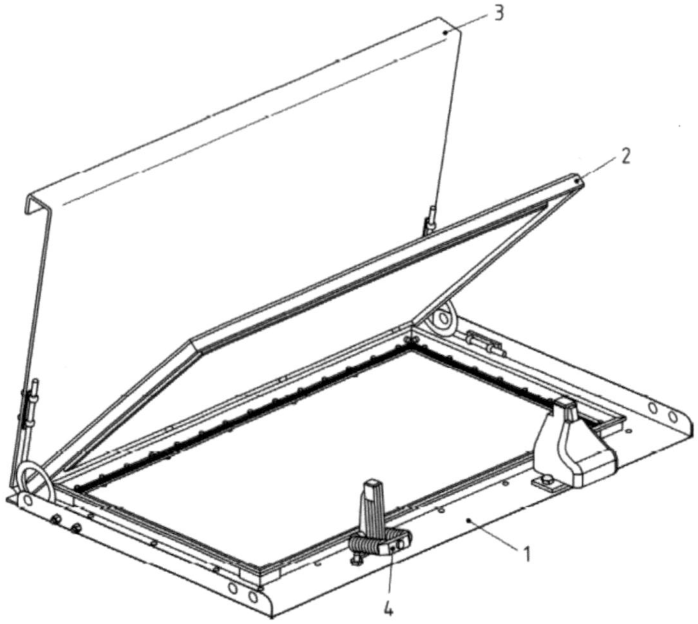

Key

1 supporting frame
2 venting element
3 baffle
4 retaining elements

Figure 17 – Explosion Door
(Source: EN 14797:2006)

Vent Closure Inspection and Maintenance Requirements

Vent closures shall be inspected at least annually [NFPA 11.4.1]. The inspection frequency depends on the environmental and service conditions to which the devices are exposed, and on documented operating experience [A.11.4.2]. The inspector shall verify [NFPA 68 11.4.4]:

- The opening is free and clear of any obstructions
- The closure is not corroded or mechanically damaged
- The closure has no buildup of deposits on the inside surfaces or between layers of the vent
- The closure shows no fatigue and has not released
- Records of inspections shall be retained for a minimum of 3 years [NFPA 68 11.7.2].
- If process material has a tendency to adhere to the vent closure, the vent closure shall be cleaned periodically to maintain vent efficiency [NFPA 68 11.9.3]

Vent Closure Heating

Vent closure operation should not be hindered by snow, ice, corrosion, debris or buildup of deposits on the inside surfaces [NFPA 68 6.5.2]. For sulfur pits, this might require heating and insulating the vent closures to keep the inside surface temperature above 250 °F to prevent sulfur from freezing on the panel.

When designing the heating system, many variables need to be considered, including ambient air temperature, wind speed, sweep air temperature (heating the incoming sweep air may be required), size of the panel, insulation thickness, etc. ControTrace heating elements may be well suited to this type of application because they can be adapted to a variety of panel mounting frame shapes and styles. Heating elements should not be placed directly on top of the vent panel (Figure 18) if they interfere with the operation or increase the panel inertia. However, heating elements may be placed on the surrounding frame to which the panel is mounted, which allows heat to be transferred to the vent panel by conduction (Figure 19). For large vent panels, the convective heat loss to the surroundings may be too substantial to keep the center of the panel above 250 °F, even with insulation [14]. Adding a vent duct to block the wind (Figure 20) may resolve this. However, vent ducts should be carefully designed because, similar to the discharge piping on a relief valve, the additional backpressure imposed by a vent duct can increase the vent area required [NFPA 68 7.4].

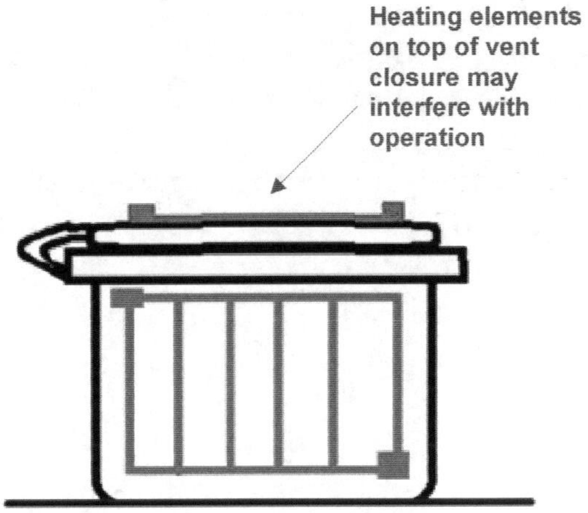

Figure 18 – Concept for Heated Explosion Door
(Figure provided by Controls Southeast Inc.)

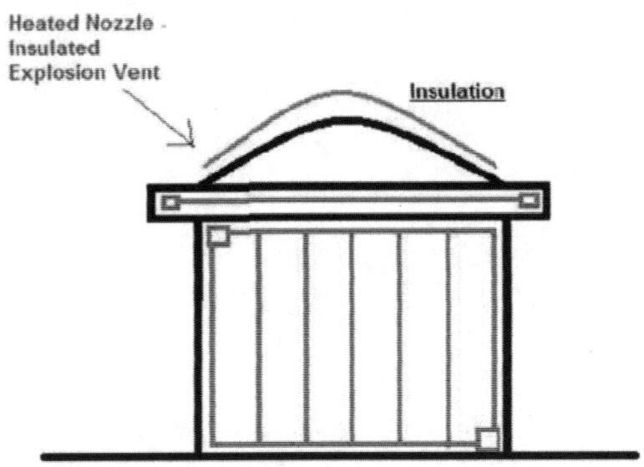

Figure 19 – Concept for Heated Rupture Membrane
(Figure provided by Controls Southeast Inc.)

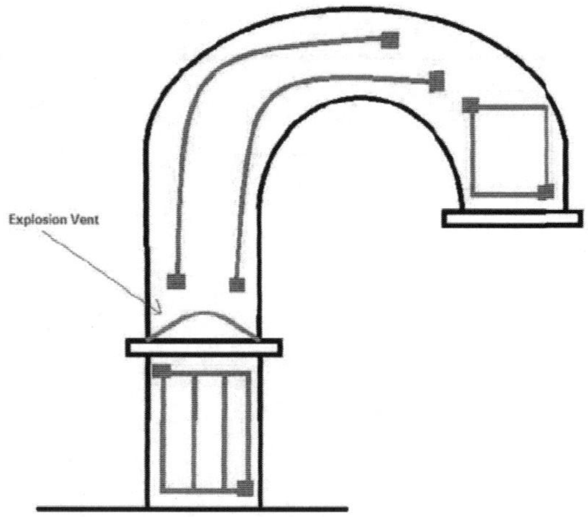

Figure 20 – Concept for Heated Rupture Membrane with Vent Duct
(Figure provided by Controls Southeast Inc.)

Conclusions

The primary conclusions derived from this study can be summarized as follows:

- There is no clear consensus within the industry on the requirement for deflagration vents on sulfur pits.
- Owners need to make deliberate, informed decisions about the need for sulfur pit deflagration vents during the design phase HAZOP, in consultation with their detailed engineering contractor, while also considering site-specific operation procedures and safety practices.
- Sulfur pit deflagration vents are challenging to implement.
- If it is deemed that other safety devices, methods and procedures cannot adequately mitigate the possibility of forming explosive conditions within the vapor space of the sulfur pit without the installation of deflagration vents, the owner may elect to install a pressure vessel rated to withstand deflagration pressure, rather than installing a concrete pit with protection devices that are difficult to design and maintain.

References

1. Johnson, J., Hatcher, N., *Hazards of Molten Sulfur Storage & Handling*, Laurance Reid Gas Conditioning Conference, February 2003.
2. SulfurExperts, *Liquid Sulfur Handling, Treating and Storage*, Sulfur Recovery Seminar, September 2005.
3. Amine Best Practices Group, *Are You Handling Sulfur Safely?*, Hydrocarbon Processing, July 2003.
4. NFPA 68, *Standard on Explosion Protection by Deflagration Venting*, 2007 Edition
5. NFPA 69, *Standard on Explosion Prevention Systems*, 2002 Edition
6. NFPA 655, *Standard for Prevention of Sulfur Fires and Explosions*, 2012 Edition
7. BS EN 14797:2006, *Explosion Venting Devices*, December 2006
8. BS EN 14994:2007, Gas Explosion Venting Protective Systems, February 2007
9. BS EN 14460:2006, Explosion Resistant Equipment, May 2006
10. B. Chase, Telephone interview. August 17, 2012. Mr. Chase is a technical expert with NFPA who provided informal interpretation of some sections of NFPA 68.
11. N. Dowling, *Explosion Limits for H_2S/CO_2/Air and H_2S/N_2/Air Mixtures*, ASRL Quarterly Bulletin 148, Vol. XLV No. 4, Jan-Mar 2009, pp. 1-2
12. D. Crowl, *Understanding Explosions*, American Institute of Chemical Engineers, New York, 2003
13. Center for Chemical Process Safety, *Guidelines for Engineering Design for Process Safety, Center for Chemical Process Safety*, American Institute of Chemical Engineers, 1993
14. D. Campbell, Interview. August 29, 2012. Mr. Campbell represents Controls Southeast Inc., makers of ControTrace heating elements, who conducted heat transfer calculations for several conceptual vent closure configurations.
15. D. Guaricci, e-mail. August 28, 2012. Mr. Guaricci represents ATEX Explosion Protection L.P., who provided photos and indicative price quotes for rupture diaphragms and explosion doors for the example sulfur pit.

CONSIDERATIONS OF TEMPERATURE MEASUREMENT IN THE CLAUS THERMAL REACTION FURNACE

STEVE CROOM

DELTA CONTROLS CORPORATION

It is necessary that the temperature in the Claus Furnace be monitored to protect it from damage, minimize maintenance, avoid costly shutdowns, and improve safety and efficiency. There are two ways to do this: the Thermocouple and the Infrared Pyrometer. The question of which is better is irrelevant, since both offer advantages as well as limitations.

For decades, thermocouples were generally thought to be unreliable in Claus furnaces, which they were. They had a tendency to fail and when they did, they could not be easily repaired or replaced until the unit was shut down. So, the thermocouple gave way to the Infrared device, which did not succumb fatally to improper installation or the harsh conditions of the furnace. And when they did malfunction, they could be serviced, by the use of an isolation valve, while the unit remained in operation.

However, the tendency for the Infrared Sensors (IR) to drift over time and the routine maintenance requirements of the IR units created a demand for something better. So, special thermocouples and infrared pyrometers were designed specifically for the Claus furnace that addressed the operating issues.

Careful examination of Claus furnace thermocouple failures revealed that problems we not so much in the design, but in the installation technique, or more accurately, the lack thereof. All thermocouple failures in Claus service are due to corrosion of the thermocouple elements. Careful examination revealed the reasons for the failures; they

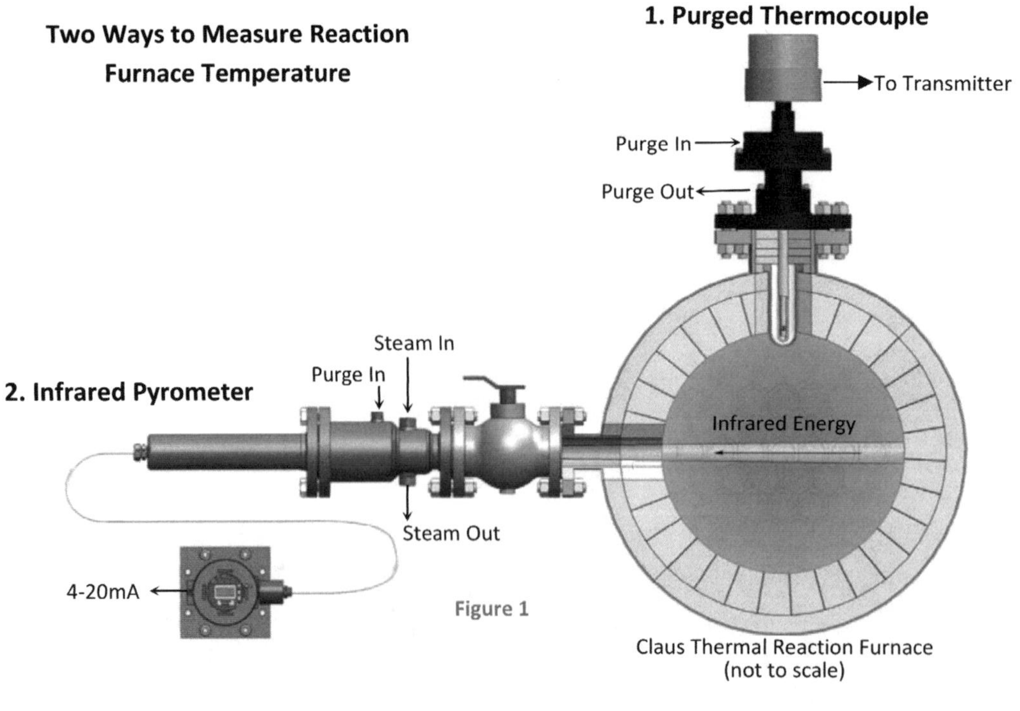

Figure 1

often were of a mechanical nature and not because of the process conditions, severe as they are. Failure was generally breakage of the thermowell, usually because of improper handling and installation, which exposed the elements to corrosion. In a couple of instances, it was observed that the thermocouple was broken during installation resulting in failure by contamination a few months after startup.

So, the solution was to design a thermocouple that did not succumb to the stresses of refractory movement and to design installation tools and fixtures that make installation simple, safe and foolproof. In addition, installation tools and proper installation technique ensure the thermocouple is positioned properly so as to provide an accurate reading of temperature at the refractory hot face.

So, why bother with the thermocouple and its apparent vulnerabilities? The reason is that it is the most accurate device to measure temperature; it is the world standard.

Let's compare them:

Thermocouple advantages:

- Relatively low cost,
- Little or no maintenance,
- High accuracy,
- Unaffected by changing gas composition.

Thermocouple drawbacks:

- May be damaged by severe thermal shock, extreme high temperature, shifting refractory, and improper installation.
- Must be continuously purged to specified pressure and flow.
- Cannot be repaired until furnace shutdown.

Infrared Pyrometer advantages:

- Mechanical isolation using the nozzle valve, permitting service or replacement while the furnace remains in service.
- Withstand severe system upsets. Normally not harmed by moderate over-temperature conditions or shifting refractory.

Infrared Pyrometer drawbacks:

- Vulnerability to sulfur condensing, building up in the nozzle bore
- Sulfur coating the viewport glass window (see Figure 2)

- (Both conditions cause an attenuation of the Infrared energy reaching the sensor, resulting in an ever-increasing inaccurate low measurement).

- Drifting sensor / electronics requires frequent recalibration. Some IR pyrometers incorporate a mechanical chopper mechanism that reduces long term reliability.

Figure 2 Sulfur coating a viewport window

So, what specifically has been done to make both of these two technologies more suitable for Claus furnaces?

For the thermocouple…

1. Installation of a secondary large outer protective thermowell that rests on the refractory and moves with it relative to the vessel shell. It is not attached to the thermocouple assembly, but fits loosely in a large hole bored through the refractory. This permits some movement and shifting of the firebrick relative to the insulating refractory layer as well as shifting insulating refractory relative to the vessel shell. In addition, this secondary well provides protection of the inner element thermowell from thermal shock.

2. Thermocouple installation fixtures provide simple and easy installation minimizing the risk of breakage. A refractory drilling kit ensures that the correct bore is created in the refractory, on center and perpendicular to the mounting flange. This ensures that the proper clearances are established permitting refractory movement without damage. Another valuable tool is an installation mounting bar kit that permits safe and easy installation in other than the recommended vertical top centerline nozzle location. In these non-vertical locations, it is difficult for the installer to insert a heavy thermocouple into the nozzle and simultaneously keep the assembly perfectly centered to avoid any side loads on the ceramic thermowell. If the installer relaxes his grip on the unit during horizontal installation, it will become broken due to a combination of gravity and the heavy flanged headworks. The mounting bar kit attaches to the vessel nozzle flange and supports the full weight of the unit, keeping the thermowell centered during insertion until it is solidly bolted in place.

3. Thermocouples must be purged to prevent contamination over the long term. No materials exist that are "gas-tight" in extremely high temperatures. And in the Claus furnace, the use of ceramic materials is necessary and at the high temperatures, certain elements of the reaction gases migrate though the wall of the ceramic thermowell. Without a purge, these corrosive elements will, in time, collect inside the thermowell and begin to corrode the precious metal thermocouple elements.

Figure 3 - Purge Panel

To prevent this, a continual purge flow at elevated pressure is maintained throughout the thermocouple assembly. Typical flow rates of approximately 11 LPH (~ 0.4 SCFH) are sufficient to protect the thermocouple without cooling the thermocouple elements any more that 1-2 degrees. Thermocouples can eventually fail if the purge and pressure are not maintained over the long term. To eliminate the purge issues, simple, easy to install purge control panels (Figure 3) are provided. In addition, the purge controls also serve as a diagnostic tool to determine the causes of any thermocouple accuracy issues.

Nitrogen is the recommended purge gas; clean dry instrument air may be suitable, but only if it contains *no* hydrocarbons; trace quantities of oil in the purge air will collect and burn inside the thermowell, leaving deposits that eventually damage the thermocouple. (see Figure 4)

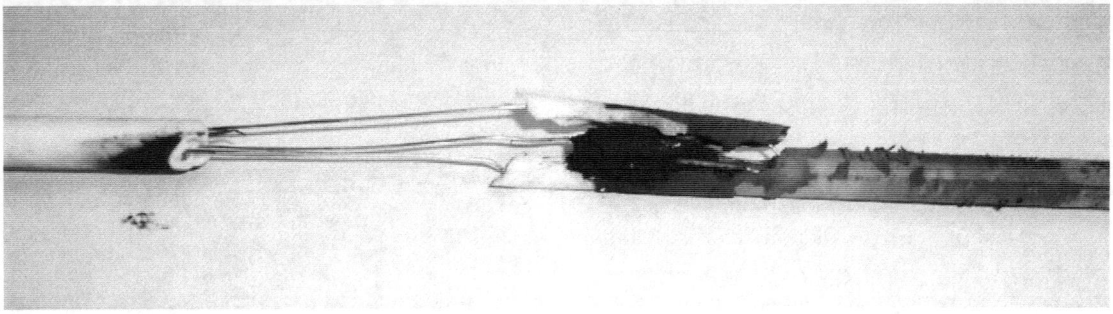

Figure 4- Damage to thermocouple element support due to burning of trace quantities of oil in the purge gas.

For the Infrared Pyrometer…

1. Implementation of a heated and insulated lens assembly eliminates the problem of sulfur buildup in the refractory bore, nozzle and block valve. The heated lens housing generates sufficient heat that when combined with the furnace heat, prevents sulfur from solidifying and obstructing the IR energy pathway to the sensor. High purge flow into the nozzle (a historic industry standard practice) should not be employed as the cooler air encourages the solidification of sulfur and corrosion of the nozzle. To prevent accumulation of dust or ash on the lens window surface, new designs incorporate a very small purge flow that is directed onto the lens window in a swirl path fashion. In addition, the low flow purge is pre-heated to avoid any cooling effect that would encourage sulfur to condense on the lens glass.

2. Early IR instruments suitable for the rigors of Claus service incorporated sensors that were subject to aging and drift, resulting in the need for frequent recalibration. This technology also employed the use of a mechanical "chopper" motor which compromised long term reliability of the instrument.

Modern Claus pyrometers use high stability semiconductor sensors that do not exhibit the aging effects of early designs. These sensors provide long term reliability and freedom from frequent calibration or maintenance. Because semiconductor sensors cannot tolerate the high temperature of the vessel nozzle, a high temperature fiber optic cable is used to connect the lens to the sensor.

Thermocouple Installation Criteria

The thermocouple performs most reliably when installed on the vessel top centerline in the vertical down position. This is because it is the easiest position to make the physical installation and minimizes the risk of breakage during insertion. Also, movement or shifting of the refractory at the top centerline is normally limited to separation of the firebrick from the castable materials and occasional separation of insulating brick from the vessel shell; both are easily tolerated by the thermocouple. There is little shear motion at the top and any that occurs is along the longitudinal axis. Mounting at the side positions of the furnace, although acceptable, makes installation more difficult and, in addition, refractory shifting may also be of some radial shear motion. This refractory movement, if severe, can cause breakage, particularly if any refractory dust/chips settle around the outer refractory thermowell and fill the space needed to permit refractory movement.

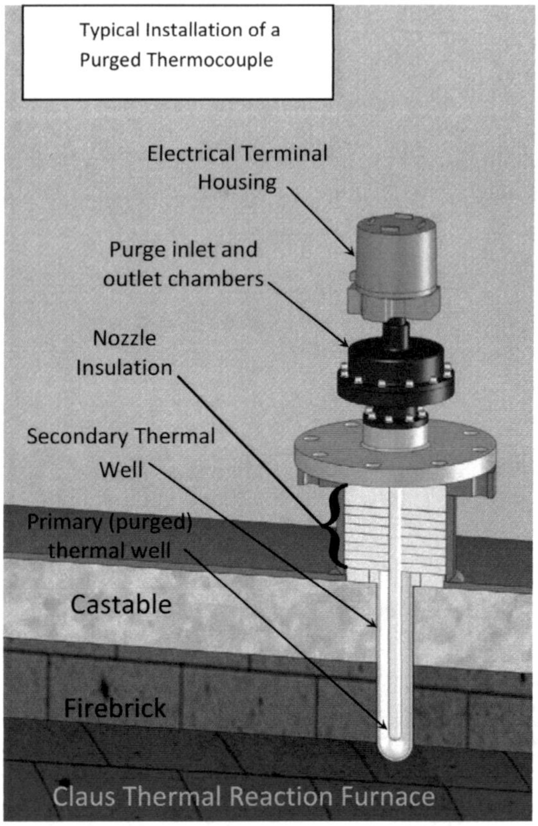

Nozzle sizes of 4" or 6" permit easy installation of nozzle insulating materials that minimize the buildup of sulfur in the nozzle, which could place stresses on the inner "gas-tight" element thermowell, again causing damage. Also larger nozzles permit refractory movement without causing the outer refractory thermowell well to press against the ID of the nozzle.

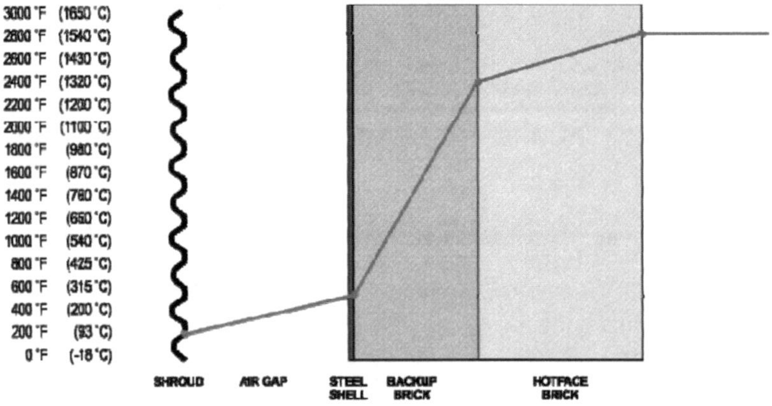

Figure 5 Temperature Profile of Refractory, Shell, and Shroud
Illustration courtesy of Thorpe Specialty Services

Nozzles should be as short as practical. Shorter nozzles maintain heat closer to that of the vessel shell, minimizing corrosion (See Figure 5) illustrating the temperature profile of refractory, shell and shroud). Taller nozzles are cooler, encouraging corrosion and the solidification of sulfur. In addition, shorter nozzles reduce the thermocouple insertion length, lowering the cost of the thermocouple. Shorter thermocouples are also easier to install and minimize the chances of physical damage during handing. It is advantageous to have the thermocouple mounting flange positioned beneath the rain shield/shroud. Nozzles should not be purged as this needlessly cools the nozzle, encouraging corrosion and the solidification of sulfur, destroying the protective capability of the specially designed nozzle insulating materials. Weld neck flanges make ideal nozzles since they have no weld beads in the ID that restrict insertion of the solid insulating rings.

The most critical detail of thermocouple installation is the creation of the bore hole through the refractory. For maximum reliability, this hole must be of the proper diameter, centered, and be perpendicular to the nozzle mounting flange. Although this seems simple, logistical issues often result in incorrect bore hole creation. A bore hole that does not meet the specification causes undue needless expense to repair the refractory for another drilling exercise, and takes valuable time. It is important to protect against castable material entering the base of the nozzle; should this occur, the nozzle depth dimension will be altered, preventing installation of the secondary outer refractory thermowell that rests on the cold face of the refractory. If material is found to have entered the nozzle above the inside surface of the vessel shell, it must be removed leaving a level surface on the refractory and flush with the inside surface of the vessel shell. The easiest and simplest means to protect against this is to use the refractory stops supplied by the thermocouple manufacturer. But, someone must take responsibility of assuring that these stops are fitted to the nozzles prior to the beginning of refractory installation.

Installing a stop on the nozzle prior to refractory installation prevents castable from entering the nozzle.

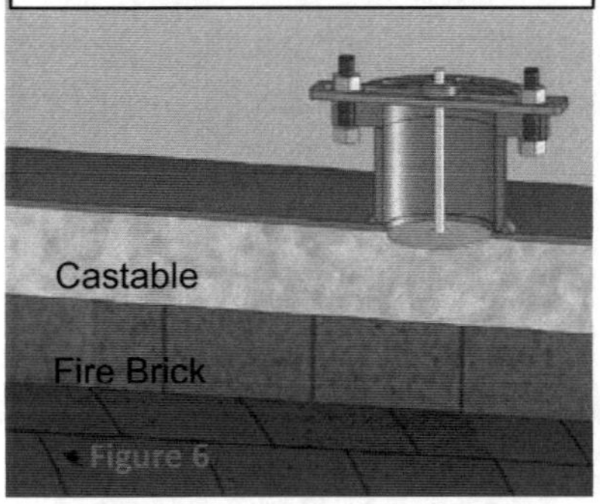

Figure 6

A thermocouple installed in a nozzle / refractory bore that does not meet specification may fail during startup due to shifting refractory and the failure will not become evident until the furnace has been in service for weeks. The drilling of the refractory cannot be performed "free style" without the proper guide fixtures; doing so usually results in failure. The easiest, quickest, safest and simplest means is to utilize the drilling kit furnished with the thermocouple.

Infrared Pyrometer Installation Criteria

The pyrometer, contrary to the thermocouple, is usually installed in a side position on the vessel. This is to provide for easy access, since it can be adjusted or serviced while the furnace remains in operation; however, keeping the nozzle location above the 9:00 / 3:00 o'clock position prevents the accumulation of particulate and molten sulfur, which could eventually block the IR energy path to the sensor. Gravity and vibration provide cleaning assistance as long as there is some downward angle of the nozzle toward the vessel centerline. Remote electronics type units may be installed in a vertical downward position, but consideration must be given to the limited length of interconnecting fiber optic cable from the lens to the electronics.

As with the thermocouple, the nozzle should be short. Again, this is to maintain heat that helps prevent corrosion and the buildup of sulfur. For IR units featuring remote electronics and steam jacketed lens assemblies, no nozzle purge should be used and the nozzle, block valve and lens should be insulated to maintain higher temperature, closer to that of the vessel shell (see Figure 7). This minimizes corrosion and discourages the deposition of solid sulfur and eliminates coating of the lens window glass, preventing the need for routine maintenance. Remote electronics type IR units require that 50 - 100# steam be fed to the steam jacketed lens housing in order to realize the benefits of immunity to sulfur build up and corrosion.

Figure 7 Pyrometer with Insulation

With integral IR units, whereby the electronics and lens are mounted directly to the nozzle, consideration must be given to the manufacturer's installation instructions regarding nozzle purging and any cooling requirements to protect the instrument. No insulation would normally be recommended for these type units.

Nozzles are usually 2" or 3". It is important to assure that the bore hole through the refractory is clear and in alignment with the lens. The bore hole should be a minimum 1.5" ID, centered and perpendicular to the mounting nozzle flange to assure a clear pathway to the interior of the vessel. Although not normally necessary, the creation of this bore hole through the refractory is most easily performed using the drilling kit furnished with the IR pyrometer.

Longitudinal Location of Temperature Measurement Devices

There is no standard recommended longitudinal placement of temperature measurement mounting nozzles in a Claus Reaction Furnace. The ideal placement is usually determined by the burner manufacturer using CFD (Computational Fluid Dynamics) to select the most appropriate locations. Consideration must also be given to physical obstructions, access to the instruments and of the refractory at the point of penetration. With

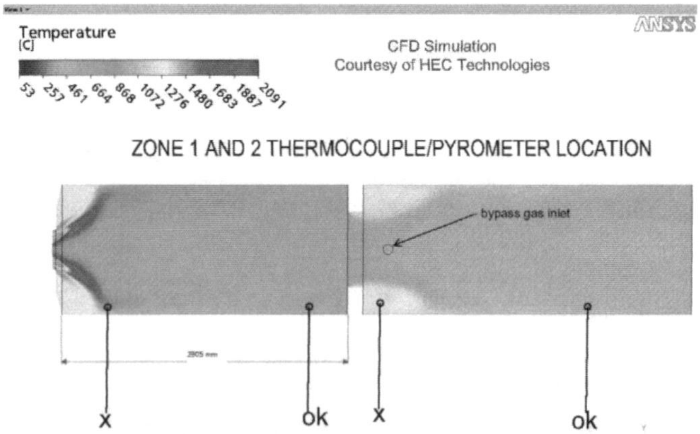

Figure 8 Furnace Temperature Profile – Uniform Combustion

no recommendation by the burner manufacturer or the benefit of CFD, placement is usually at a point about two-thirds the distance from the burner to a checker wall / choke ring and another about two-thirds the distance from the checker wall / choke ring to the WHB tube sheet. (See Figure 8). Additionally, CFD profiles may reveal other possible problem areas, such as burner design, feed streams or choke ring / checker wall effectiveness, any of which could result in premature failure, shut down and costly remediation (See Figure 9).

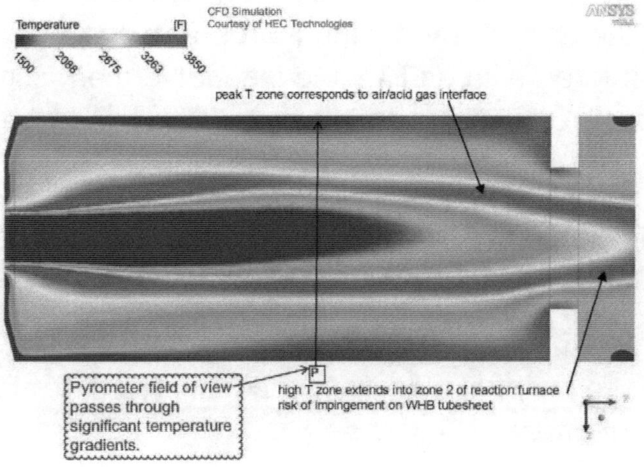

Figure 9 CFD Simulation of poor burner design. Pyrometer or T/C sees relatively cool refractory even though hot gases impinge on WHB tube sheet.

Refractory Dry-Out Temperature Measurement

The curing of refractory is critical to long term refractory performance. Failure to adhere to the refractory contractor's recommendations may cause premature refractory failure and invalidate warranties. Accurate measurement of the critical and relatively low temperatures during dry-out is a challenge. Generally, IR pyrometers and thermocouples that are suitable for measuring reaction temperatures are unable to accurately measure the critical low temperatures. But, there are a couple of ways to accomplish this.

One is to select the proper thermocouple element type. Type "S" or "R" are generally suitable for Claus service to a maximum temperature of 1500°C and provide usable accuracy to as low as 100°C, useful in measuring the refractory dryout profile. To maximize immunity from damage from excessive temperatures, Type "B" thermocouple elements are employed (the most commonly used in Claus service) and will perform to 1800°C without damage. However, the Type "B" is not suitable for measuring the low "dryout" temperatures. To utilize the Type "B" and also have low temperature measurement capability, an auxiliary element (usually a Type "K" element) can be fitted as a "third" element making the unit suitable for accurate measurement of the low end dryout temperature. Although this is a convenient method, the single drawback is that when the reaction temperature exceeds that of the dry-out thermocouple element's max temperature (about 1400°C), it will melt away, making it a "one use only" "dry-out" thermocouple. So, if it becomes necessary to shut down for additional refractory work and subsequent dryout, this "third" thermocouple would no longer be available.

A more suitable method of dryout would entail the use of the insertable "calibration" thermocouple that is an option offered with Infrared Pyrometers. This provides an excellent means to monitor this dry out profile. However, it requires that the IR unit remain out of service until the refractory dry-out cycle is complete, but the IR pyrometer normally is of little use at low temperatures anyway. Upon reaching the end of the dry out cycle and before going on line, the insertable thermocouple is withdrawn and the IR device returned to its operating position and placed in service.

A point of interest... some refineries have opted to prohibit opening of any Claus Furnace port during operation, including that of utilizing thermocouples and cleaning devices fitted with a mounting gland and an operable block valve.

Pyrometers installed to monitor the reaction gas temperature in lieu of the refractory hot face may not effectively read the gas temperature of fuel gas or natural gas during the dry-out phase.

By far, the most effective means to obtain ideal refractory monitoring is to install multiple "K" thermocouple elements at critical locations inside the furnace. This minimizes the chance of improper readings due to a single element placement, which could be in an unsuitable location

Use of Both IR and Thermocouple

Use of both technologies provides redundancy and protection from common-cause failures; i.e., conditions that affect one technology will likely not affect the other technology.

Since thermocouples are the accuracy standard, they provide a constant means for verification of IR units in actual service.

Thermocouples succumb to extreme over-temperature conditions (1825°C melting point of "B" type), loss of long term purge, severe shifting refractory or severe thermal shock. None of these conditions would normally have any detrimental effect on IR units.

IR units, as mentioned above, may fail because of other problematic issues. Setting aside the newer remote electronic units designed to eliminate the problems of blockage of the IR sight path due to build up of sulfur and coating of the lens glass, these are common problems. Although these issues may be corrected by opening the nozzles and cleaning them out, it is a costly procedure. With IR units that suffer these problems or that require frequent calibration due to sensor drift, the thermocouple can be used to determine when IR maintenance should be performed.

This cleaning maintenance work on the IR can only be performed however, when the nozzle block valve is operable and not frozen in the open position. Means to prevent this are the selection of a proper fireproof valve and keeping it at elevated temperature. Also, periodically operating the valve handle may prevent it from freezing.

Differences in Measurement Reading between IR Pyrometers and Thermocouples

It has been frequently observed that there is a discrepancy between the temperatures reported by thermocouples versus pyrometers. In these instances, there are three possible causes:

1. The instruments are reporting the actual temperatures which are different.
2. The thermocouple is wrong.
3. The pyrometer is wrong.

Of course, discrepancies can also be caused by any combination of the above.

Actual Temperature Differences

First, it is important to understand that no two points in the furnace are at exactly the same temperature. Reaction gases start out at near ambient temperature at the burner and rise in temperature as they react while traveling down the furnace. At the same time, heat is being lost through the wall of the furnace. High gas velocities, turbulence, stratification and radiant heating tend to smooth out the temperature gradients somewhat, but temperature differences of several hundred degrees at different points along the furnace length are not uncommon. These differences can change with gas flow rate and gas composition.

It is also important to understand what is being measured. A thermocouple reports the actual temperature at the thermocouple junction. An infrared pyrometer measures the infrared energy emanating from a small region on the refractory wall; about 20% of this energy is radiated by the wall that the pyrometer is focused on, the other 80% of the energy is reflected energy that was radiated by other parts of the furnace. Thus the infrared measurement is a kind of weighted average of the temperature in that region of the furnace. (Those familiar with infrared temperature measurement may be tempted to think that the emissivity adjustment on the pyrometer can allow the instrument to read only the temperature of the wall at the focus point, but that concept only applies where a hot object is surrounded by a cooler environment – in that case there is virtually no reflected energy. Inside a reaction furnace, the reflected energy is abundant and cannot be distinguished from radiated energy.) The result is that if there are significant temperature gradients within a furnace, then a perfectly calibrated pyrometer could read a different temperature than a thermocouple, even if the pyrometer were focused directly on the thermocouple's thermowell.

In practice, there is usually less than 25°C difference between two instruments located close together. If the measurements are within this range, it is probably about as good as can be reasonably expected.

Thermocouple Errors

Thermocouples cannot read erroneously high. If the thermocouple is reporting a higher than expected temperature, strong consideration must be given to the possibility that expectations, rather than the thermocouple, are in error. One reason for a high temperature reading is a direct flame impingement on the thermocouple. In this case, the thermocouple is not in error, but is reporting the high temperature to which it is exposed.

Thermocouples can become inaccurate by reading lower than the actual temperature. This is usually caused by contamination or corrosion. Contamination is caused by lack of proper purging of the thermocouple, either due to a problem in the purge system, or else due to breakage of the ceramic thermocouple element well.

Most Claus furnace thermocouples are equipped with a second 'reference' thermocouple that is internally embedded in ceramic and not directly exposed to the purge gas inside the thermowell. If these two separate thermocouple elements in a given thermocouple unit are consistently reading within approx 25°C of each other, it may be assumed that they are reading accurately.

Thermocouples can also read low if the purge flow rate is significantly higher than the recommended purge rate (approx 0.4 scfh or 11 LPH). This low flow rate cools the thermocouple by only about a degrees or two, but cooling rapidly increases as the purge rate is increased. Excess purge cooling can easily be tested by temporarily turning off the purge and observing a rise in the indicated temperature. (Briefly turning off the purge will not harm the thermocouple.)

Of course, as with any instrument, the temperature transmitter connected to the thermocouple may be improperly programmed or spanned.

Infrared Pyrometer Errors

IR pyrometers are factory calibrated by pointing them into a 'black-body' calibration environment. This type of device very precisely models the IR emission of a furnace at uniform temperature. As discussed above, actual Claus reaction furnaces are not completely uniform in temperature, so temperature reported by a pyrometer will always be a weighted average of the furnace temperature. The measured wavelengths are carefully chosen to be such that the reaction and reactant gases are transparent, so normally the gas composition and flame temperature will have little effect on the IR measurement.

The most common problem with pyrometers has to do with material coming between the furnace and the sensor and occluding the infrared energy. Usually, pyrometer errors come on gradually whereby the IR pyrometer slowly decreases its signal relative to the thermocouple(s). Usually, the cause is from a very thin layer of sulfur accumulating on the lens viewport glass. As this becomes more and more opaque, the IR energy reaching the sensor is attenuated, causing a lower and lower reading over time. Another may be sulfur building up in the base of the nozzle or in the bore through the refractory, again, partly blocking the view path or the IR. This problem is minimized and often prevented by keeping the viewport at a temperature above the freezing point of sulfur. This is done with the use of heated lens assemblies, pre-heated low flow lens purge (only). And by insulating the nozzle, block valve and lens assembly, buildup of sulfur in the optical path will be eliminated. High flow purging of the nozzle cools the nozzle and encourages the solidification of sulfur; it is not recommended.

Older IR designs for Claus service often experience drifting of the accuracy of the sensor / electronics, requiring frequent recalibration. This is a historic issue with lead-based sensors; modern semiconductor sensors usually perform for many years without calibration.

Although uncommon, vibration or refractory shifting as the furnace heats rises, particularly upon initial start up, can cause a misalignment of refractory to the sight path of the IR lens, partially obscuring the “view”. Most modern pyrometers provide for visually checking of the boresight of the instrument to detect misalignment.

Mitigating Measurement Discrepancies

If the difference in measurements between two nearby instruments is relatively small, (perhaps 30°C, although that number is somewhat arbitrary) things are probably doing as well as can be expected. Larger differences may require some diagnostics as mentioned above to determine which instrument(s) may be in error. Discrepancies that increase over time are a particular warning that something is amiss.

After verifying that the optical path is not occluded, the operator may determine that the pyrometer needs to be recalibrated. The best method to calibrate a pyrometer is with another pyrometer. The process pyrometer is temporarily removed and replaced with a portable pyrometer that has matching optical characteristics. After reading the furnace temperature, the process pyrometer is placed back in service and adjusted to match the portable pyrometer reading. Note that the portable pyrometer should be one that is specifically designed for the job. A general-purpose IR gun typically has too wide of a viewing angle to accurately see down a valve, nozzle, and borehole into the furnace. And its calibration does not take into account any attenuation in the viewing window.

An easier, but less accurate method of calibration is to set the IR device to match the reported reading of the nearest permanent thermocouple, once it is established that the thermocouple(s) are operating properly.

The use of insertion thermocouples at the IR port does not necessarily make an accurate calibration, although it is the best method if the above methods are not available. When using the insertion thermocouple, it is wise to turn off any high flow nozzle purge, as this has a cooling effect. And, the insertion thermocouple will likely be destroyed if introduced to a temperature higher than 1400°C, which would melt the "K" type element. Higher temperatures would also cause physical damage to the insertion thermocouple's stainless steel sheath.

Recommendations:

- Use both thermocouple and IR technology to reduce susceptibility to common-cause failures.
- Install both in different regions of the furnace
- Use thermocouples designed specifically for Claus service and ensure that proper purge specifications are carefully followed.
- Consider IR pyrometer designs that are designed for Claus service and which are heated to protect against sulfur build-up in the optical path.
- Follow the details of installation instructions and use the proper tools and fixtures for installation.
- Span Instruments to read higher than the maximum expected operating temperatures.

UNDERSTANDING TODAY'S LIMITING FACTORS ON REACTION FURNACE LININGS

by

Thorpe Specialty Services Corporation
www.ThorpeSSC.com

Andy Piper – VP/Sales & Marketing Manager
Jeff Proctor, P.E. – Technical Services Manager

Prepared For Presentation at the 2013 AIChE Spring National Meeting
April 28 – May 2, 2013
16th Topical Conference on Refinery Processing
Sulfur Management Session
San Antonio, Texas

ABSTRACT

SRU operators are experiencing increasing demands for improved reliability due to significant changes in emission regulations and penalties, more severe operating conditions, high cost of repairs and the consequential impact a reaction furnace failure can have on other operating units in refineries. Management is finally paying more attention to these issues, however, before the SRU problems can be properly addressed, they first must be fully understood.

Our paper outlines the common problems plaguing SRU linings today, including a lack of understanding of refractories in general and the unique requirements for SRU applications, as well as the consequences of the technical and commercial decisions made by licensing companies, project teams and owner/operating companies. We will show examples of common failures throughout the industry and then illustrate some of the complex technical aspects that need to be understood and tradeoffs that must be considered in the design process. The ultimate goal of the design/implementation process is to "balance" many competing factors in order to maximize the performance, reliability and life of the lining system; not for theoretical steady-state operating conditions, but for the real world conditions that are imposed on the refractory linings.

INTRODUCTION

If you are reading this paper then the chances are fairly high that you or your company has suffered the pain common to refractory issues in reaction furnaces. This is not a pleasant or an inexpensive experience; nor is it isolated to only you or your company. We see the same issues regarding premature refractory failure repeated time and again all over the world. The painful impact to the plant may include the following:

- Premature &/or frequent maintenance repairs or lining replacements required between or during planned outages
- Longer downtimes due to extra refractory work requirements
- Extended startup times or even failures due to dryout issues
- Shell breeches and the safety & environmental issues associated with this extreme failure
- Loss of production capability of related units during unexpected or extended reaction furnace outages

Typical Extreme Refractory Failures

In other words - less runtime, less reliability, higher costs and worse, less profit. Many operating companies today are only too aware of these issues but are unaware of how to correct them.

There are reasons the same failures occur repeatedly and there are ways to significantly improve these situations. As with most problems, the devil is in the details. In this case, the devil is in even understanding what details. This is especially true for refractories as they are normally considered a commodity that anyone can design, supply &/or install; making refractories almost an afterthought.

In actuality, SRU linings are complex structures that require careful consideration. They are not only the first line of defense to protect the vessel shells, but are the only defense. In our opinion, these units operate under more severe service conditions than any other unit in a refinery. The consequences of "getting it wrong" up front then brings in the added costs and frustrations of dealing with substandard design, materials of construction, or workmanship for the entire working life of the SRU or until lining improvements are implemented.

We recognize that under certain conditions, all thermal reactors are capable of melting any refractory made today. Having in the past supplied these units on a turnkey basis and having performed countless emergency repairs for customers due to unexpected failures, we base our lining recommendations on an Optimized lining approach. Typically we find that due to cost pressures &/or lack of awareness of the operating potential, linings are designed to the theoretical normal operating conditions. We are aware of the particular difficulties with these units due to startups and shutdowns, natural gas firing, O_2 enrichment, hydrocarbon slugs, operational errors, thermal striations or complete unit overheating from poor burner mixing or damaged burner components. We consider these to be the real world operating conditions for thermal reactor service; it is not a matter of "if" but "when" a thermal event stresses the lining. Every unit should be evaluated for its size, geometry, thermal profiles, etc. in an effort to optimize the lining to better endure these events.

We will begin with the commercial or project structure issues involved in most new projects and then outline some of the important technical issues that need to be considered to optimize the refractory performance.

PROJECT CONSIDERATIONS

Many of the problems with today's refractory lining systems actually begin in the project phase. While it is sometimes politically expedient or convenient to identify only one thing as the root cause of a problem, we find in refractory that there are normally numerous contributory issues combining to eventually result in a failure or need for repair, sooner than is necessary. Below we will discuss some of the typical problems we find in different projects which we believe have contributed to various refractory issues.

Project Structure – How a project is structured and how the team is rewarded can have a significant impact on how decisions are made and therefore on the final end product and its performance. The general refractory guidelines are normally established by the licensor early in a project and are used to determine vessel and duct sizing, etc. Unfortunately, due to the extreme variations in size, geometry, and operating requirements for these units, such guidelines are often inadequate. Even worse, some specifications require spending excessive dollars to satisfy unnecessary requirements that bring no true value to the lining, while totally disregarding more important considerations. Getting refractory knowledgeable people involved early in the project is dependent on how the project is structured and is essential in mitigating problems that can develop "down the road". If an EPC is awarded the full turnkey project, the customer normally will not challenge the design basis, believing the EPC has the refractory knowledge as well as being reluctant to impose any requirement that could void a performance warranty. If the Project has broken the refractory components into pieces between a licensor, mechanical erector(s), various equipment OEM's, vessel fabricators and local installers, the communication path and responsibility trails are often too long to effectively assure optimal lining designs and installations. Too often inappropriate companies are involved in the refractory process.

Problems associated with the project structure are typically the result of the team underappreciating the complexity and importance of the refractory system as they focus primarily on a low cost or convenient solution. From the standpoint of assuring the success and reliability of the lining, we believe that the refractory design, supply and installation should not be fragmented and broken apart, and the refractory company should be involved at the very beginning of the project as the vessels and ducts are sized.

Lack of Refractory Knowledge – Several factors contribute to the lack of an industry wide, thorough understanding of the diverse and complex issues associated with sulfur reaction furnaces today. First of all, it is important to recognize that owners, manufacturers, designers and installers have become more and more dependent upon improved monolithic refractory lining systems in recent decades. The significant improvement over the last 25 years in monolithic product formulation, installation knowledge and equipment, has contributed to the replacement of many brick applications with monolithic linings throughout the industrial markets. There has been a corresponding loss of understanding of brick systems. This is particularly true in refineries where either monolithic materials or ceramic fiber based products are used in most other units outside the SRU. Since it is generally accepted (and we agree) that brick linings are the best choice for thermal reactors there has arisen a "disconnect". We have a preferred product form that has experienced a declining user base.

With the loss of brick design and installation skills, it is easy to understand why the EPC/Licensor/OEM's, for which refractories are only a small component to the entire SRU process, are not aware of the details involved in this critical, specialized part of the unit. While we have not worked or visited with every EPC/Licensor/OEM, we have worked or visited with many. As yet, we have not found one with a true refractory design engineer on staff. We normally find an engineer assigned to follow that company's refractory use and specifications. The lack of proper refractory design knowledge combined with their companies having to compete for projects on a price basis, results in a lack of desire to consider improved methods. This lack of internal knowledge leads to the use of outdated specifications ultimately allowing the use of suboptimal materials and systems. Most of the specifications focus primarily on the individual materials to be used without sufficient regard for how those materials can be successfully integrated for the particular application to meet performance expectations. It is the systems' performance that is critical, not the individual materials.

Lack of Customer Feedback – In an attempt to defend the EPC/Licensor/OEM's from some of our comments above, it should be noted that many are very good, quality conscious companies and individuals operating on the firm belief they are providing the best linings possible. Part of the problem between belief and reality is that most times customers do not give proper feedback to the original project participants. We have seen numerous occasions where failures occurred well within the warranty period but the plant proceeded to perform repairs without contacting the original responsible parties. This makes for a more economical repair compared to getting the original parties notified, mobilized to inspect, time to evaluate the failure, etc…all before beginning the actual repairs. Plants typically take this approach as lost production is far more expensive than just performing the repairs and getting the unit back up. The reasons for not contacting the responsible parties may change, but the results do not change; a continuation of poor designs or processes from project to project by the original suppliers who are unaware they had a problem.

WHAT NEEDS TO CHANGE?

Higher Priority on the Refractory Process – To optimize lining life and performance it is strongly suggested to think about refractories sooner in the project rather than later. If reliability is important, the refractory linings should be considered as an integrated system requiring careful consideration through all phases of the entire project instead of passing this important function down the project food chain. Ideally the lining design, material supply and installation would be of single source supply and responsibility in order to prevent miscommunications and misunderstandings. As these functions are normally broken apart and each step bid out independently, it may require restructuring the project to allow for a single-source evaluation and award. Overall, since these linings should be brick construction, the evaluation process should focus on companies that emphasize and are active in industries and units involving complicated brick construction, but also including active participation in brick design and construction in SRU linings. The demanding nature of these units makes it imperative that awards be based on the knowledge and ability of the refractory company rather than the low dollar proposal or to a "Good Ole Boy" network system.

Better Understanding of Reality – First of all, again, the burner systems in these units, whether for low temperature gas plant operations or for refinery related units with or without O2 enrichment, can (and do) melt any refractory made today. Designing for only the theoretical operating range should not be acceptable. Optimizing the system to better survive those upset conditions is the only way to get the best reliability and longest refractory life with the lowest life cycle cost.

Demand an Engineered System – One of the biggest misconceptions in the marketplace is that if it is on a drawing, then the lining is engineered. If so, one can definitely argue the quality and depth of the engineering on some of the drawings utilized. Many of these should be labeled "Bill of Material" or "Refractory Placement" drawings which would be more accurate. Material science does apply to refractories just as it does to the steel components. There are physical material properties that when understood against the operational demands of these systems, can be applied to thermally and mechanically balance the system to optimize the lining performance. An engineered system will address and control many aspects of the system which goes through many changes on its way to a potential of say 3000°F. Much of the true engineering of a self-supported brick system is performed behind the scene and is not obvious to the untrained person looking at the drawing. To illustrate the point, we will discuss a few of the concerns that need to be addressed to optimize a lining system.

- Thermal Balance Of The Entire System – While the refractory lining system is designed to protect the steel shell from high temperature corrosion and metal failure, it is the exterior weather shroud (often inappropriately called a rain shroud) that protects the shell from low temperature corrosion. Both systems, in unison, are designed to protect the vessel shell from high and low temperature corrosion. To those unaware, it seems natural to separate these two items (the refractory and the external shroud) in the initial construction. Many times the two systems are designed by completely different companies (assuming an external shroud was even used). However, the external shroud can have a significant effect not only on the vessel shell temperature, but also on the thermal profile through the refractory lining. This thermal profile can have a significant effect on the ability of the refractory lining components to perform under the extreme stresses and temperatures to which they are exposed. These systems should be designed by the same engineer familiar with the dynamics of both systems and the interaction between these systems.

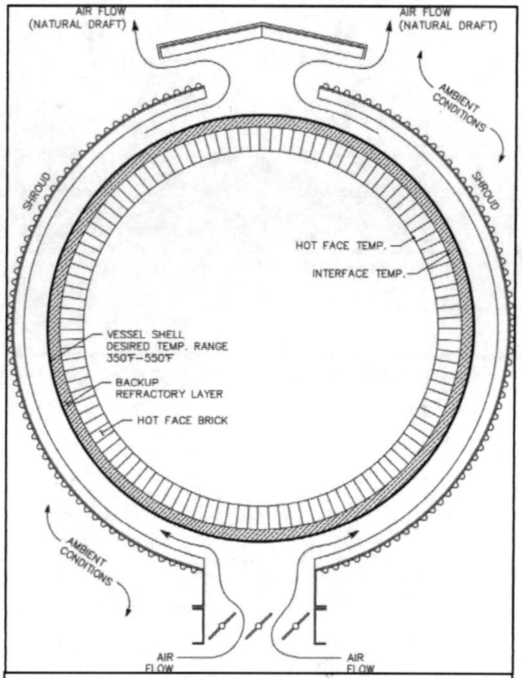

Full Weather Protection "Shroud"

- Proper Lining Thickness – Determining the ideal lining thickness to optimize performance is a complicated process. The importance of this step is often overlooked. Many times, generalized rules of thumb or a similar sized unit is used from a past job that appears to have been successful and as long as a simple thermal calculation seems to show a proper shell temperature; this process stops. However, lining thickness is a function of many factors, some of which are; vessel diameter, thermal profile thru the system, mortar quality and design, brick shape, physical properties of lining materials, etc. Determining proper lining thickness is usually an iterative process as the design engineer balances the effects of the complete thermal system to stresses in the refractory lining and how to control or moderate those effects within the properties of the materials utilized.

- Hot Load Strengths – For unsupported (unanchored) brick linings, consideration of the hot load strength of the refractory products is critical and is the beginning point for designing these systems. Generally the unrestrained thermal expansion of the lining will be approximately 3 times as much as the vessel shell. When heat is applied, the lining goes into severe compression. If the thermal profile thru the lining is excessive, it will plastically deform under the stress resulting in what is referred to as "Hot Load Deformation". So essentially, hot load deformation occurs when the temperature of the material exceeds the point at which it will be able to resist the compressive stress to which it is subjected. While there can be other contributory issues, the evidence of hot load deformation is seen when units are inspected after a service campaign in the form of open joints overhead, sagging brick, or complete collapse of rings overhead.

Classic examples of high temperature deformation

Minimizing deformation is a critical component of an overall design. Most specifications do not even address these issues or are misdirected by attempting to make references to specific manufacturers' brand names (or equals) or other physical properties that may have no bearing on the design. One example is specifications for alumina content. The error is assuming that higher alumina content automatically translates into higher hot load strengths. This is sometimes a result of looking at the melting temperature of the individual brick product and not understanding that materials weaken before they melt; i.e., these properties are not directly related to each other. A severe, high temperature, long duration testing regimen clearly shows there are significant differences in these supposedly competitive high alumina bricks and also show, for example, that the highest alumina (99%) bricks are not good for typical usage in reaction furnaces. Field experience has also proven this to be true. Direct testing for the needed properties should be required.

- Brick Keying Action – Due to lining movement as a result of thermal expansion/contraction, it is essential that the individual brick shapes have the necessary "keying action" to keep them from slipping out of position. One must remember that gravity is pulling down on the overhead brick 24/7. If there is not enough keying action or taper on a brick, it will slip out. Should this happen, the entire ring will become unkeyed and the likely result is more brick falling down creating a larger hot spot and typically a shell breech.

Example of brick slippage due to insufficient keying action.

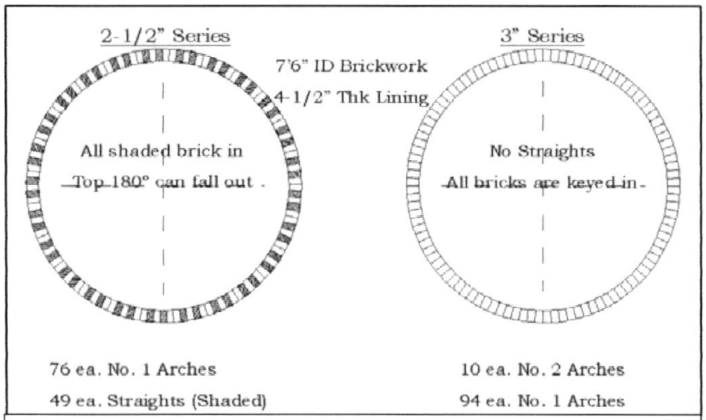

Two sample ring combinations from standard industry manufacturers' guides. The tables do not indicate what combinations are appropriate for the applications.

Utilizing standard brick ring tables will provide a design engineer many options to simply turn a circle, but will not tell them which combinations are best for the application. Choosing the correct brick shapes with the proper amount of keying action requires an experienced designer familiar with the brick shape options, their benefits and disadvantages combined with the need of the specific application.

- Controlling Brick Movement – We have already described the significant thermal movement expected with these products. Unrestrained, brick will continue to move as the unit is thermally cycled with the end result being large open joints, hot spots or collapse or failure of the immediate area of the lining. On the other hand, an excessive amount of restraint will overstress the lining and the vessel shell resulting in possible damage to both. Controlling brick movement requires knowledge and experience to utilize the correct balance between restraining brick movement by use of the selective brick shapes in the correct places while relieving the expected stresses with properly placed and designed expansion relief.

Drawing on left illustrates improperly oriented brick on the dished head and the result of unrestrained brick on the bullnose ring (center picture). Drawing on right illustrates properly oriented head brick resulting in an anchored bullnose ring with expansion relief designed into the outlet throat.

Summary

Improved reliability and performance is possible for these units, even for full time oxygen enriched, high temperature operations. First one must realize the importance of the refractory linings and then raise and maintain the visibility (the coordination) of all aspects involving refractory. Once this is accomplished we believe that by following three simple keys, reliable linings are possible for these severe operating units. Detail attention at every step of the project, start to finish. We call these the **Three Keys to Reliability**:

1. **Engineered Lining Design** – Qualifying a Designer cannot be based on how old they are or how many drawings they have generated. Careful consideration must be given to:
   - Knowledge and familiarity of the operational atmosphere and capabilities of a thermal reactor; normal and upset; real world conditions
   - Understanding the needs of the refractory system to survive the normal and upset conditions, as much as possible.
   - Must be completely familiar with complicated brick construction specifically for SRU's.
   - Able to perform the complicated thermal calculations resulting from the shroud effect and operational swings under different outside ambient conditions
   - Familiar with the information important to the installers of the lining system
   - Clear understanding of the critical material properties needed for the specific lining design

2. **Proper Material Selection** – This should not include brand named products (or equals) as the only criteria. The appropriate material properties critical to meeting the needs of the system must be properly tested for before shipment. For the hotface brick, this should include a regimen of high temperature testing for extended periods of time to verify suitability for service per the design. Additional standard quality control testing should be performed and related to those of the testing regimen to verify consistency for the balance of the manufactured brick. For the other backup and/or monolithic materials, specific critical properties should be determined and included in the manufacturer's requirements.

3. **Experienced Installation Crews** – Not all refractory installers are knowledgeable and skilled at complicated brick construction for SRU reaction furnaces. In today's world of high monolithic refractory use, there are fewer skilled craftsmen than in the past. Careful attention should be paid to the skill level of the installers, the use of appropriate equipment and installation procedures to assure that only experienced crews are involved in the installation of these linings.

(27a) Life Cycle Assessment and Life Cycle Cost of the Pyrolysis of Switchgrass to Use As an Energy Source in a Conventional Power Plant

Nuttapol Lerkkasemsan, Chemical Engineering, Virginia Polytechnic Institute and State University, Blacksburg, VA

Biomass is expected to become an important energy source. This paper considers both life cycle assessment and life cycle cost of the pyrolysis of switchgrass to use as an energy source in a conventional power plant. Specifically the greenhouse gas (GHG) emission for power generation was investigated through life cycle assessment. The process consisted of cultivation, harvesting, transportation, storage, pyrolysis, transportation and power generation. Here pyrolysis oil is converted to electric power through co-combustion in conventional fossil fuel power plants. The conventional power plants which are considered in this work are diesel engine power plant, natural gas turbine power plant, coal-fired steam-cycle power plant and oil-fired steam-cycle power plant. Several scenarios were conducted to determine the effect of selected design variables on the production of pyrolysis oil and type of conventional power plants. The set of design variables consisted of land fraction, land shape, the distance needed to transport switchgrass to the power plant, the distance needed to transport pyrolysis oil to where, and the pyrolysis plant capacity.

The results show that with an existing average agriculture land fraction in the United State which is 44.1 percent of all the United State land area, the cost of electricity from pyrolysis of switchgrass is lowest at $1.2 per gallon with a capacity of 5000 ton of switchgrass per day of pyrolysis plant. However, power generation from switchgrass required huge land especially for an existing average agriculture land fraction which is 0.441. In order to provide enough switchgrass for a pyrolysis plant capacity of 5000 ton per day, a large land area with a size of 1706.11 square miles is required; note that the land area is bigger than the state of Rhode Island. Using natural gas turbine power plant for electricity generation, the price of electricity can go as low as 7.70 cent/kwh; contrast this with the average electric price of 10 cent/kwhin the United States. Based on 5000 tons of switchgrass per day, a natural gas turbine power plant can produce 5011.86 million kilowatthours, which is enough to supply the state of Hawaii. The main advantage in using a pyrolysis plant is the negative GHG emission from the process. This conclusion came from Life cycle assessment analysis of power generation using pyrolysis oil, which showed that it was environmentally friendly and would therefore reduce the GHG emission to the atmosphere. If and when carbon credit becomes significant, pyrolysis oil could be even more competitive.

(27b) Natural Gas as a Transportation Energy Resource: Life Cycle Considerations for Personal Mobility Options

Christian M. Lastoskie, Civil & Environmental Engineering, University of Michigan, Ann Arbor, MI

Presently the U.S. is experiencing a significant boom in domestic natural gas production due to the application of hydraulic fracturing and horizontal drilling to Marcellus and other formations with an abundance of shale gas. The combination of decade-low natural gas prices and increased restrictions on coal utilization for electric power has led to a displacement of coal-fired generation with electricity from natural gas.

With the prospect of the United States undergoing a transition to being a net energy exporter within the next decade, it is timely to consider scenarios wherein a displacement of petroleum as the predominant domestic transportation energy resource occurs, with steadily increasing utilization of cheap abundant natural gas as the source fuel for personal mobility. In this presentation, we consider three options for natural gas-derived personal mobility: compressed natural gas automobiles, battery electric vehicles, and hydrogen fuel cell cars. The life cycle environmental impacts are compared for the operation of these three types of vehicles, including the fuel processing infrastructure required to deliver the respective fuel for each vehicle. We consider the impact of potential future constraints on carbon emissions on the environmental footprint of each mode of natural gas-powered personal mobility, on a per mile driven basis.

(27c) Environmental Controls in Coal-Based Electrical Energy Production

Noah Meeks, Research & Environmental Affairs, Southern Company Services, Inc., Birmingham, AL

Over the past half-century, the coal-fired electric power plant has evolved from a simple machine to a complex chemical process plant. While the heart of the industry remains the same: turning turbines with coal-generated steam, the treatment of coal combustion emissions has grown considerably and more processes are under development. There is a significant role for chemical engineers to play in meeting these current and future needs. This tutorial will cover the environmental impacts and pollution control unit operations of coal-fired electric power plants, both current and future. The four main unit ops are selective catalytic reduction of NOx, electrostatic precipitation, fabric filtration ("baghouse"), and flue gas desulfurization ("scrubber"). These incorporate a number of chemical engineering principles including separation processes and reactor design. In addition, new compliance drivers are pushing additional pollution controls for mercury, which is typically done as an adsorption process. Finally, concerns over global climate change have resulted in extensive research into carbon dioxide capture and sequestration using absorption processes. Also water and wastewater impacts from the power plants will be covered.

From Research to Practice: A Story of Mathematically Predictive Actionable Process Safety Leading Indicator Index

AICHE - 2013
Abstract ID: 294440
Submitted to: ENVIRONMENTAL DIVISION

Author Contact Details

Speaker: Joe Stough, **VP Innovation & Technology**
Company: IHS
Email: joseph.stough@ihs.com
Phone: +1 713 499 4288

Author Bio:
Joe Stough is the visionary and strategist behind the IHS Operational Excellence Forum methodology. With a professional background in large-scale database software design a graduate degree in Applied Statistics from the University California, Stough founded Syntex Management Systems and led the company in providing solutions that successfully combine the disciplines of enterprise-level
business software design and advanced statistical analysis to dramatically impact Enterprise Risk on a global basis. Syntex is now a part of IHS, Inc., where on an annual basis, Stough is continuing to refine his findings in the context of the Operational Excellence Forum.

ABSTRACT:

For the past several years, a group of global Energy companies, including nearly all of the super majors, has been working on a research initiative with a common mission to find an actionable leading indicator index to drive Safety and eventually overall Operational Excellence performance.

This group has been collaborating to provide a large multi-year data set, which contains millions of data records from events such as incidents, investigations, near misses, audits, observations, assessments, and many other routine field-level activities. Since 2008, a rigorous statistical analysis process has been iteratively applied to this data to identify the mix of leading metrics which most effectively predict Safety performance outcomes.

This paper will review the leading indicator research findings drawn from analyzing this large Energy industry data set. It will reveal the key components of an actionable leading index which has been found to be uniquely strong in organizations who continue to sustain top-level Safety lagging performance.

From the analytical research and careful review with this consortium of industry QHSE executives and Subject Matter Experts, the index that was found to best predict the Safety outcome performance of an organization included measurements of (a) proactive event reporting (Reporting Culture), (b) the discipline and consistency in execution of QHSE business processes, (c) the timeliness in closure of important corrective / preventive actions, and finally (d) the responsiveness of supervisors and line managers.

The paper will provide an overview of the analytical process that produced these results and will outline alternatives for applying these insights on leadership performance dashboards. Although the leading indicator research and analytics activities continue annually to strengthen the effectiveness of the leading indices, companies are applying the index to provide actionable leading metrics on executive scorecards. We will outline a sample process for implementing practical, "actionable" leading metrics in each key area to drive line manager / leadership behaviour and ultimately improve QHSE performance outcomes.

Overview and Background: Researching and Solving the Problem

An organization attaining true Operational Excellence has combined best-of-class business and financial performance with zero loss of production up-time, asset integrity, the environment, and safety. However, the frequency and severity of operational losses are significantly above zero for most corporations. Contributing to this is the lack of a standardized measurement of operational loss to enable performance comparisons and benchmarking.

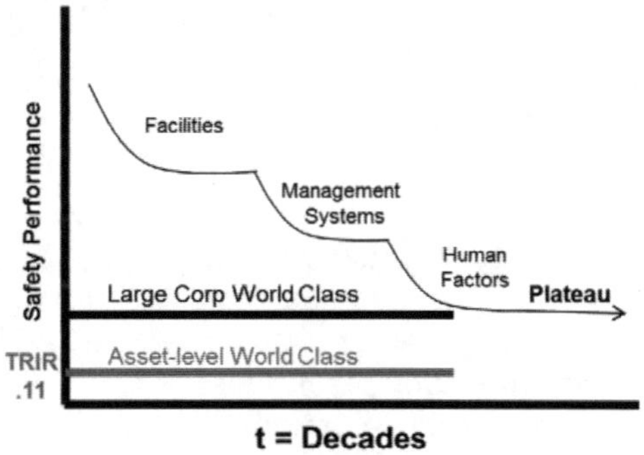

Although it is difficult to standardize the measurements of production up-time, asset integrity, and environmental outcomes across varying types of assets, the one area that does enable a relatively standard view of performance is **safety**. And since many industry executives hold the belief that the safest operating assets are also the best run operations, our research views the Operational Excellence problem through the lens of safety performance.

Figure 1: The Problem
Over the past few decades, the industry as a whole has seen safety incident frequency rates trend toward zero. But two key problems remain:

1. Even for the world-class performers, this trend has reached a plateau; and
2. The rate of high-severity injuries and fatalities has NOT followed the trend toward zero.

With these issues in mind, the focus is not only to drive Total Recordable Injury Rates (TRIR) to zero but more importantly to drive **severity-Weighted** Total Recordable Injury Rates (WTRIR) to zero. Whereas this goal of zero is perceived by most to be lofty at the aggregate or corporate-level, our research data set revealed operating assets (as depicted in Figure 1) which have sustained performance very close to zero.

In addition, the research data set revealed that the top 10% of assets as measured by WTRIR had the following interesting characteristics in comparison to the bottom 10%:

- WTRIR was roughly two orders of magnitude better
- Rate of near misses reported was 78% higher
- Incidents were reported 3 times more promptly
- Action items were authorized by leaders over twice as fast
- Rate of completed action items was 79% higher

Since these assets are sustaining near zero losses in the same risky environment with a similar level of safety exposures, this indicates the possibility of a much higher degree of Operational Excellence than is routinely achieved in industry.

A Path to Solving the Problem

With the top-performers raising the bar, a key challenge is to discover what the leaders, the workforce, and the overall organizations of these assets are doing differently. And the answer needs to come in a form that is measurable and actionable so that it can be repeated and applied to lower performing assets to continuously drive toward zero loss.

Specifically, the question that needs to be answered is: "What are the actionable, measurable factors which yield the near zero operational loss performance that is being achieved by these top performers?"

Using a severity-weighted safety incident frequency rate as the standardized outcome measure of operational losses, the research studies assets with performance varying from near zero to orders of magnitude worse to understand what creates high performing assets. The inputs in this research include data from workforce involvement in proactive activities such as audits, inspections, management system assessments, and observations as well as reactive events such as incidents, near misses, investigations and corrective actions.

By applying a rigorous statistical analysis process, the research reveals correlations between multiple leading factors and severity-weighted safety outcome performance. The focus is to discover those leading metrics which are not only predictive of outcome performance, but also are the most practical for use by top leadership as actionable, routinely measurable management controls. The process is continuously repeated with the aim of creating the optimum set of leading metrics which can be combined into a single, predictive multi-variable leading index.

The intended use of this index is two-fold:

1. To provide operating executives with actionable management controls for stewarding the process of continuously improving Operational Excellence performance; and
2. To deliver benchmarking of statistically proven leading indicators which reveal performance gaps and enable measurable goals and annual performance improvement plans.

Additional Obstacle: Management Addiction to Lagging Indicators

A primary obstacle to realizing the benefits of the above approach is top-level management's historical addiction to lagging indicators. Key Performance Indicators (KPI's) are management metrics used by business leaders to measure operational performance and to compel leadership to make performance improving operational changes. Lagging KPI's for QHSE are calculated from the occurrences of medium to high consequence loss incidents and provide a historical view of outcome performance. An often-referenced metaphor infers that using lagging indicators to drive performance is like "driving by looking in the rear view mirror". But in our research, it is extremely rare at the top-executive level to see anything other than lagging metrics of Safety performance used to judge performance.

Figure 2: Lagging KPI's from loss events

Per this depiction of the "Heinrich triangle", reporting data from incident events of the highest severity is routinely governed by a corporate standard (and often regulated) - thus rendering lagging data widely applicable and measurable across the enterprise.

In most companies, managers still use TRIR or a similar lagging metric as the primary KPI for judging an operation's Safety performance. One reason for this dominance is the practicality of having a standard in producing a normalized performance metric which can deliver an "apples to apples" comparison of these lagging KPI's across the enterprise (as described above). In addition to being routinely measurable, lagging measures are believed by management to be more quantifiable, tangible, objective, and generally more reliable than any leading KPI's proposed by the business.

Criteria of Leading KPI's as Management Controls

For business leaders to overcome this addiction and enable a leading KPI to truly rival lagging

KPI's as the preferred metric on a corporate-wide scale, the leading KPI must meet the following criteria to become a management control:

Actionable

To drive leadership behaviour, the leading KPI must offer the operational leader with a practical solution which can be acted upon to affect operational changes.

Objective

Surveys and opinion based assessments are too subjective. The best leading KPI's involve objective calcutions drawn from detailed activity or measurable events.

Normalized

TRIR enables and "apples to apples" comparison. Similarly, a leading KPI must be normalized to enable management comparisons across operating units.

Routinely measurable

The most effective and accurate leading KPI's are calculated from measuring leadership and worker involvement in routine operational activity on a relatively frequent (i.e. monthly) basis – as opposed to annual opinion-based surveys.

Believable and predictive

To attain the full commitment of field-level leaders, the leaders must believe that outcome performance can be affected with proper changes to performance in the leading KPI. Data and math can be used to assist leaders in developing this belief.

Throughout the aforementioned research and statistical process, the above criteria were applied to assure the resulting index of leading KPI's would likely be a cure. The rest of this paper will describe the research and business practices designed to overcome management's lagging indicator addiction.

Research Hypothesis

Per the premise described above, businesses are involved in the research to find the keys to preventing operational losses. These companies believe that an operational loss is a function of an organization's inability, either via failures in systems or behaviors, to effectively mitigate the risks involved in the loss. Therefore, the following is the research hypothesis:

"Operating units that are the best at reducing risks will sustain the best lagging Safety performance."

Risk Reduction Cycle (RRC) Database

An important characteristic of the research initiative is the availability of data pertaining to an organization's efforts, business processes to identify and **reduce the risks** to operational losses. The companies involved in the research have implemented a common enterprise-wide database application to consolidate data from business practices that fit the "risk reduction cycle" pattern. These companies are able to draw data from the database to observe the execution of these routine operational activities and are minimizing the effort of gathering leading data along the way. The following risk reduction process pattern of work practice activities applies to many QHSE business processes found in routine operations.

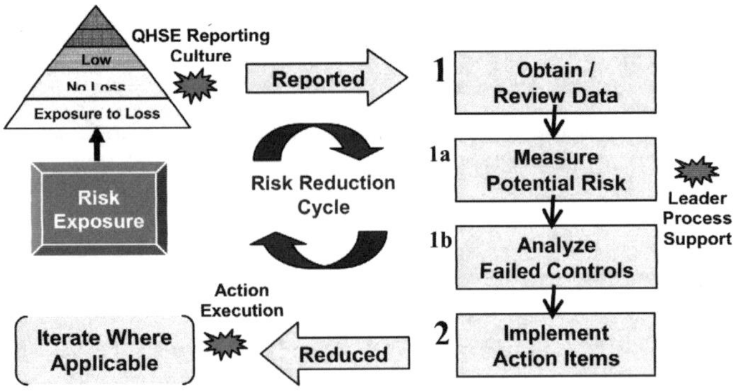

Figure 3: Risk Reduction Cycle pattern

As depicted, there are 2 key elements which are common to all risk reduction activities: (1) obtaining awareness via "reported" events and (2) implementing corrections to reduce the risk – i.e. "reporting and fixing" reported items.

Through researching operating facilities in the Energy industry, hundreds of event types have been discovered as risk reduction events – and thus contribute to the data set. The list ranges from multiple different types of incidents, near misses, and investigations to the many different types of inspections, observations, audits, etc...

The Core Elements of the Risk Reduction Cycle

The following describes how each element of the risk reduction cycle generally maps to the key steps in work practices resulting from both reactive (incident-based) events and proactive (assessment-based) events.

1. Obtain / review data (event reporting)
If you don't know about the risk, you can't reduce it. This element maps to the initial incident and assessment-based process steps whereby events are reported and people are assigned to contribute expertise / input and to review / approve the report.

1a. Measure potential risk (risk score)
For incident events, this is an optional / advanced practice of classifying "how bad it could have been" by using a risk matrix. Risk assessments, PHA's, and various other proactive processes include this as a key process step where the risk level is formally scored.

1b. Identify failed controls (root cause analysis / investigations)
For incident events, this element maps to the investigation process. Since many companies only investigate high-severity or high-risk events, this step is optional as well. An advanced practice is to apply an informal investigation to classify root causes and map those causes to Management System elements for ALL incident events – including near misses. For assessment-based processes, this element often represents the core purpose of the event – i.e. to identify the areas where risk control needs improvement.

2. Implement / repair controls (action item management)
For incident events and proactive events alike, this element maps to the most important step in all risk reduction processes – the execution of tasks to repair broken controls or implement new ones to ultimately reduce exposure to risk.

The research involves defining and calculating metrics across each of the above aspects of risk reduction process execution. Per the hypothesis, the intent is to identify the appropriate mix of metrics which both effectively measure an organization's risk reduction capability and are predictive of safety performance.

Researching a World-wide Data Set of Risk Reduction Processes

The companies involved in the research project provided their data on a scale that spans thousands of operating facilities in over 100 countries. Engaged in this initiative are world-wide operations from multiple Energy companies which have world-class QHSE performance... as well as leading companies in Oil Field Services, Chemicals, Manufacturing, and other industries where managing operational risks is a material concern.

Although most of the companies referenced in this study deploy unique company-specific programs for managing QHSE performance, they share one thing in common. Each of them is applying a common technology solution to collect and analyze data from a myriad of risk reduction process events such as incident investigations, near miss reports, management system audits, risk assessments, assurance reviews, behavioral observations, field-level inspection programs, hazard analysis, and many other processes. As depicted in the below diagram, the intended scope of this database is to be the organization's collection of all events that result in risk reduction actions.

Figure 4: Risk reduction cycle events and actions database

On the surface, these organizations have simply been using a common "mechanism" to manage their own unique set of QHSE risk reduction processes and ultimately analyze the resulting data. However, at a deeper level, these companies are not only collecting data resulting from the event occurrences (e.g. incident reports, injury details, spill quantities, near miss types, root causes, audit results, assessment scores, inspection findings, etc.) but also the "work practice behaviors" reflecting the organization's tendencies in executing the processes behind such events (e.g. mean-times between completion of "critical" process steps, rate of leadership involvement, ratio of near-misses reported to high-consequence incidents, etc.). Applying statistical analysis to this vast data set has revealed the measurable organizational tendencies which are unique to top performers. The ultimate goal is to convince top business leaders to use these types of leading measurements.

From Research to Practice

Going forward in this paper, we will unveil some of the math-grounded process-based metrics which were discovered via the research initiative. Then, we review the critical role that each of these measurable areas play in the mission to proactively drive Safety performance. We'll end by showing the best approaches for applying these leading indicators on Key Performance Indicator (KPI) dashboards through an iterative measurement process to shift leadership focus, continuously improve QHSE performance, and attain Operational Excellence.

The Research Initiative: A Statistically Significant Data Set

Over the last several years, a collective data set of millions of records of data covering over a million direct and contractor employees of customer companies has been accumulated. As conveyed in the below diagram, the risk reduction cycle database manages both lagging and leading data. This has afforded the opportunity to study the relationships between lagging and leading data elements – with the purpose of finding statistically proven leading indicators.

The Opportunity to Correlate

The data set includes lagging metrics derived from high-consequence incidents as well as multiple sources of potential leading metrics including (a) leading events such as near-misses, low-consequence incidents, assessments, observations, action items, etc. and (b) the organizational tendencies in managing the business processes associated to such events.

NOTE: Although as depicted in Figure 4 low-consequence incidents and near misses are part of the incident event database, the organizational behaviours associated to reporting and managing such "non-mandatory" events are categorized in Figure 5 as leading areas of measurement.

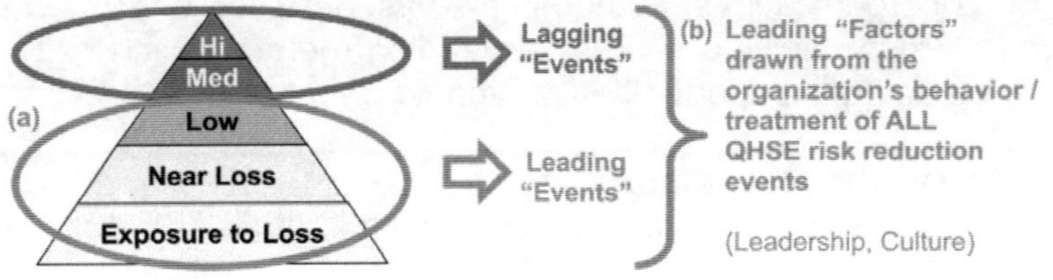

Figure 5: Sources of Data

With lagging and leading measurements coexisting in the same centralized database, a unique opportunity exists to practically, rigorously study correlations and identify math-proven leading indicators.

(a) A Database of Risk Reduction Events and Actions

As previously described, a risk reduction event is any event that starts with the (lagging or leading) discovery of a risk exposure and ends with the assignment and completion of action items to strengthen protections and ultimately reduce risk exposure. Examples of "lagging" events contained in the data set include near misses, incidents, and associated causal investigations from safety, security, environmental, quality, reliability (maintenance), and other types of loss events. Also included are "leading" events such as audits, inspections, behavioural observations, management system assessments, and over 400 other types of assessment-based events. Collectively millions of incidents, assessments, and action items have been accumulated over the past several years – establishing a rich, statistically significant data set for the research initiative.

(b) Data to Measure "how organizations manage risk reduction events"

The goal of these business process software tools is not only to collect the detailed data from risk reduction events but also more importantly to manage the full process lifecycle from initiation (reporting) of events through each key step in processing the event all the way to completion of the final corrective action. The research applies statistical analysis to this vast set of data to study the organizational tendencies in how the various parts of the organization are managing their roles in the risk reduction process.

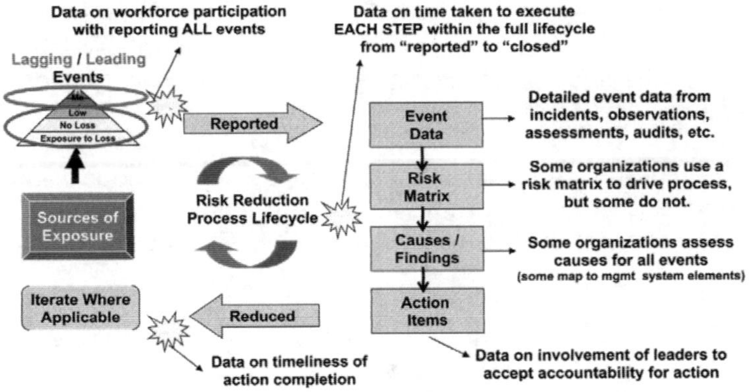

Figure 6: The Process-based Data

The diagram to the right depicts the many organizational touch points throughout the full "risk reduction process lifecycle" from which the metrics are being drawn. Through collaboration with Subject Matter Experts (SME's) from participating companies, over 200 different metrics have been defined in the research thus far – ranging from measures of reporting culture to various types of leadership and process-based metrics to measures of action item performance.

Research Premise: Find Opportunities for Improvement

The premise of the research initiative is to study relationships between the ordinary lagging metrics and the multitude of leading metrics as described above. The objective is to reveal measurable differences in the organizational factors which convey how organizations treat risk reduction events. These differences are reflected by substantial variation in metric scores when calculated across a group of organizations. Although the vast data set (described above) includes many different types of data elements for study, it is the leading metrics which expose such differences that represent the truly meaningful opportunities for improvement.

Getting to the Meaningful Data

The result of an enterprise-wide implementation of such a risk reduction process database is a practical means of collecting and analyzing data spawned from the event processes within the scope of the project. After establishing such a platform, some company executives have asked the following question:

"How do we get the meaningful data to enable our managers to lead performance improvement?"

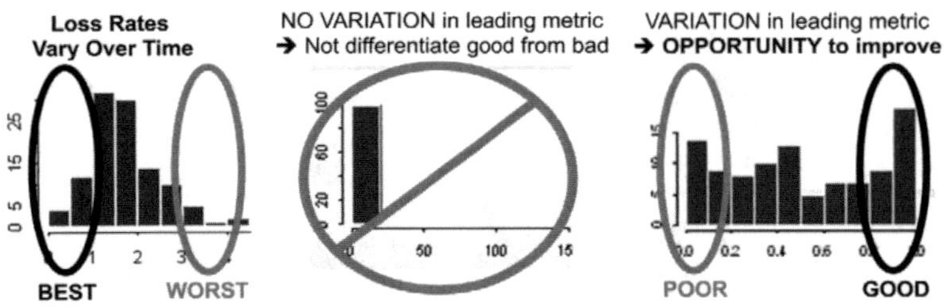

Figure 7: Meaningful Data

The best leading metrics are the ones for which the top performers (in terms of loss rates) are doing well and the worst performers are doing poorly. Because this variation exists in loss performance, a meaningful leading metric must show variation (e.g. example metric to far right) to represent an opportunity to improve.

Answering the Question: "What are the BEST performers doing differently?"

As conveyed in the above diagram, the histogram to the far right shows a metric with substantial variation – i.e. a material number of organizations have scores in each major segment of the performance spectrum. In this example metric (near misses as % of total incident events), organizations with a score near zero show up on the left side of the histogram and those with a score near 100 show up on the right side – i.e. some good performers and some bad.

For the best leading metrics, the majority of those organizations which show up on the "good" side of the metric's histogram are also present on the list of the BEST performers in terms of loss rates. These metrics represent the unique measurable characteristics of top performers and are the target of the research initiative.

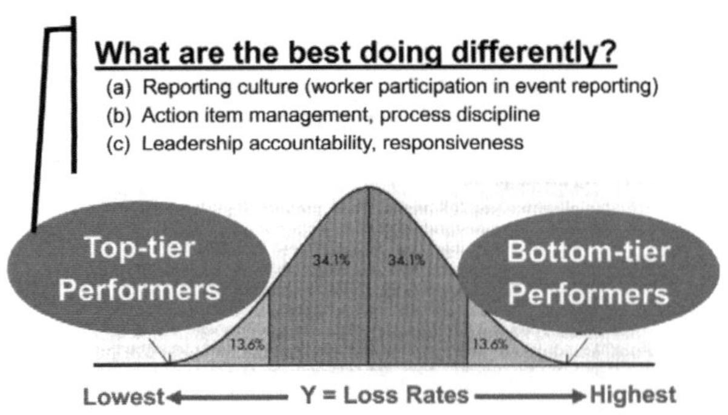

Figure 8: Finding the unique characteristics of the top performers

As conveyed in the diagram to the left, the research and benchmarking initiative has revealed a list of factors that are collectively predictive of performance and meet all of the previously listed criteria.

These factors were revealed after many repeated cycles of statistical analysis and review by SME's. The final list of the most meaningful metrics was then calculated in quartiles. From there indices were created to establish a leading performance measurement for each of the three factors. Finally the indices were combined into an overarching index to provide a single score which was found to be most predictive of WTRIR.

The Ultimate Leading Indicator: A Composite Safety Performance Index

The composite index is made up of the three dimensions: ***Reporting Culture Composite Index***, ***Action Item Composite Index*** and ***Leadership / Process Composite Index***. Each dimension is calculated from multiple metrics to form a unique measurement that results from the "tension" between the combined metrics.

Index from "tension" between metrics

For example, by combining a measure of the volume of action creation with a measure of discipline in on-time completion, there is tension between volume and disciplined execution – i.e. it is easier to complete a very short list of actions on-time than it is to carry such disciplined execution across a much longer list of action items.

Reporting Culture Composite Index = "culture of reporting and fixing"

This index results from the tension between the volume of "voluntary" event reporting and the organization's tendency to include actions as follow-up to reported events. In order to perform well in this index, an organization must not only sustain a substantial rate of near miss reporting (normalized by work hours) but also must follow-up reported events with actions. An organization must be strong in BOTH areas to be a good performer in the Reporting Culture Index.

Action Item Composite Index = "rate of action with timely execution"

This index results from the tension between the volume of actions generated and the organization's tendency to complete actions in a timely manner per the planned due date. In order to perform well in this index, an organization must BOTH sustain a relatively high rate of action AND continue to complete actions in a timely manner.

Leadership / Process Composite Index = "responsive, disciplined leadership"

This index results from the combination of supervisor / leadership involvement in two key steps in the risk reduction process lifecycle – i.e. the inital response to a reported event and the initial authorization of actions derived from events. To be a top performer in this area, an organization must have low lag times in BOTH of these areas.

The Importance of Reporting Culture

A common view of reporting culture is that it is an organization's total event reporting volume. Per the above index components, it is evident that event reporting "volume" has a critical affect on both of the first 2 above indices. Event reporting volume directly affects the Reporting Culture Index and indirectly affects the Action Item Index in that total action volume is minimized if reported events are minimized. Thus two-thirds of the total safety performance index discovered through the research project is affected by the volume of reported events – i.e. Reporting Culture.

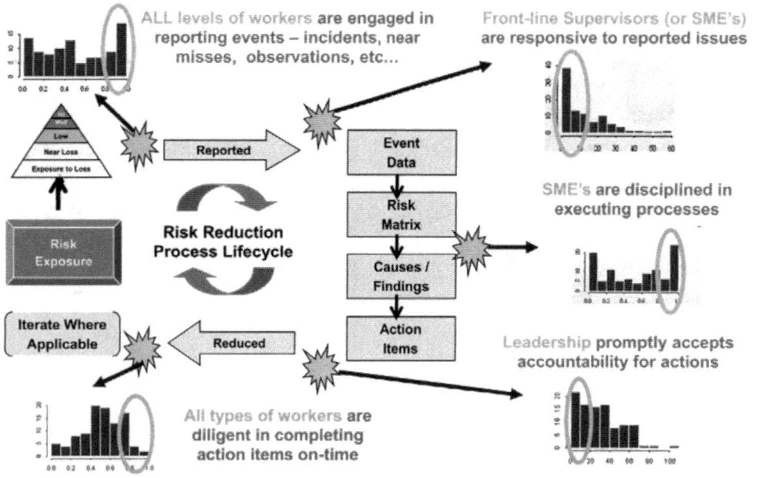

Figure 9: The unique tendencies of top performing organizations

Reading the histograms (to the right) in a clockwise manner not only portrays a list of math-grounded leading metrics but also tells a practical and believable story about organizational tendencies.

By studying the statistical relationships between the many leading metrics and WTRIR, it was revealed that organizations which perform well in each of these areas were much more likely to be top-level QHSE performers. Since these types of leading metrics are contained in the same database tool, they are applied as KPI's via the same tool to continuously improve the factors that lead to better performance outcomes.

Applying the Research Findings in Practice

Some companies have corporate databases which are justified on the ability "to produce better pareto charts of event data". But if these tools could directly DRIVE better QHSE performance, the true return on investing in these programs would be achieved. Many companies are preparing analytical reports for which some do include "meaningful" leading metrics. But are they getting the commitment of operations leaders to use those reporting tools to directly monitor and improve (i.e. measure and act upon) the math-proven factors that result in better performance? Leaderhip commitment to implement such leading KPI's as management controls is critical to drive improvements in QHSE outcome performance. Helping to get this commitment is a key purpose of the analytical research and the overall KPI measurement strategy.

Figure 10: Leadership as "Users" to continuously monitor and improve

To improve, leaders must be the KEY USERS of the continuous improvement solution.

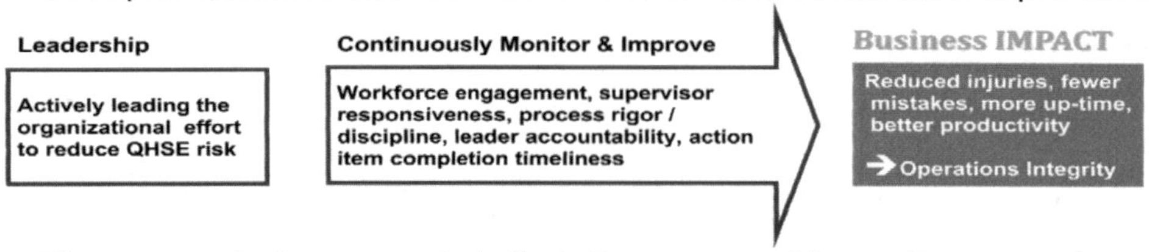

The research has revealed that the measurable math-proven keys to improvement are "organizational" in nature – i.e. workforce engagement, leadership accountability, etc.. Active management sponsorship is required to truly lead the required improvements in these areas. So, to achieve continuous improvement, it is imperative to get the data and math to help obtain management buy-in as well as the iterative benchmarking measurement tools to sustain their focus year over year.

Managed KPI Scorecard			
	Recordable Rate (Lagging)	Safety Triangle Ratio (as %)	% High Actions On Time
Div 1	2.50	37%	67%
Div 2	3.10	22%	62%
Div 3	1.70	83%	95%
Div 4	1.30	75%	77%
Div 5	0.95	93%	75%

Figure 11: Measure leading KPI's to DRIVE performance

What "leaders" measure and act upon gets improved. To obtain such active response from operations leadership, the KPI's must be carefully selected. Only those metrics which resonate at all levels of business management are included on the "managed KPI" scorecard. An example of such broadly applicable metrics is provided to the right. The managed metrics are typically calculated on a 12 month rolling basis and cascaded through multiple levels of the organization to assure improvements are realized enterprise-wide. But these only represent a sub-set of the leading indicators which have potential to affect performance outcomes.

Executive Analysis ONLY			
Predictive Index	Report Culture	Action Execution	Leader Response
4.79	2.10	24%	11%
5.25	1.40	17%	41%
17.10	7.50	54%	57%
7.10	4.10	72%	3%
22.0	18.30	46%	21%
5.20	6.10	64%	12%
10.24	6.58	45%	24%
4.10	1.50	93%	8%
9.70	6.30	71%	31%

Figure 12: Preparing top leadership to evolve the "managed" KPI's & continuously improve

To evolve management's focus year over year, another required component entails an additional set of metrics included only for analysis by members of the top-level executive steering committee. This example (to the right) displays a sample list of the math-grounded metrics discovered through the research. Although many organizations may not quite be ready to hold managers accountable for such a list of metrics, these additional benchmark KPI's may reveal valuable trends and help to prepare the top leadership to consider such metrics for inclusion as "managed" KPI's in the subsequent measurement cycle – establishing a continuous improvement process.

Iterating with Benchmarks to Continuously Improve

The traffic light metrics are derived from benchmark calculations. Companies may choose from various types of benchmark calculations during design of the KPI reports. Internal benchmarks provide comparisons to other organizational units' performance within the same company. And external benchmarks provide comparisons to the aggregate scores per other companies using the same database tools and involved in the collective benchmarking efforts. Additionally, a discrete scale may be used to establish the traffic light scoring effect.

Division/Site	TRIR	WTRIR	Preventive Index	Reporting Culture	Action Execution	Leader Response
Org Unit 1	0.60	0.60	1	1	1	3
Org Unit 2	0.44	0.44	1	2	1	2
Org Unit 3	0.47	0.47	2	2	2	4
Org Unit 4	0.76	0.76	2	4	2	2
Org Unit 5	0.00	0.00	1	1	1	2
Org Unit 6	0.41	0.41	2	2	3	2
Org Unit 7	0.78	0.78	2	2	3	2
Org Unit 8	0.42	0.42	3	4	2	2
Org Unit 9	0.23	0.23	1	1	1	2
Org Unit 10	0.91	0.91	3	2	3	4
Org Unit 11	0.51	0.51	2	1	2	2
Org Unit 12	1.80	1.80	3	1	2	2
Org Unit 13	0.55	1.64	3	3	3	2
Org Unit 14	1.42	1.42	3	1	2	4
Org Unit 15	1.81	1.81	4	4	3	3
Org Unit 16	0.63	2.31	3	4	2	3
Org Unit 17	1.00	4.98	4	2	4	4

Figure 13: Example annual benchmark report

Each year the research members receive a benchmarking report at the operating unit level. The purpose of the report is to identify performance gaps and improvement opportunities in the areas measured by the index of leading measurements.

By updating these statistics periodically, a continuous improvement platform is established to leverage the knowledge gained across all member companies. As more companies get involved in the benchmarking program and more business leaders are engaged in improving the math-proven leading indicators, the statistical inferences are continually sharpened and the overall community of collaborating companies benefits with improved QHSE performance results.

Summation

Through applying structured statistical processes and subject matter expert interpretations to a large statistically significant data set, it has been discovered that top-performing organizations (in terms of Safety outcomes) are the ones which have their leadership and front-line workers engaged in a collective effort to reduce operational risks.

This discovery may not be a surprise but rather a validation of existing beliefs. However by deriving these facts from a system which manages daily operational activities (e.g. incidents, near misses, investigations, observations, inspections, action items, etc.), a new opportunity is afforded operations managers to continuously measure and thus improve these organizational factors and ultimately to gain control of Operational Excellence and Safety performance outcomes.

In conclusion, the research discoveries regarding the leadership and cultural dynamics of top-performing organizations are certainly not new to the Energy industry. However the ability to institute management controls which enable all levels of leadership to measurably improve these factors on a routine, daily basis is likely NEW to most companies. But for those companies who are measuring and improving broadly applicable QHSE process-based KPI's and are engaged in the collective benchmarking programs to continuously improve, this ability has become a key part of their ongoing strategy to achieve and sustain world-class QHSE performance.

(31a) Marketing

Jon Haughton, Health & Nutrition, Corn Products International, Westchester, IL and Meagan Lewis, UOP, IL

This is a tutorial-like presentation of the MBA-In-A-Day session. An invited speaker will give an overview of the title subject, who could be a faculty member from a regional business school or an expert practitioner.

The session will be of interest to those who want to learn about MBA programs and content: Those who may be considering enrolling in an MBA program, those who want to understand the type of information that is covered, and those who would benefit from having an overview of a typical MBA program. The speakers will direct their presentations to the interests of chemical engineers. The session will be highly of interest to young professionals: The typical MBA student has six years of experience and is 29 years old, and students with engineering or science backgrounds make up 35 to 40% of MBA programs.

(31b) Strategy and Economics

John M. Forgac, BP Naperville Campus, BP Products North America, Naperville, IL, Meagan Lewis, UOP, IL and Jon Haughton, Health & Nutrition, Corn Products International, Westchester, IL

This is a tutorial-like presentation of the MBA-In-A-Day session. An invited speaker will give an overview of the title subject, who could be a faculty member from a regional business school or an expert practitioner.

The session will be of interest to those who want to learn about MBA programs and content: Those who may be considering enrolling in an MBA program, those who want to understand the type of information that is covered, and those who would benefit from having an overview of a typical MBA program. The speakers will direct their presentations to the interests of chemical engineers. The session will be highly of interest to young professionals: The typical MBA student has six years of experience and is 29 years old, and students with engineering or science backgrounds make up 35 to 40% of MBA programs.

(31c) Operations, Supply Chain, and Risk Management

Jon Haughton, Health & Nutrition, Corn Products International, Westchester, IL, Meagan Lewis, UOP, IL and John M. Forgac, BP Naperville Campus, BP Products North America, Naperville, IL

This is a tutorial-like presentation of the MBA-In-A-Day session. An invited speaker will give an overview of the title subject, who could be a faculty member from a regional business school or an expert practitioner.

The session will be of interest to those who want to learn about MBA programs and content: Those who may be considering enrolling in an MBA program, those who want to understand the type of information that is covered, and those who would benefit from having an overview of a typical MBA program. The speakers will direct their presentations to the interests of chemical engineers. The session will be highly of interest to young professionals: The typical MBA student has six years of experience and is 29 years old, and students with engineering or science backgrounds make up 35 to 40% of MBA programs.

(31d) Leadership, People Management, and Organizations

John M. Forgac, BP Naperville Campus, BP Products North America, Naperville, IL, Meagan Lewis, UOP, IL and Jon Haughton, Health & Nutrition, Corn Products International, Westchester, IL

This is a tutorial-like presentation of the MBA-In-A-Day session. An invited speaker will give an overview of the title subject, who could be a faculty member from a regional business school or an expert practitioner.

The session will be of interest to those who want to learn about MBA programs and content: Those who may be considering enrolling in an MBA program, those who want to understand the type of information that is covered, and those who would benefit from having an overview of a typical MBA program. The speakers will direct their presentations to the interests of chemical engineers. The session will be highly of interest to young professionals: The typical MBA student has six years of experience and is 29 years old, and students with engineering or science backgrounds make up 35 to 40% of MBA programs.

(45a) Pilot and Full-Scale Application of Technologies for Mercury Control

Noah Meeks, Research & Environmental Affairs, Southern Company Services, Inc., Birmingham, AL

With the final MATS rule going into effect in 2015, electric utilities must now consider mercury control as a needed technology in their emissions train. This talk wiill cover recent full-scale injections of activated carbon and alkali injection, as well as pilot-scale studies of non-halogenated carbons or non-carbon sorbents. We will also discuss recent mixed matrix membrane pilot studies. **(Most recent data will be included in the abstract closer to the meeting)

(45b) Acid Tunneling Vs Acidizing : A Case Study of Rudatain Oilfield-Kuwait

Adel Elsharkawy, Petroleum Engineering, Kuwait University, Kuwait, Kuwait and Mohamed Al-Hadher, Pet Eng, Kuwait Oil Company, Safat, Kuwait

Acid Tunneling Vs Acidizing : A Case Study of Rudatain Oilfield-Kuwait

By Mohammed Al-Hadher and Adel Elsharkawy

Kuwait University, College of Engineering & Petroleum

Po Box 5969 Safat 13060 Kuwait

Abstract

Acidizing is one of the most important techniques used in oil industry in order to enhance the reservoirs and increase oil production. There are three types of acidizing techniques which are matrix acid stimulation, acid fracturing, and acid tunneling. In this paper a comparison these types of acidizing is discussed. The case study of RA-0007 acid tunneling completed in Kuwait Oil Company is presented. It represents the first dendritic with extensive worm holes created by acid tunneling for maximum reservoir contact. The whole effect is aptly demonstrated by the change in PI of the well before and after the acid tunneling as mentioned before. Analysis of the well designed and implemented job showed that two tunnels were created successfully in MA-C formation through a two days rig-less operation. The results indicated an improvement in the well productivity index (PI) which had been increased from 1.3 to 3.87 bpd/Psi.

(45e) Sustained Casing Pressure Modeling and Well Integrity

Tony Rocha-Valadez, Mary Kay O'Connor Process Safety Center, Artie McFerrin Department of Chemical Engineering, Texas A&M University, College Station, TX, A. Rashid Hasan, Harold Vance Department of Petroleum Engineering, Texas A&M University, College Station, TX, C. Shah Kabir, EPT, Hess Corporation, Houston, TX and Sam M. Mannan, Artie McFerrin Department of Chemical Engineering, Mary Kay O'Connor Process Safety Center, Texas A&M University, College Station, TX

SUSTAINED CASING PRESSURE MODELING AND WELL INTEGRITY

T. Rocha-Valadez[1*], A.R. Hasan[2], C.S. Kabir [3] and S.M. Mannan[1]

[1]Mary Kay O'Connor Process Safety Center, Artie McFerrin Department of Chemical Engineering, Texas A&M University, College Station, Texas 77843-3122, US., email: *tonyrochav@tamu.edu, mannan@tamu.edu.

[2] Harold Vance Department of Petroleum Engineering, Texas A&M University, College Station, Texas 77843-3122, US, email: rashid.hasan@pe.tamu.edu.

[3]Hess Corporation, 1501 McKinney Street, Houston, TX 77010, US, email: skabir@hess.com.

With an increasing demand for hydrocarbons and other sources of energy, the methods for delivering such energy needs have become complex. In the case of the oil and gas industry, exploration and production have gone to deeper waters and deeper horizons than ever before, resulting in drilling several miles into the earth's core. Deeper well penetration means encountering increasingly high formation pressures and temperatures. For example, in deepwater drilling, well construction and production management deal with significant heat transfer between the produced fluids and its surroundings. The high-rate production and the attendant heat transfer demand close attention in terms of operational safety and potential environmental impact. One phenomenon that occurs in producing wells is sustained casing pressure (SCP), which presents a risk to both process safety and the environment.

In oilfield terminology, the casing is a steel pipe cemented to the surrounding formation. Multiple concentric casings are involved in a typical well construction. The casing immediately surrounding the production tubing is known as the production casing, which is most vulnerable to thermal and mechanical stresses and fluid leakage. The cement sheaths provide zonal isolation, support the casing, and protect it from corrosive fluids. Ideally, pressures monitored at the wellhead for multiple casings should change minimally after being bled off through a needle valve at the surface. When the pressure rebuilds after being bled off, that casing is said to exhibit SCP [1].

SCP occurs because of stresses that the system experiences owing to fluids production up the tubing. Besides the normal source of SCP, a poor cement job may allow hydrocarbon migration from the producing or other horizons and accumulation into one or more of the casings. The risk of not attaining well integrity can range from the activation of rupture discs to the blowout of hydrocarbons. Because of the safety and environmental dangers of SCP, regulations require remedial treatments for a casing string that exhibits significant SCP problems. Unfortunately, available methodologies to test for SCP are often qualitative and empirical in nature [2, 3], rendering them inapplicable in many situations. No practical systematic testing procedure has emerged to quantify SCP.

In this paper, we present a model for SCP that is rooted in the transport processes of the system. The model assumes gas leaking through the cement barrier, and then migrating up the annular-liquid column. It accounts for the fluid-temperature profile along the wellbore and liquid compressibility. Some robust assumptions about the transport processes allow us to write the governing equation as a first-order, linear-differential equation that is solved with appropriate boundary conditions. The resulting algebraic expression for the casing pressure rise as a function of time paves the way for all applications. A comparative study of the model's performance with field data suggests excellent agreement for wells where the casing pressure increase is owes gas influx into the annulus. The proposed model could form the basis for standardization of SCP testing. This study also offers some test procedures to assess SCP-related safety concerns.

References

1. Bourgoyne Jr., A.T.J., S.L. Scott, and W. Manowski, ***A Review of Sustained Casing Pressure(SCP) Ocurring on the OCS***, 2000, Final Report Submiteted to MMS.

2. Xu, R. and A.K. Wojtanowicz, ***Diagnosis of Sustained Casing Pressure from Bleed-off/Buildup Testing Patterns.*** SPE, 2001: p. Productions and Operations Symposium.

3. Wojtanowicz, A.K., S. Nishikawa, and R. Xu, ***Diagnosis and Remediation of Sustained Casing Pressure in Wells***, July 2001, MMS.

(45f) Evaluation of Particle Growth in an Electrochemical Fluidised Bed Reactor

Emmanuel Ehirim, Rivers State University of Science and Technology, Port Harcourt, Nigeria

The process of electrodeposition within electrochemical reactor results in the increase of particle sizes, which increases the volume of the bed. This work entails a mathematical model to describe the effect of this phenomenon on the behaviour of a three-dimensional fluidised bed electrode operating under the condition of electron transfer. The model incorporates the pseudo-stationary behaviour in the reactor since the bed increases in volume due to the particle size growth. The Orthogonal Collocation method was used to solve the developed differential equations. The overpotential distribution and the reaction rate within the reactor have been calculated. The principal results are the effects of the electron transfer operation on the superficial displacement with time. The effects of total applied current, electrode geometry and the bed porosity on the behaviour of the system have been analysed. The result shows a very gradual increase in the bed height.

(45g) Debottlenecking Gas Dryers and Desiccant Adsorbers

Benjamin A. Schmitt, Hydrocarbon Processing Industry (HPI) - Business Unit, Johnson Screens, Inc., New Brighton, MN and Michael Ekholm, Engineering, Johnson Screens, Inc., New Brighton, MN

Debottlenecking Gas Dryers and Desiccant Adsorbers

Though Maximization of the Useable Bed Volume with Johnson Screens' SSG™

Benjamin Schmitt, PE

Senior Mechanical Engineer, Technology

Michael Ekholm, PE

Applications and Product Design Manager

Johnson Screens Inc. St. Paul, MN

AIChE Spring Meeting, April 2013

13th Topical Conference

San Antonio, TX

Introduction:

The following presentation of the Shaped Support Grid (SSG™) releases additional supporting data and updates to the initial presentation given by Robert G. Norell at the 2010 AIChE Spring Conference.

Limitations on ethylene dryers and desiccant adsorbers can create bottlenecks in the production process which can lead to a need for increased dryer and adsorber capacity. Capital intensive investments may be necessary in order to eliminate the bottleneck which may include the addition of an additional vessel. With limits on the plot plan, this may not be a workable solution. Adding to the problem is the amount of time consumed in engineering, planning and construction, which can have negative effects on the operation of the ethylene unit. Increased desiccant volume and operational efficiency can increase the unit's production with minimal capital expenditures. Johnson Screens', patent pending, SSG™ can add the increased process efficiency and cycle life necessary to extend the operating capability of the unit.

Extending the operation, though debottlenecking, can be achieved by maximizing the time before breakthrough of the media bed. There are three primary functions that maximize the breakthrough time:

Proper gas distribution in both normal operation and regeneration conditions which maximizes the bed utilization.

Increased molecular sieve or desiccant bed volume are achieved through the elimination of traditional bed supports comprised of support grids and beams. The support interface for the screen configuration also allows for thermal expansion typically present in vessels with a thermal cycle as part of the process. The design allows for thermal cycling without the risks of seal or structural failure inherent in other solutions.

Minimizing the bed volume consumed by the distribution and support system increases the net desiccant volume lost when utilizing alternative bed supports.

Johnson Screens ***SSG*** is designed to provide the industry with an alternative, cost effective, approach to increasing capacity. Supporting data and preliminary case studies will be depicted to quantify the benefits that are achievable through the implementation of the ***SSG***.

Conclusions:

The use of a Johnson Screens ***SSG*** provides a reliable immediate solution to reduce capacity bottlenecks. The ***SSG*** increases the bed volume, when compared to traditional bed support arrangements comprised of grids and beams. The increase in desiccant volume will allow for increases in feed flow rate and/or adsorbtion cycle time. The ***SSG*** maximizes the flow distribution during desiccant regeneration, maximizing the utilization of the complete cross section of the vessel. This solution can be implemented in both new and existing vessels, providing consistency in design between planned increases in performance of current vessels and future expansions.

The modular design of the ***SSG*** allows for easy installation and maintenance. Connection of the ***SSG*** to the vessel is performed within the outlet nozzle, eliminating all welding on the pressure vessel and eliminating any need for recertification of the pressure vessel after installation. The design allows for ease of dismantling for inspection of the vessel head during routine maintenance and inspections. The low profile will also reduce interference with any required media exchanges. The design package of the ***SSG*** provides the end user with a holistic solution that fulfills debottlenecking needs with limited capital expenditures.

Biography of Speaker:

Benjamin Schmitt is a Senior Mechanical Engineer in Johnson Screens' Hydrocarbon Processing Industry Technology Group. Ben graduated from North Dakota State University in 2004 with a Bachelor of Science degree in Mechanical Engineering. He currently working toward his Master's in Business Administration from Hamline University. Ben is member of AIChE and is a registered professional engineer in the state of Minnesota.

AMACS
PROCESS TOWER INTERNALS

AIChE 2013 Spring Meeting Poster board

Design Considerations for effective liquid removal In Suction Drums

Presented By:

Kanti Patel, PE

Compressor down again?

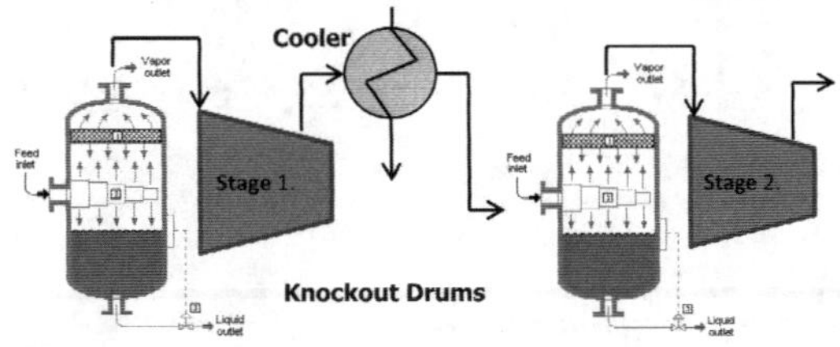

Is liquid carryover the culprit?

<u>We have heard it all before</u>:

-Premature compressor shutdown.

-Production time lost.

-High yearly compressor maintenance costs.

-Impeller damage due to pitting corrosion.

-Corrosion due to salt deposits.

AMACS
PROCESS TOWER INTERNALS

Common Design Problems with Suction Drums which promotes liquid carryover

1. Choosing the correct mist eliminator (M.E. equipment)
2. Knowing where to install the M.E. in a suction drum.
3. Consider a high mist loading in the gas when selecting your M.E.
4. Fouling considerations.
5. M.E. disengaging space, bad gas distribution & liquid slug in the feed.
6. One micron and great droplets, versus submicron mist removal.
7. High viscosity liquids.
8. Dislocated M.E. due to pressure surges.
9. Through put exceeds original design capacity.
10. Capacity range too large for suction drum.

AMACS
PROCESS TOWER INTERNALS

Choosing the correct Mist eliminator equipment

- Proper sizing

+ Use Souders-Brown equation to find optimum velocity.

$$\mathbf{Vd = K} \cdot \sqrt{\frac{Pl}{Pg} - 1}$$

+ Select the correct vapor velocity factor K. based on the operating pressure, and type of equipment.
  - Lower K-factor 0.01 for every 100 psi
  - K factor to select when using:

Mesh Pad	K= 0.1 - 0.55
Chevron Vane	K= 0.45 - 0.65
MultiPocket® Vane	K= 0.75 – 1.0

AMACS
PROCESS TOWER INTERNALS

Choosing the correct Mist eliminator equipment cont.

- Selection based on droplet size.

Table 2. Droplet sizes (water in air) typically captured with 99.9% efficiency by mist eliminator elements of various types

Element type	Size range microns
Fiber candles or panels	0.1 and larger
Mesh with co-knit yarn	2.0 μm and larger
0.006-inch knitted mesh	5.0 μm and larger
0.011-inch knitted mesh	10 μm and larger
AMACS MultiPocket ® Vane	10 μm and larger
Conventional vane arrays	15 μm and larger

AMACS
PROCESS TOWER INTERNALS

Where to install the M.E. in the suction drum

- Improperly installed M.E. can cause severe liquid carryover problems.
- General accepted spacing guidelines.

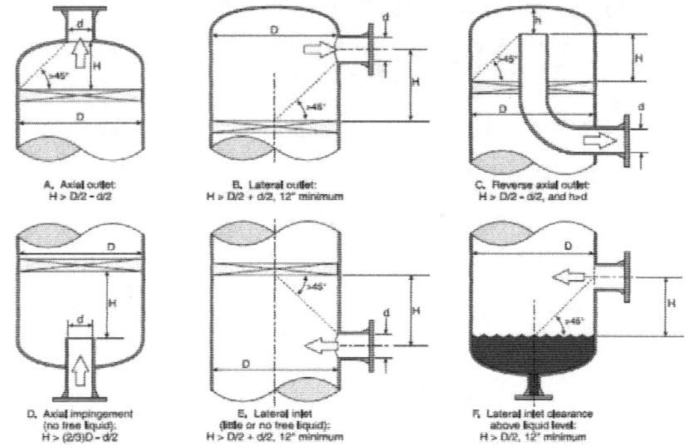

AMACS
PROCESS TOWER INTERNALS

Consider high mist loading when selecting your M.E. equipment

+ High mist loading in the gas stream can flood your M.E. equipment and cause liquid carryover.

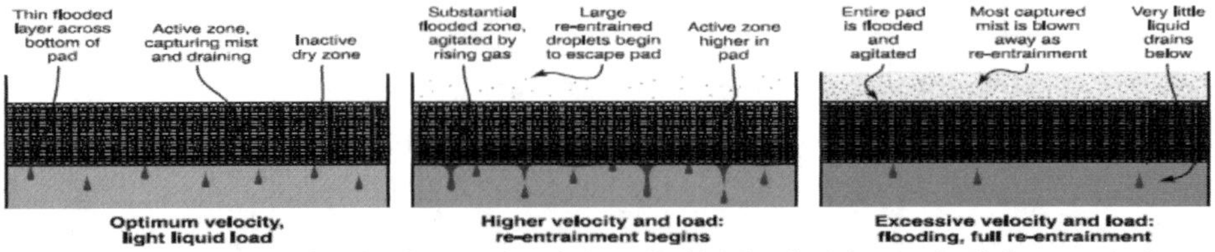

Figure 4. Envisioning mesh pad performance degradation as liquid and gas loads increase (vertical cross-sections)

AMACS
PROCESS TOWER INTERNALS

Consider high mist loading when selecting your M.E. equipment cont.

+ Limited drainage capabilities
  - mesh M.E. <0.5 gal/FT2/min
  - Vane type M.E. <2 gal/FT2/min

+ Select a combination of vane & mesh mist eliminators

AMACS
PROCESS TOWER INTERNALS

Fouling Applications

+ Fouling of mesh mist eliminator will cause liquid carryover.

+ For mild fouling install a water spray system below the mist eliminator for occasional inline cleaning (based on Delta-P)

+ For consistent fouling problems, install a Vane mist eliminator upstream of the mesh mist eliminator.
  - Vanes remove solids and bulk liquids therefore, protecting the downstream mesh mist eliminator

AMACS
PROCESS TOWER INTERNALS

Fouling Applications cont.

+ For consistent fouling problems where the solids are "sticky" install a water spray system below the vane mist eliminator.

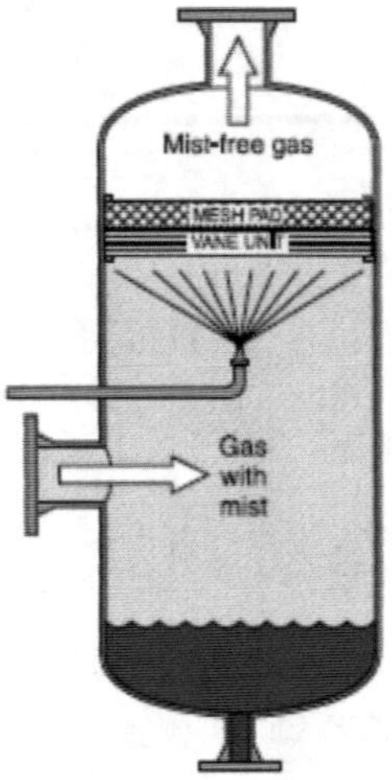

AMACS
PROCESS TOWER INTERNALS

Bad Gas distribution, limited disengaging space and liquid slugs.

+ Bad gas distribution, limited disengaging space in a suction drum can cause severe liquid carryover.
+ CFD analysis can assist in locating high velocity areas.

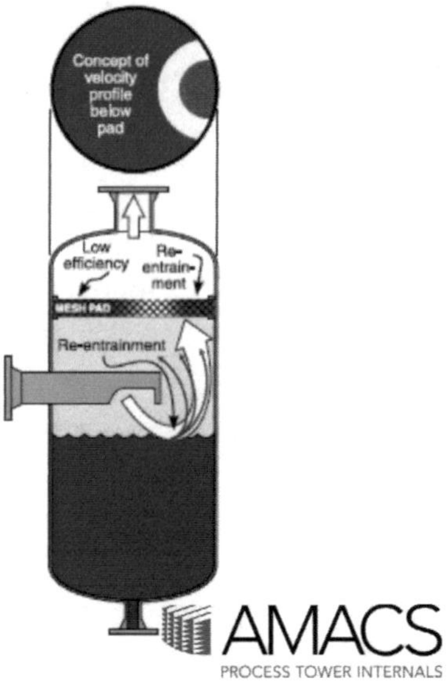

Bad Gas distribution, limited disengaging space and liquid slugs cont.

+ Use inlet diffuser to force uniform gas distribution i suction drum.

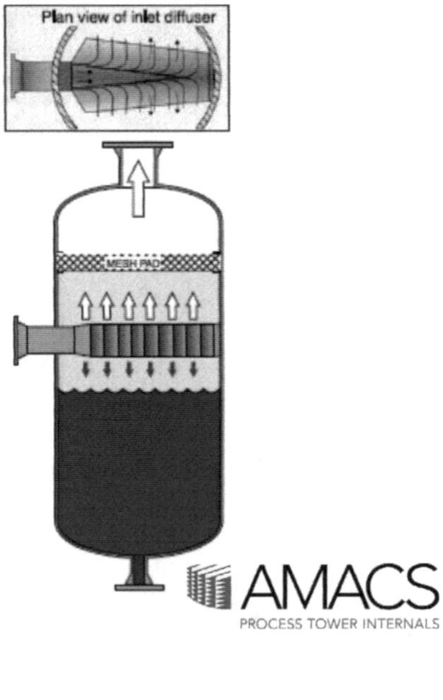

Inlet Feed Device

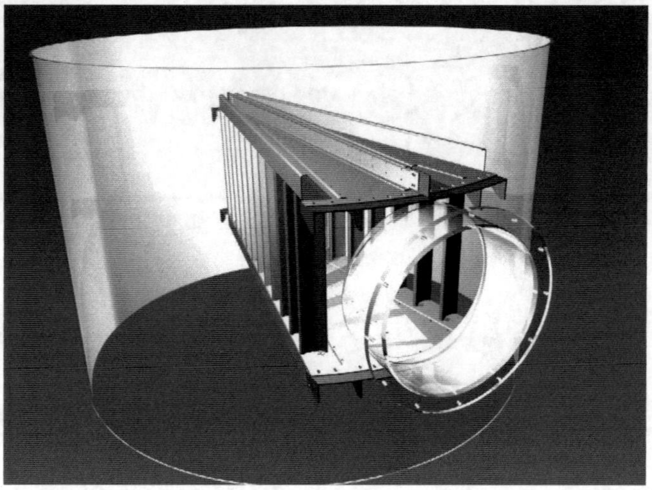

AMACS
PROCESS TOWER INTERNALS

Inlet Feed Device

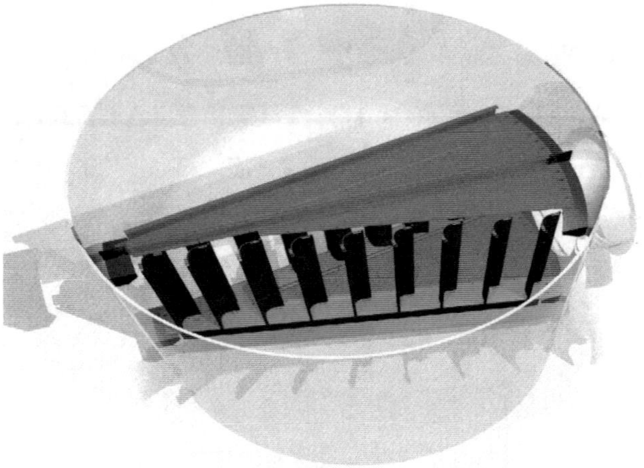

AMACS
PROCESS TOWER INTERNALS

Bad Gas distribution, limited disengaging space and liquid slugs cont.

+ Use a perforated distribution plate downstream of the M.E. for additional distribution assistance.

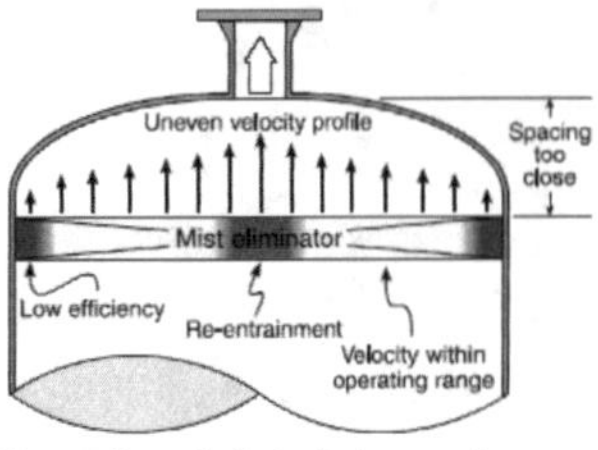

Figure 7. Example of mist eliminator performance degradation due to uneven velocity profile

+ Inlet diffusers also handle incoming liquid slug which otherwise can cause flooding of the Mist eliminator.

AMACS
PROCESS TOWER INTERNALS

Know what mist size droplets You are dealing with

+ Know if your mist is > 1 micron or submicron.
+ When gas travels from A to B, condensation in the pipeline will cause a fine mist (fog or submicron droplets)

  Examples:
  - Incoming gas from an offsite location
  - Fuel gas to furnaces/heaters with low NOx burners

AMACS
PROCESS TOWER INTERNALS

Know what mist size droplets You are dealing with cont.

+ Submicron mist is not captured by conventional M.E.

+ Submicron mist requires a coalescing filter, typically
installed in a two stage filter separator.

AMACS
PROCESS TOWER INTERNALS

Know what mist size droplets You are dealing with cont.

<u>1^{c} stage;</u> submicron droplets are coalesced to > 10 micron droplets.

<u>2^{c} stage;</u> greater than > 10 micron droplets are ejected from the coalescing filter and captured by a vane, which captures droplets > 8 micron.

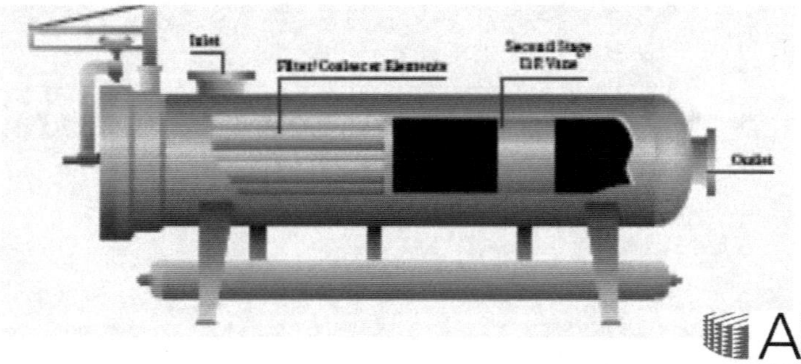

AMACS
PROCESS TOWER INTERNALS

High viscosity liquid as mist

+ Viscous liquids drain slower out of the M.E. and could cause localized flooding.
+ Use a mesh mist eliminator with an open drainage layer on the bottom to improve liquid drainage.
+ Use a lower K- value (lower face viscosity) for the M.E. To allow additional drainage time.
+ Use a vane type M.E. upstream of the mesh mist eliminator, to remove the bulk of the viscous liquid, so the M.E. can effectively handle the residual liquids.

Dislocated M.E. due to System pressure surges

+ M.E. sections can be dislodged into the downstream equipment, which then creates massive carryover.
+ Standard M.E. equipment often gets installed and secured to the support ring with tie wire.

Dislocated M.E. due to System pressure surges cont.

+ Where system pressure surges are common, there are more secure ways to install the M.E.

  - dual support rings
  - through bolts
  - install a pressure relief door

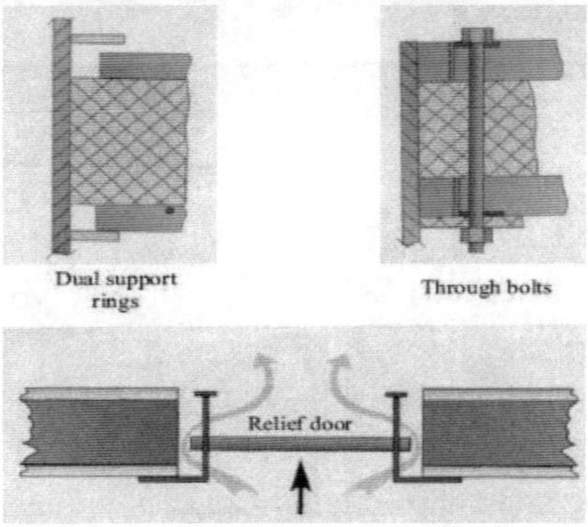

AMACS
PROCESS TOWER INTERNALS

When throughput exceeds original capacity

+ There are various solutions to increase capacity in an existing vessel.

  - Patented High capacit'
    mist eliminator 0% - 1(

AMACS
PROCESS TOWER INTERNALS

When throughput exceeds original capacity cont.

-Patented MISTFIX™
cylindrical mist eliminator 0% - 25%

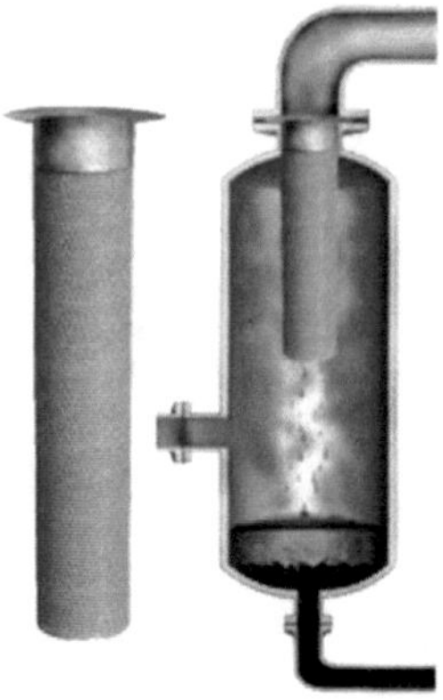

When throughput exceeds original capacity cont.

- High capacity Patented MultiPocket® vane. 0%-50%

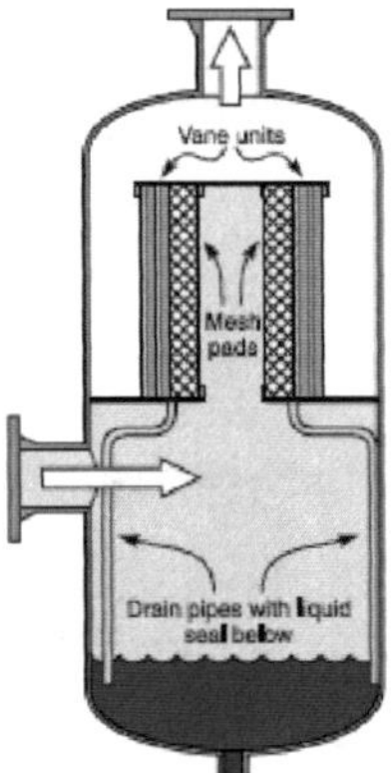

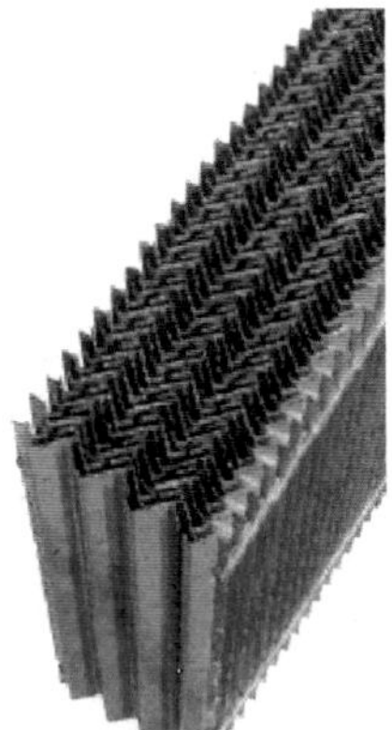

AMACS
PROCESS TOWER INTERNALS

When throughput exceeds original capacity cont.

+ MistFix™ is installed through the gas outlet, all other M.E. equipment comes in sections, installed through the manway.
+ With increased capacity comes increased bad gas distribution.
  - Use an inlet diffuser.
+ To avoid welding new support rings and housing for the high capacity new equipment, use double expansion rings, a compression fit.

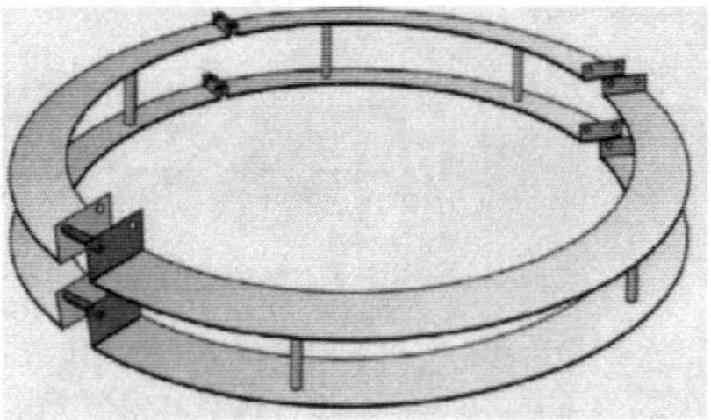

AMACS
PROCESS TOWER INTERNALS

Support Rings

AMACS
PROCESS TOWER INTERNALS

Support Rings

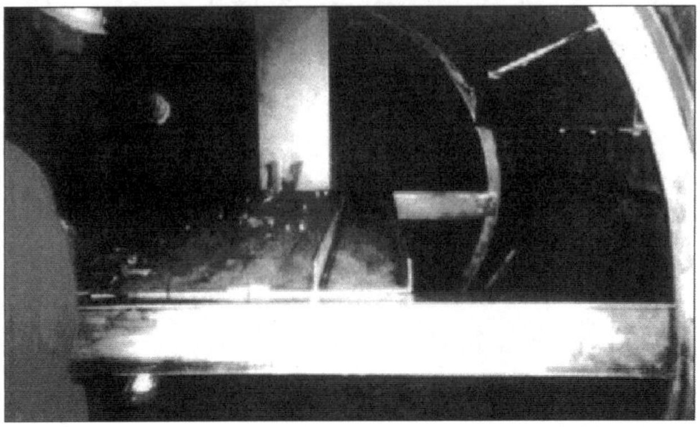

When the capacity range is too large for a suction drum

+ Typical M.E. equipment will operate between

  30% – 110% capacity.

+ When a suction drum often operates below 30% liquid carryover may occur.

When the capacity range is too large for a suction drum Cont.

To facilitate a capacity range of 10% - 110% consider a 2 phase M.E. in the same drum.

- one that operates 30% - 1
- second that operates between 10% - 30%

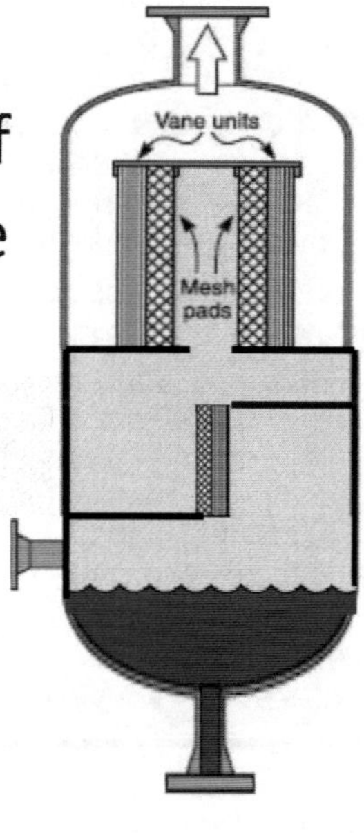

(45i) Renewable Feedstock for Steam Crackers: Catalytic Upgrading of Crude Tall Oil (CTO) Into Bio-Naphtha

Jinto Anthonykutty, Juha Linnekoski, Antero Laitinen and Ali Harlin, VTT Technical Research Centre of Finland, Espoo, Finland

Renewable feedstocks from bio resources are better alternatives to petroleum derived naphtha for the production of olefins and aromatics by steam cracking or reforming.[1, 2] Mostly, these renewable feedstocks (bio-naphtha) rich in n-alkanes are produced by catalytic upgrading methods via hydrotreating/hydroprocessing; in which oxygenates are removed from bio-oil by various deoxygenation processes. At present, hydrotreated vegetable (HVO) oil is produced in many countries on industrial scale using existing petroleum refinery methods[3]. However, employing higher priced vegetable oils could be a major concern in these processes and the main challenge is therefore finding an inexpensive and non-food chain affecting feedstock. Tall oil, obtained as a by-product from Kraft pulping process, is abundant and cheap;and hence, meets the criteria of a low cost starting material for upgrading[4]. Moreover, the utilization of tall oil in conventional oil refineries could serve as an eco-friendly method to convert bio-wastes into bio-naphtha[1].

In the present research approach, catalytic upgrading (hydrotreating) of crude tall oil (CTO) into bio-naphtha suitable for a steam cracker is discussed. CTO, which contains mainly fatty acids (42.9 wt. %) and resin acids (25.7 wt. %) was obtained from Stora-Enso Kaukopää (Finland) mill and used as received. A commercial NiMo catalyst was employed and the experiments were conducted in a continuous, down flow, fixed bed reactor at different process conditions (WHSV 1-3h^{-1}, Pressure 5 MPa and Temperature 350 - 450 ^oC). After the stabilization period (6 h.), the liquid sample was collected and separated into organic and aqueous phases.The product distribution in organic phase was analyzed with a Shimadzu GC/MS equipped with a Zebron MS 30m × 0.25 mm × 0.25μm column.

This current approach reports that CTO produces mainly paraffins, cycloalkanes and aromatics during upgrading process. Paraffinic hydrocarbons mainly comprise n-octadecane (20-25 wt. %) and n-heptadecane (15-20 wt. %), are obtained in maximum yield (55- 60 wt. %) in low temperature reactions. The yield of paraffinic hydrocarbons increased as a function of increasing space time in low temperature reactions. 18-Norabietane and other mono-, di- and tri- cycloalkanes are obtained from resin acids in CTO, and these cyclic structures are converted to aromatics such as naphthalene and phenanthrene in high temperature reactions (>400 ^oC). Stigmastane, apparently produced from sterols in CTO are also present as a major three ring component in the product stream. Additionally, gaseous products obtained and the yield increased as a function of temperature. The trend observed with the distribution of gaseous products reveal that decarboxylation and cracking reactions are favored at higher temperatures, while hydrogenation and hydrodeoxygenation (HDO) are suppressed.

The elemental composition of CTO shows that it contains relatively high sulfur content (1800 mg/kg), could be useful to keep the catalyst's activity with time on stream (TOS). However, the considerable amount (87 mg/kg) of alkali metals (Na, K, Ca and P) in CTO might be critical for the catalyst activity with TOS, will be discussed in this contribution.

The product composition obtained with CTO is assessed in comparison with compositions and characteristics of refinery naphtha feeds, and will be discussed in detail. Furthermore, this contribution will discuss the achieved deoxygenation rate and success criteria for the product composition of upgraded CTO.

References

1. Gornay, J.; Coniglio, L.; Billaud, F.; Wild, G. Steam Cracking and Steam Reforming of Waste Cooking Oil in a Tubular Stainless Steel Reactor with Wall Effects. ***energy&fuels***, **2009**, 23(11), 5663-5676.
2. Pyl, S.P.; Schietekat, C.M.; Reyniers, M.F.; Abhari, R.; Marin, G.B.; Van Geem, K.M. Biomass to olefins: Cracking of renewable naphtha. ***Chemical Engineering Journal***, **2011**, 176-177, 178-187.
3. Sunde, K.; Brekke, A.; Solberg, B. Environmental Impacts and Costs of Hydrotreated Vegetable Oils, Transesterified Lipids and Woody BTL- A Review. ***Energies***, **2011**, 4(6), 845-877.
4. Coll, R.; Udas, S.; Jacoby, W.A. Conversion of the Rosin Acid Fraction of Crude Tall Oil into Fuels and Chemicals. ***energy&fuels***, **2001**, 15, 1166-1172.

AIChE Paper Number 45j
High Efficiency Ethylene Cracking Furnaces

Rajaram Ramesh

Product Development Manager - Ethylene
Technip
Netherlands

Marco W. M. van Goethem, PhD

Product Development Manager – Modeling and Testing
Technip
Netherlands

25th Ethylene Producers Conference
AIChE 2013 Spring National Meeting
San Antonio, Texas, April 29 – May 3, 2013

Abstract

Environmental and economic pressures generate a driving force to produce products more efficiently. A direct effect of more efficient processes is that the emission of green house gases, particularly carbon dioxide decreases. In households nowadays "high efficiency furnaces" are available with efficiencies in the range of 94-97%. This paper describes the outcome of a feasibility study on applying the high efficiency furnace concept to steam cracking furnaces. The high efficiency is achieved by enhanced heat recovery from the flue gases before they are released to the atmosphere at ca. 65°C.

Technip studied the application of this concept for cracking furnaces based on gas as well as liquid feedstocks. Natural gas was used as basis for the energy price. Since energy prices have a strong impact on project economics, the evaluations were done for an EU average natural gas price and a US natural gas price.

Two alternative heat sinks have been considered: preheating of combustion air and preheating of cold hydrocarbon feed. Both are viable for liquid cracking furnaces. In the case of gas cracking furnaces feed preheating results in a pinch and this makes it not viable.

Capital costs are based on Technip's cost database and quotations from air preheater vendors. Special attention has been paid to the material selection of the affected convection section banks.

This study shows that high efficiency ethylene cracking furnaces are economically viable. For Europe the study indicates an attractive payback period. In the US the payback period is longer due to the lower energy prices. If energy prices increase in the future, the concept's economic viability will become better. Inclusion of CO2 credits will make the payback period more attractive.

Furnace Efficiency

In this paper furnace efficiency has been calculated as net thermal efficiency as interpreted by API560[1]. Furnace efficiency is determined by drawing an envelope around the furnace and including in it the radiant section, convection section and any heat exchanger equipment that recover heat from the flue gas (air preheater, economizer, steam superheater, etc.). The efficiency is the ratio of the useful heat absorbed to the fired heat of the fuel and any other external heat inputs. Ambient temperature is a reference temperature for efficiency calculations and any heat inflow or outflow above this reference is taken into consideration as a heat input or heat loss.
Furnace efficiency can be determined as follows:

$$\frac{heat\ absorption\ in\ radiant\ section\ +\ convection\ section}{(firing\ rate\ \times\ fuel\ heating\ value) + sensible\ heat\ of\ fuel\ and\ air\ above\ ambient\ temp. + other\ external\ heat\ inputs}$$

Alternately, $\text{Furnace efficiency} = 100\% - \text{Total heat loss}$, where:

$$\text{Total heat loss} = \frac{stack\ loss\ +\ radiation\ heat\ loss}{(firing\ rate\ \times\ fuel\ heating\ value) + sensible\ heat\ of\ fuel\ and\ air\ above\ ambient\ temp. + any\ external\ heat\ inputs}$$

In a furnace where fuel and combustion air enter the furnace at ambient temperature and there are no other external heat inputs, the first expression can be simplified to the following:

$$Furnace\ efficiency = \frac{heat\ absorption\ in\ radiant\ section\ +\ convection\ section}{fuel\ firing\ rate\ \times\ fuel\ heating\ value}$$

Please note that net thermal efficiency, as defined by API 560, is based on the lower heating value of the fuel.

Study objectives

Technip undertook this study with the goal to reduce fuel firing in cracking furnaces by at least two percent. Although this figure does not look very ambitious, it must be remembered that since efficiencies in current designs are already in the mid nineties (upper part of the technology S-curve), further improvements will not be easy and can only be incremental.

Since stack loss is the main source of inefficiency (radiation heat loss is comparatively smaller and more difficult / expensive to reduce) it was decided to make stack loss reduction as the main focus of the study.

By a relatively straight forward calculation it can be shown that the two percent efficiency rise target would require the flue gas to be cooled to approximately sixty or seventy degrees Celsius.

The main challenge in reducing stack temperature to this level is finding an appropriate heat sink for the recovered heat. In a cracking furnace there are two suitable heat sinks, combustion air and hydrocarbon feed, and both were studied. The following sections discuss the results for liquid furnaces and gas furnaces.

Basis of Study

We have done the study on a naphtha furnace and an ethane furnace. We have chosen these two feedstocks since a large number of furnaces operate on these two feeds and we can draw conclusions that can be applied approximately to other liquid and gas feeds. The study has been performed on the following bases:

- The reference liquid and gas furnace designs used in this study are based on actual furnaces from Technip's past projects. The furnaces were simulated using Technip's proprietary tool SPYRO which is used in the design and simulation of cracking furnaces.
- The budgetary capital cost need to achieve the target furnace efficiency was calculated as an incremental cost. It was estimated from vendor quotes and Technip's internal cost database.
- The cost estimates are for a new project, not a revamp, and are for a West European location.
- Operating costs are based on cost of base utilities and derived utilities. Base utilities are natural gas and demineralized water. These are taken from market data.
- Since natural gas prices are very different in different parts of the world, the operating costs and economic analysis have been done for European and United States gas prices.
- EU natural gas price was fixed at €25.1 /MWh of higher heating value (HHV) which is equivalent to €28.0 /MWh of lower heating value (LHV). The prices are based on EUROSTAT data for 1st half of 2010 and were obtained from the publications of the Department of Energy & Climate Change in the UK. They correspond to the EU-15 average for Large Industrial Consumers (i.e. 2.13 pence/kWh HHV).
- US Natural gas prices were obtained from the US Department of Energy. The average value was 6 \$/MM BTU (HHV) or 17 €/MWh (LHV) for 1st half of 2010. The average HHV used is 1029 BTU per cubic foot or 1.029 MM BTU per thousand cubic feet.
- Cost of demineralized water was assumed at € 2.00 /ton.
- The prices of other utilities (boiler feed water, steam and electricity) have been calculated from those of the two base utilities. Electricity was priced assuming 35% overall efficiency of power generation.
- The economic calculations do not include any CO2 credit for the fuel saved.
- Assumed exchange rate: 1 US dollar = 0.75 euros
- The plant is assumed to operate 8400 hours per year

The resultant utility prices used in the study are summarized below.

UTILITY COSTS

		EU basis	US basis	
FUEL as NATURAL GAS	Base	**28.0**	**17.0**	€ / MWh
DEMINERALISED WATER	Base	2.0	1.5	€ / ton
VHP STEAM (net value)	Derived	23.7	14.4	€ / ton
MP STEAM (net value)	Derived	13.9	8.5	€ / ton
LP STEAM	Derived	9.7	5.9	€ / ton
ELECTRICAL POWER	Derived	80.1	48.6	€ / MWh

Study Results

Naphtha Furnace using Air Preheat

Combustion air can be preheated in a plate type air preheater (APH). These are cross flow type exchangers and have low pressure drops. Combustion air is at a lower temperature than any of the process or utilities streams entering the convection section; therefore, from the standpoint of efficiency, APHs are ideally suited to be the last heat recovery section before the stack.

Fig 1 shows the Base Case Naphtha furnace while Fig 3 shows the same furnace with an APH. Please also refer to Table 1 for a performance comparison of the APH design with the Base Case naphtha furnace. The following points are of interest:

1) The overall efficiency of the furnace will improve from 94.4 to 97.0% for the APH option. This is mainly due to the lower stack temperature but also partly due to the smaller excess air, which is explained below.

2) This option will require a forced draft combustion air fan to overcome the pressure drop in the APH. As a result the furnace operation will change from induced draft mode to balanced draft. The availability of air in forced draft will enable more accurate control of the excess air and it will be possible to operate the furnace with lower excess air (5% instead of 8%). This leads to a lower flue gas flow, a lower stack loss and therefore better efficiency.

3) There will be a drop of 4.8% in fuel firing which is greater than the efficiency increase. This is because of the lower absorbed heat. The furnace is designed and controlled in such a way as to supply all the required process heat and to absorb the surplus heat into utilities to the extent possible. When the furnace operates more efficiently and the heat demanded by the process media remains unchanged, as in the APH case, there will be less surplus heat available for VHP steam. As a result the VHP steam generation will go down from 79.7 to 76.0 t/h.

4) The heat transfer surface in some parts of the APH will be below the water dew point of the flue gas (around 60°C). This can lead to water condensation and create two possible issues: 1) accumulation of water in the APH and 2) corrosion due to the presence of liquid water, oxygen and carbon dioxide.

   Two measures have been included in the design to counter these problems:
   - Placement of unit such that flue gas flows downward to aid easy draining
   - Use of corten steel to resist corrosion

5) We have evaluated the impact of the air preheat option on NOx emissions using API 535[2)]. The rise in air temperature and the drop in excess air will have opposing effects on the NOx production. The net effect is that the NOx concentration will be close to the Base Case.

   If a DeNOx unit is present, as is often the case in the US, the APH option may help to slightly extend the life of the DeNOx catalyst. This is because the total NOx load will be smaller than the Base Case due to the smaller flue gas flow.

Table 2 summarizes the results of the economic analysis of this option. We have used payback period to compare different options. Payback period is the ratio of the incremental capital cost above the Base Case CAPEX divided by the savings in yearly operating cost.

1) The capital cost (CAPEX) of the APH option includes the cost of the APH, the combustion air fan, extra ducting and structural steel. The operating cost (OPEX) savings is the savings in furnace fuel but this must be offset against the value of the lost VHP steam production (which notionally must be made up in a utility boiler elsewhere) and the additional electricity for the two fans.

2) It can be seen that the payback period is three years based on European gas prices and five years based on American gas prices.

Naphtha Furnace using Feed preheat

Naphtha is normally stored at ambient temperature and is preheated to 60°C using quench water before it is routed to the furnaces. Most cracking furnaces are equipped with a Feed Preheater (FPH) bank in the convection section. If flue gas is utilized to preheat naphtha from ambient conditions the stack temperature can be lowered to the desired range of sixty to seventy degrees Celsius. Till now designers and operators have not taken full advantage of this mainly because of the higher CAPEX and fear of dew point corrosion.

One of the consequences of this feature will be that the quench water duty previously absorbed by the naphtha will now shift to the air coolers in the quench water circuit. The additional electrical power that the air cooler fans will require will be about 10 KW per furnace.

Please refer to Table 3 for an overview of the performance of a naphtha furnace with and without an extended feed preheater. Please note that the Base Case naphtha furnace is the same as the one considered in the APH option.

1) The overall efficiency of the furnace can be improved from 94.4 to 96.0% for this option.

2) Unlike the APH option a combustion air fan will **not** be required. The furnace will continue to operate in induced draft mode.

3) The fuel firing and the VHP steam generation will drop marginally.

4) The upper tube rows of the FPH will be below the water dew point of the flue gas and some condensation will occur. As a precaution we have considered that the top four rows will be of stainless steel.

5) The flue gas NOx concentration will be unaffected by this option. However the total hourly NOx emissions and load on the DeNOx unit, if any, will decrease to the same extent as the fuel firing, which is around 0.6%.

Table 4 summarizes the results of the economic analysis of this option.

1) The capital cost (CAPEX) of this option covers the incremental cost of the extended FPH, extra ducting and structural steel. The OPEX savings is the reduction in furnace fuel after adjusting for the reduction in VHP steam production and the additional electricity for the ID fan and the Quench Water Air Cooler fans.

2) In the Base Case the naphtha feed piping between the external feed preheater and the convection section is insulated for heat conservation. In the FPH option naphtha will be fed at ambient temperature to the furnace and so these lines do not have to be insulated. This will yield some capital cost saving but this has not been taken into account in our analysis.

3) The payback period excluding CO2 credit will be three and a half years with European natural gas and five and a half years with US natural gas.

Ethane Furnace using Air Preheat

Fig 2 shows the Base Case Ethane furnace that was used in the study. The APH arrangement on an ethane furnace will be quite similar to the one on the naphtha furnace. Table 5 shows the impact that the APH would have on the furnace performance. The main points are as follows:

1) The overall efficiency of the furnace will improve from 91.7 to 96.5% for the APH option. This is mainly due to the lower stack temperature but also partly due to the smaller excess air.

2) In this furnace also the APH will require a forced draft combustion air fan. This will allow the excess air to be reduced from 8% to 5% which will result in lower stack losses and a better efficiency.

3) The fuel firing will go down by 11.0%, from 8.2 to 7.3 t/h. The absorbed heat will drop from 116.8 to 110.6 MW. As a result of these the VHP steam generation will go down from 97.0 to 88.8 t/h.

4) A portion of the APH will operate below the flue gas water dew point. Consequently the unit must be designed to prevent water accumulation and corrosion.

5) The impact on NOx was evaluated here also. The rise in air temperature will have a greater impact on NOx than the fall in excess air. We estimate that as a result the flue gas NOx concentration will increase slightly (around two percent). This can be a cause for concern if the furnace does not have a DeNOx unit and is operating close to the limits imposed by authorities. However since the total firing will go down, the hourly NOx production will drop. Plants can consider discussing with authorities to obtain a higher allowable NOx limit,

   If the furnace has a DeNOx unit, the chemical load on the unit will reduce, which should help to obtain a longer catalyst life.

The operating cost savings and payback for this option have been summarized in Table 6.

1) The capital cost includes the cost of the APH, the combustion air fan, extra ducting and structural steel. The operating cost savings is the savings in fuel firing with an adjustment for the value of the reduced VHP steam generation and the extra electricity for the ID fan and FD fan.

2) The payback period with natural gas priced at European rates will be one and a half years. The corresponding figure for US gas prices will be three years.

Ethane Furnace using Feed Preheat

Using feed at ambient temperature does not significantly reduce the stack temperature in a gas furnace. This is because the feed preheat curve has a steeper slope than the flue gas cooling curve, causing a pinch on the hot side of the Feed Preheater. Fig 4 shows the heating and cooling curves of ethane, boiler feed water and flue gas in the convection section. The hatched shape on the plot shows the maximum additional heat duty that can be recovered into feed, over and above the Base Case. The plot clearly shows that when the ethane feed temperature is lowered from the base value of 55°C to 30°C, the stack temperature will drop from 170°C to about 150°C. The flue gas cannot be cooled any more than this modest amount.

For this reason the feed preheat option was not investigated any further.

Conclusions and Summary

The key results from the different study options are listed together below.

Cracking furnace feed	Naphtha		Naphtha		Ethane	
Cold sink	Combustion air		Hydrocarbon feed		Combustion air	
	EU nat. gas	US nat. gas	EU nat. gas	US nat. gas	EU nat. gas	US nat. gas
OPEX savings	€ 0.60 mill	€ 0.36 mill	€ 0.17 mill	€ 0.10 mill	€ 1.27 mill	€ 0.77 mill
Payback time	3 years	5 years	3.5 years	5.5 years	1.5 years	3 years
Salient features of each option	Small number of additional equipment		No additional equipment		Small number of additional equipment	
	Higher CAPEX		Lower CAPEX		Higher CAPEX	
	More OPEX savings		Less OPEX savings		Best OPEX savings	
	Slight effect on NOx concentration		No effect on NOx		Around 2% higher NOx concentration	

All three scenarios have attractive payback periods and will be worth considering in high energy cost countries such as Europe and Asia. In US and other energy rich nations, the economic incentive will be much lower. In the US, where ethane cracking is common and increasing even more, improvement of furnace efficiency with air preheat can be worth considering as the payback time is less than 3 years.

If the options are compared against each other, in our opinion the Air preheating option has more potential, irrespective of feedstock.

The Feed Preheating option in liquid furnaces is cheaper but it also gives proportionally modest savings. Moreover, as earlier mentioned, the feed preheat duty is being shifted from one waste heat source (quench water) to another (flue gas). This option will not lead to a net improvement or cash savings in the plant as a whole unless there is a shortage of quench water and there are other, better uses for the saved quench water duty

Acknowledgements

The authors wish to express their thanks to their colleagues Mr Ashok Laghate, Ms Susan Cooper, Mr Sachin Singare and others in Technip Benelux who contributed to this study.

References

1) API 560: Fired Heaters for General Refinery Service, 4th edition

2) API 535: Burners for Fired Heaters in General Refinery Service, 2nd edition

Table 1
Naphtha Furnace with Combustion Air Preheating – Performance Summary

		Base Case	Air Preheat Case
Naphtha feed	t/h	60.8	60.8
Ambient conditions	-	15°C, 85% R.H.	15°C, 85% R.H.
Air temp. to burners	°C	15	77
Stack temperature	°C	108	61
Excess air	%	8	5
Absorbed heat	MW	113.2	110.4
Overall efficiency	%	94.4	97.0
Fuel flow	t/h	8.4	8.0
Fired heat	MW	119.9	113.9
VHP steam production	t/h	79.66	76.01
NOx concentration		Base	Close to Base

Table 2
Economics of Naphtha Furnace with Combustion Air Preheating

		Air Preheat Case (EU location)	Air Preheat Case (US location)
Savings in fuel firing	million €/yr	1.42	0.86
Cost of lost VHP steam capacity	million €/yr	0.73	0.44
Cost of additional electricity	million €/yr	0.10	0.06
Net OPEX saving	million €/yr	0.60	0.36
Payback	Years	3	5

Table 3
Naphtha Furnace with Feed Preheating – Performance Summary

		Base Case	Feed Preheat Case
Naphtha feed	t/h	60.8	60.8
Feed temperature	-	60°C	30°C
Combustion air temp.	°C	15	15
Stack temperature	°C	108	71
Excess air	%	8	8
Absorbed heat	MW	113.2	114.3
Overall efficiency	%	94.4	96.0
Fuel flow	t/h	8.4	8.35
Fired heat	MW	119.9	119.1
VHP steam production	t/h	79.66	79.63
NOx concentration		Base	Base

Table 4
Economics of Naphtha Furnace with Feed Preheating

		Feed Preheat Case (EU location)	Feed Preheat Case (US location)
Savings in fuel firing	million €/yr	0.19	0.12
Cost of lost VHP steam capacity	million €/yr	< 0.01	< 0.01
Cost of additional electricity	million €/yr	0.01	< 0.01
Net OPEX saving	million €/yr	0.17	0.10
Payback	Years	3.5	5.5

Table 5
Ethane Furnace with Air Preheating – Performance Summary

		Base Case	Air Preheat Case
Ethane feed	t/h	51.9	51.9
Ambient conditions	-	15°C, 85% R.H.	15°C, 85% R.H.
Air temp. to burners	°C	15	129
Stack temperature	°C	169	64
Excess air	%	10	5
Absorbed heat	MW	116.8	110.6
Overall efficiency	%	91.7	96.5
Fuel flow	t/h	8.2	7.3
Fired heat	MW	127.3	114.5
VHP steam production	t/h	97.0	88.75
NOx concentration		Base	Base + 2%

Table 6
Economics of Ethane Furnace with Air Preheating

		Air Preheat Case (EU location)	Air Preheat Case (US location)
Savings in fuel firing	million €/yr	3.01	1.83
Cost of lost VHP steam capacity	million €/yr	1,64	1.00
Cost of additional electricity	million €/yr	0.10	0.06
Net OPEX saving	million €/yr	1.27	0.77
Payback	Years	1.5	3

Fig 1
Naphtha cracking furnace

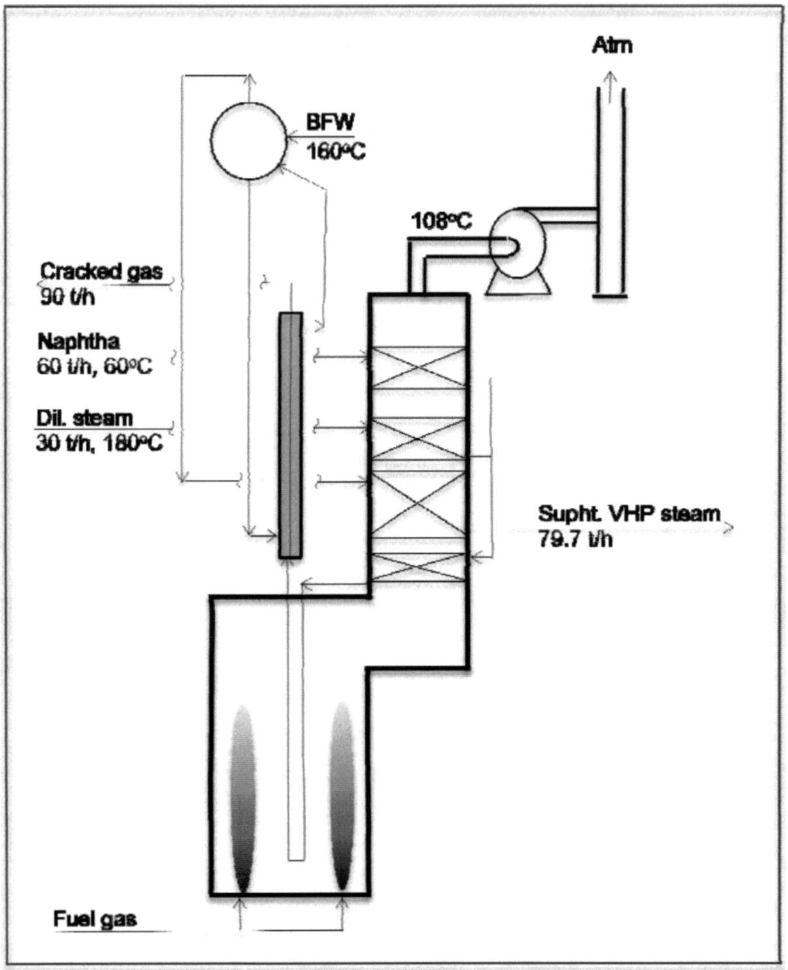

Fig 2
Ethane cracking furnace

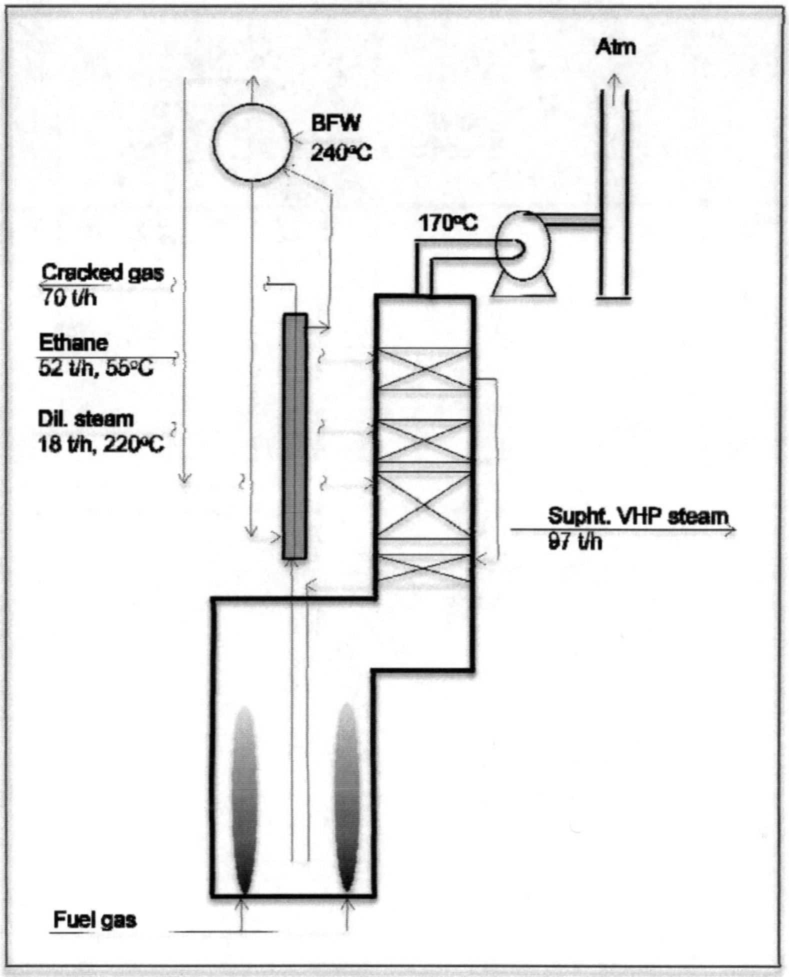

Fig 3

Naphtha cracking furnace – Air Preheater option

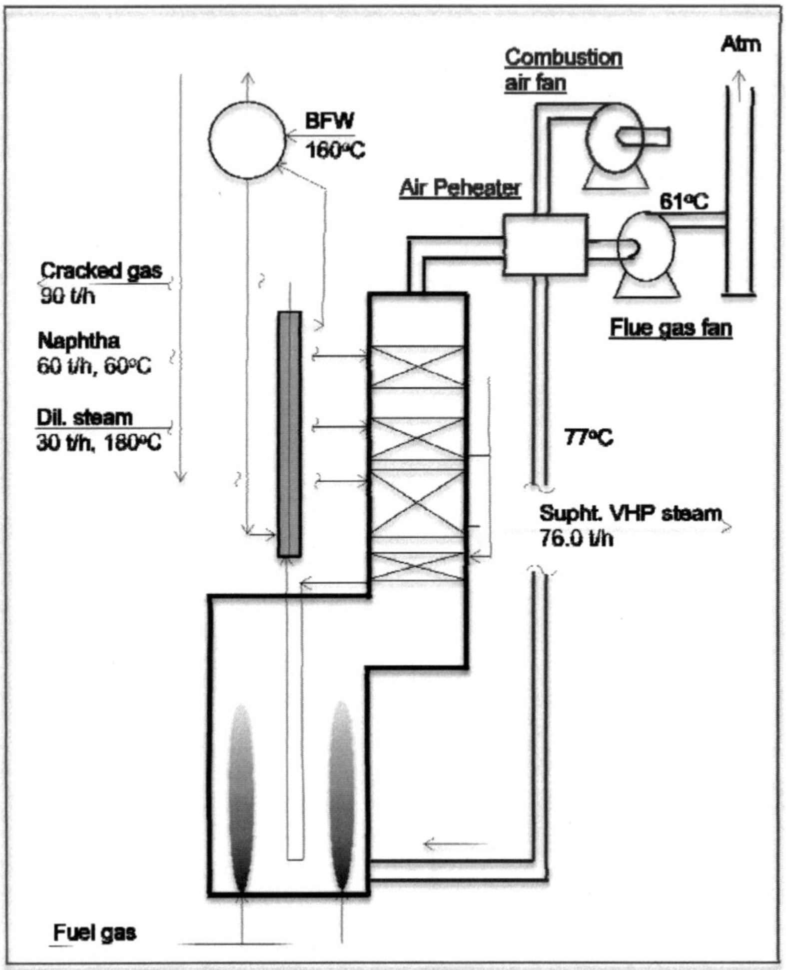

Fig 4

Ethane cracking furnace – Feed Preheater option

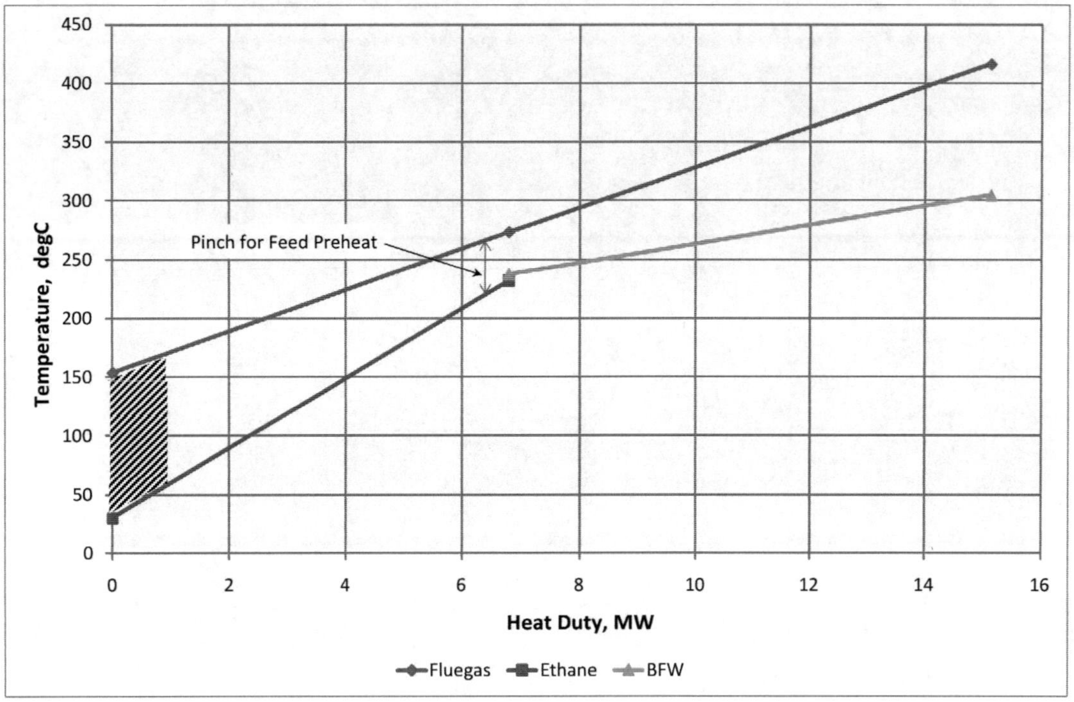

Modernize Your Control Functionality
Easy Steps to Start Process Optimization

13 AIChE
Spring Meeting, San Antonio, TX

Presenters

- Laurie Ben
- James Beall

EMERSON
Process Management

Agenda

- Introduction
- Examples – Best Practices & Lessons Learned
  - Split Range Duty
  - Smith Predictor
  - Gain Scheduling/Adaptive Control
  - Signal Characterization
- Summary

Introduction

- Modernizing Control System – Decision Done Check!
- 'Do I' or 'Don't I' take advantage of new technology
  - Implemented older cumbersome techniques?
  - Complex custom strategies that handle automation challenges?
  - Hard to implement & maintain?
- Standard features that are field proven, practical, and easy to implement
  - Easier to implement & maintain
  - More efficient and effective control strategies
  - Achieve improvements in process performance
  - Lays the foundation for optimization

Agenda

- Introduction
- Examples – Best Practices & Lessons Learned
  - Split Range Duty
    - Control Challenge
    - Legacy Solution
    - Newer Technology Solution
    - Summary
  - Smith Predictor
  - Gain Scheduling/Adaptive Control
  - Signal Characterization
- Summary

Split Range Duty – Control Challenge

- Use steam and hot oil as energy source for distillation column for temperature controller.
- Hot oil is "free energy" and should be maximized
- Control Scenarios
  - Maximize hot oil and supplement with steam
  - Base load steam (if meter bad) and supplement with hot oil
  - Base load hot oil flow (if bad meter) and supplement with steam
- Constraints
  - Minimum and maximum hot oil valve position and flow rate
  - Minimum steam flow rate

Split Range Duty – Legacy Solution

– Old legacy system implemented this feature using a velocity PID with incremental output and a “micro expert rules system”

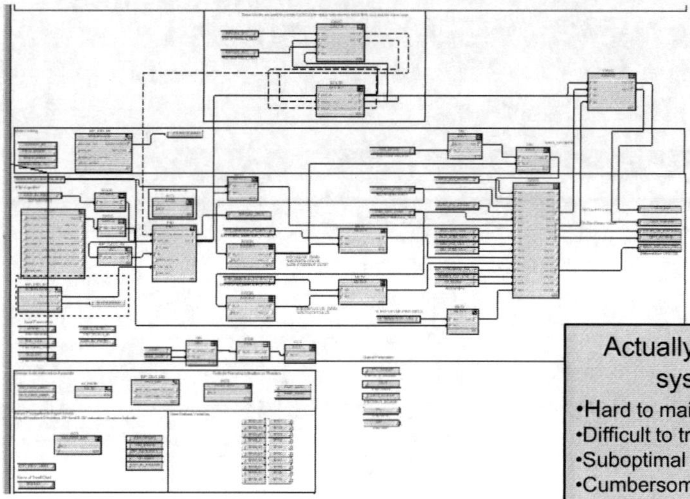

Actually Duplicated in New system, but still…

- Hard to maintain and modify
- Difficult to troubleshoot
- Suboptimal performance
- Cumbersome/Kludge/Ugly

Split Range Duty – New Technology Solution

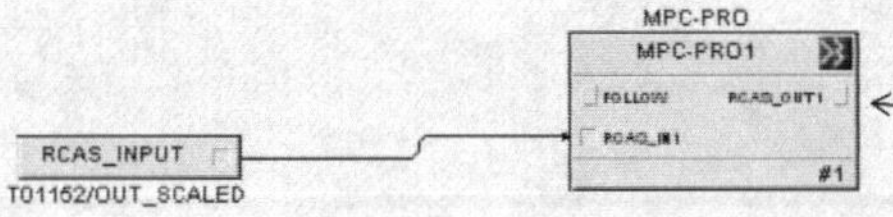

Embedded MPC Function Block provides all control functionality! EASY!!!

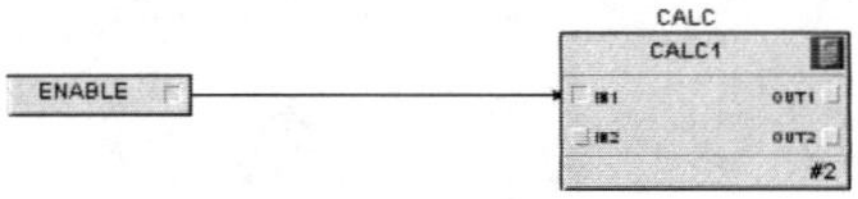

Simple calculations to calculate limits

Split Range Duty – New Technology Solution

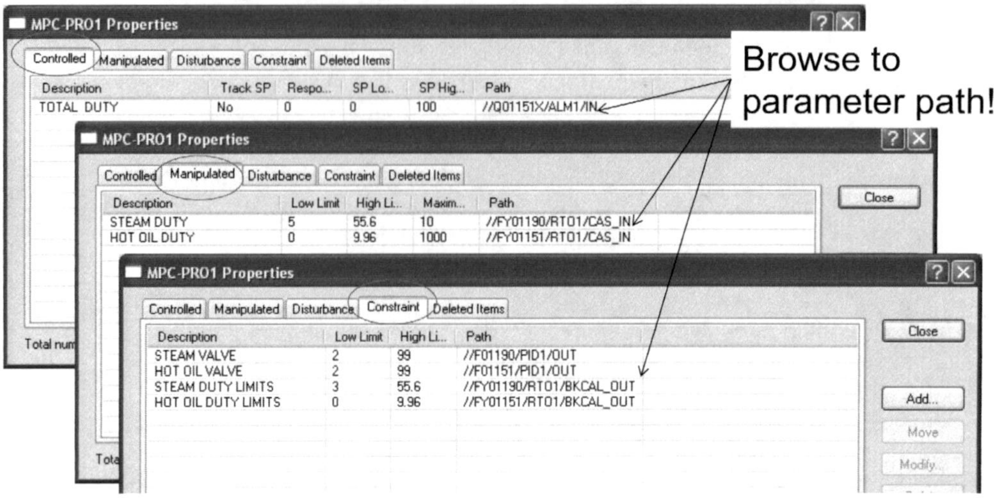

Split Range Duty – New Technology Solution

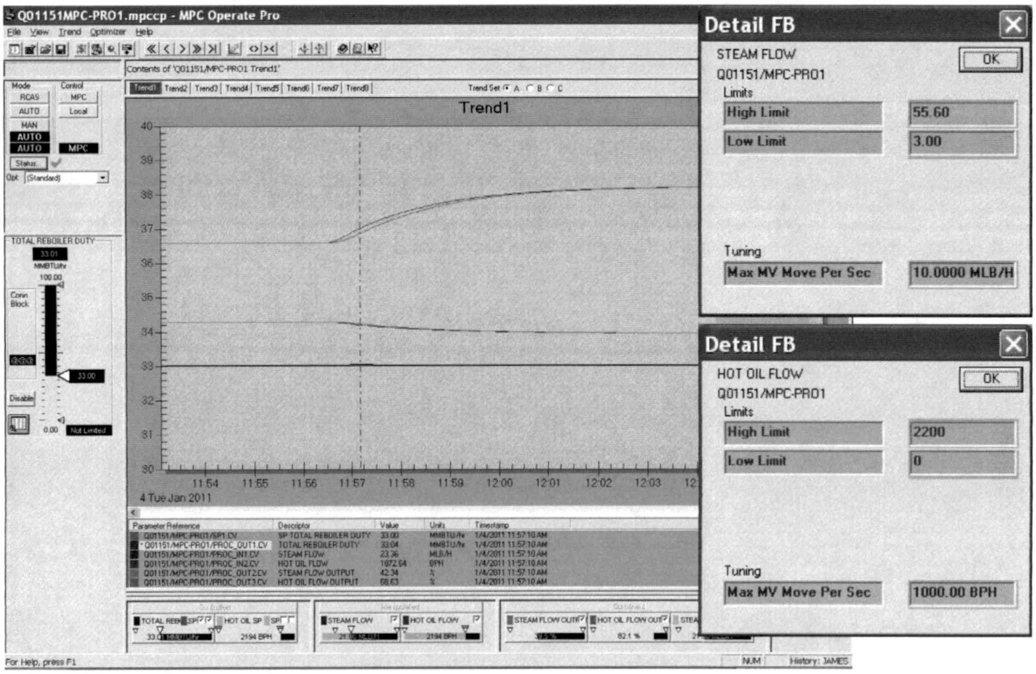

Summary Split Range Duty Control

Embedded MPC (New Technology) vs Legacy Solution

- Better performing solution
  - Bumpless transfers and constraint control
- Easier to Maintain
  - Change limits and settings
  - Simpler implementation
  - Easy to understand, train others, document
- Easily expands to handle future constraints
- Reduce energy costs!

Agenda

- Introduction
- Examples – Best Practices & Lessons Learned
  - Split Range Duty
  - Smith Predictor
    - Control Challenge
    - Legacy Solution
    - Newer Technology Solution
    - Summary
  - Gain Scheduling/Adaptive Control
  - Signal Characterization
- Summary

Smith-Predictor PID – Control Challenge

- Control of processes that are "dead time dominant" (dead time > time constant)
- Typically custom configured
  - some legacy control systems have preconfigured options, e.g. PROVOX
- Algorithm uses a process model
  - (gain, dead time and time constant)
- Challenge: as process dynamics change from model used, performance degrades rapidly

Smith-Predictor PID – Legacy Solution

- Traditional Smith-Predictor

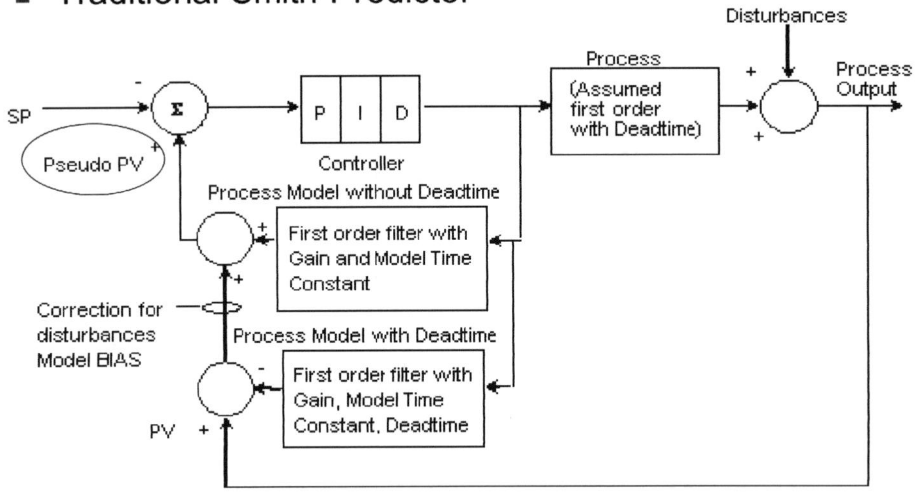

•High maintenance when process dynamics change
•Performance degrades rapidly as process dynamics change from configured model

Smith-Predictor PID – New Technology Solution #1

Easy to implement the Smith-Predictor but…

High Maintenance when process dynamics change

Performance degrades rapidly as process dynamics change from configured model

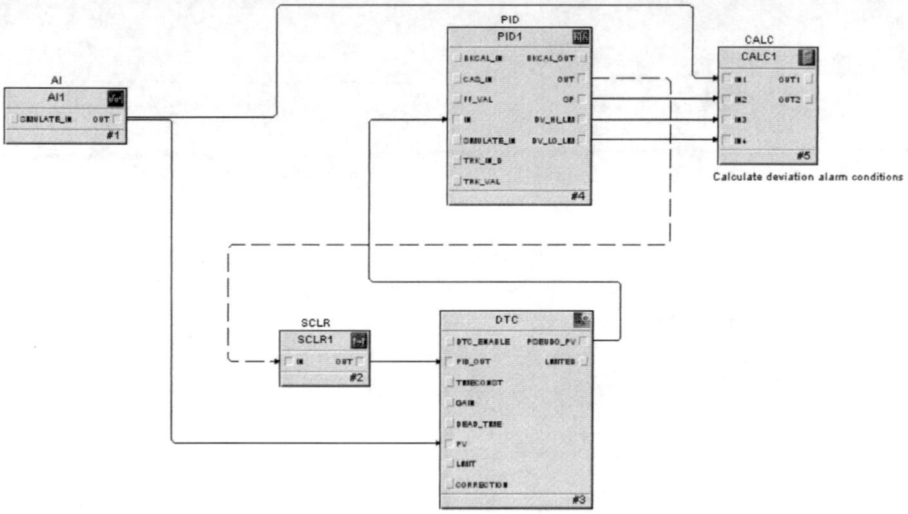

Smith-Predictor PID – New Technology Solution #2

- Embedded MPC Solution

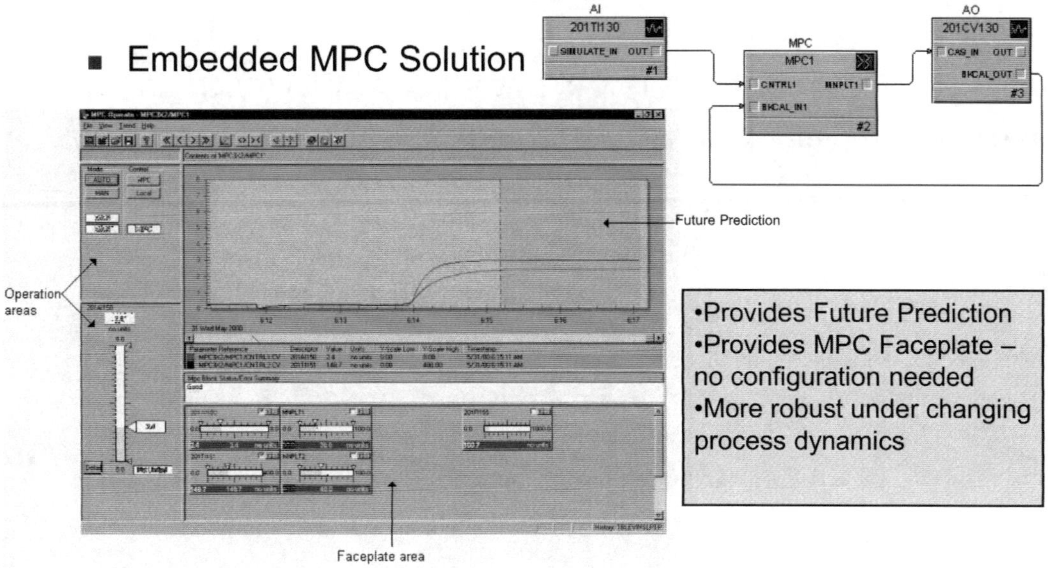

•Provides Future Prediction
•Provides MPC Faceplate – no configuration needed
•More robust under changing process dynamics

Smith-Predictor PID – New Technology Solution #2

- Embedded MPC Solution

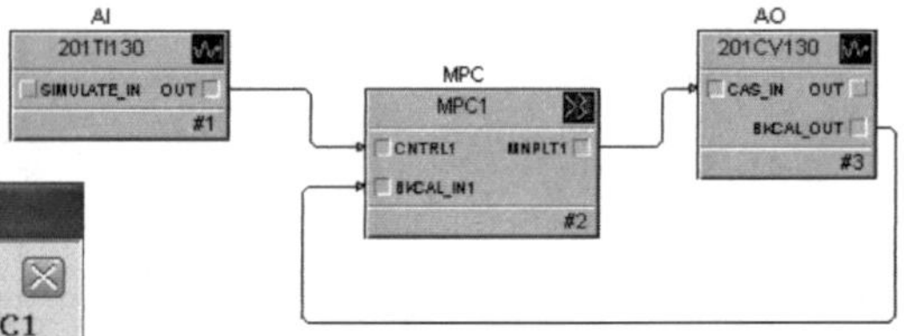

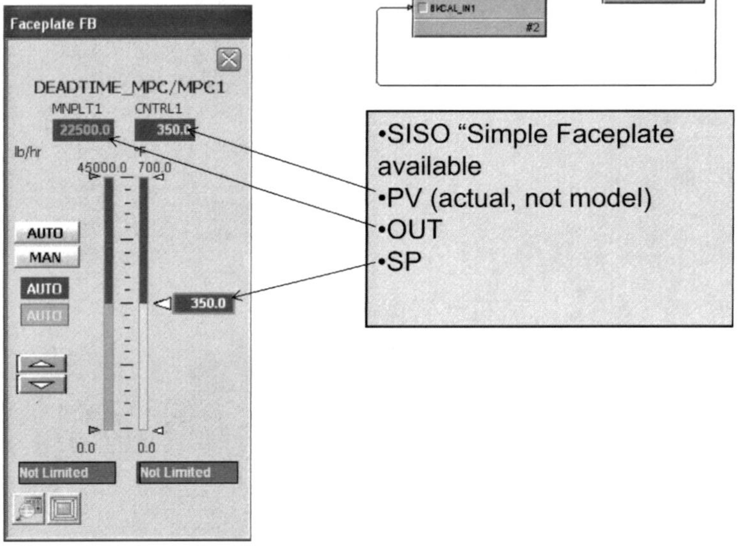

Summary Smith-Predictor PID

New Technology Solutions

- Standard Smith Predictor (#1) and Embedded MPC (#2)
- Both easy to implement and understand!

New Technology Solution #2 (Embedded MPC)

- More robust solution under a changing process model
- Provides visual display of the "future" PV
- Save money through better control performance

Agenda

- Introduction
- Examples – Best Practices & Lessons Learned
  - Split Range Duty
  - Smith Predictor
  - Gain Scheduling/Adaptive Control
    - Control Challenge
    - Legacy Solution
    - New Technology Solution
    - Summary
  - Signal Characterization
- Summary

Gain Scheduling – Control Challenge

- Legacy systems have limited capability to modify tuning based on process conditions
- Others had add-on application software, e.g. PROVOX "Gain Scheduler" system provided 3 sets of Gain, Reset, and Rate applied in 3 separate regions of a "state variable"
  - State variable = PV, OUT or "other"
  - Transition between tuning sets interpolated over "dead band"
- Beware: some suspend loop execution when tuning parameters change, so don't change the tuning every scan

Gain Scheduling – Legacy Solution

- Custom code
- Multiple points, registers, displays
- Restricted (slower) update rate

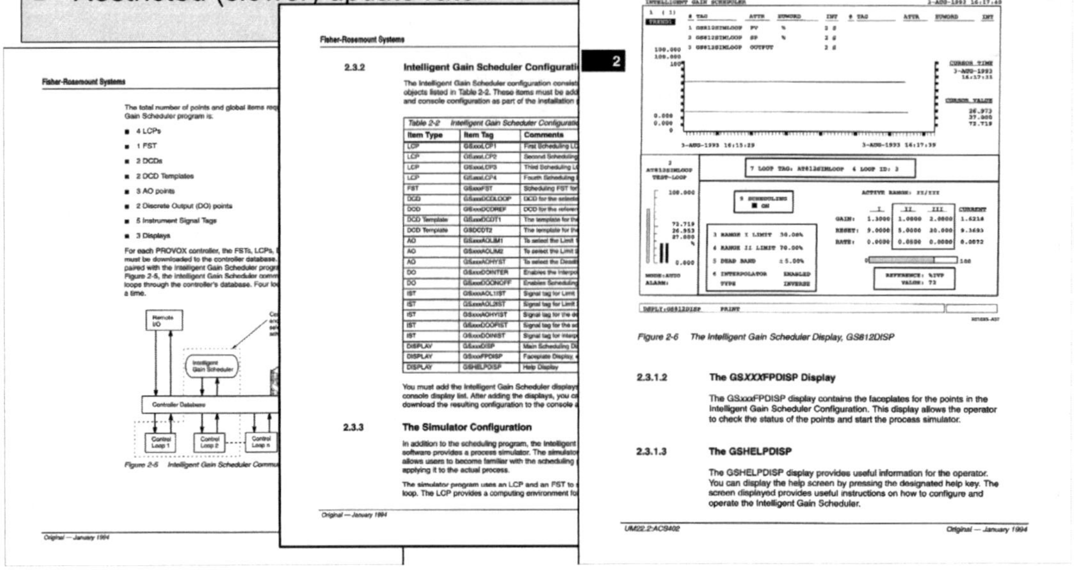

Gain Scheduling – New Technology Solution #1

Standard Modules

- Standard configuration
- Preconfigured faceplate & detailed display
- Easy to implement
- Tuning updates at loop execution

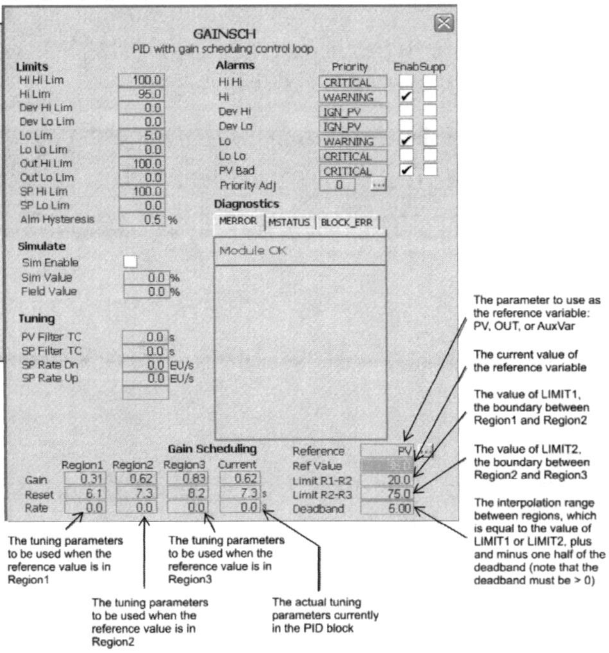

Gain Scheduling – New Technology Solution #2

New Technology "Adaptive Control"

- "Learns" process models from normal operating moves
- Allows up to 5 tuning regions of the "state variable"
- Partial Adaptive Control
  - Use "approved model" only & selected tuning rule in each region
  - Lambda PI or IMC PID; "Closed loop time constant" or "Lambda Factor".
- Optional Full Adaptive Control
  - continuous "Learning" of process model in each region
  - updates tuning while in each region per selected tuning rule
  - set allowable deviation (%) from approved model in each region

New Technology – Adaptive Control

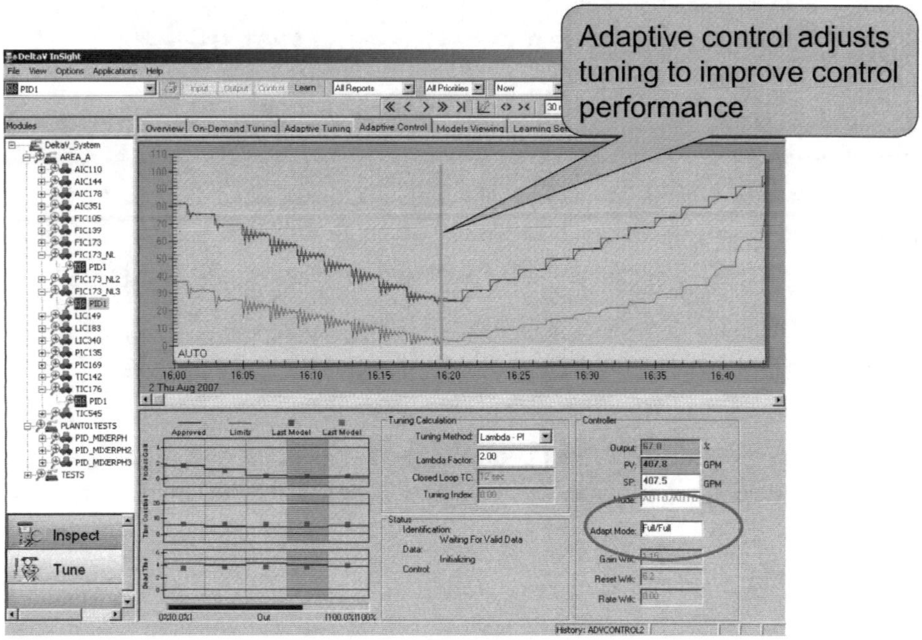

Summary Gain Scheduling

New Technology Solution(s) vs. Legacy Solution

- Gain Scheduling easy to implement and understand!
  - Standard configuration
  - Preconfigured faceplate & detailed display
  - Tuning updates at loop execution
- Adaptive Control provides additional benefit and ease of implementation
  - Learns models during normal operation – NO STEP TESTs!
  - More Regions – Up to 5
  - Automatic tuning updates when process model changes
- Improved control performance saves money!

Summary Gain Scheduling

New Technology Solution(s) vs. Legacy Solution

- Gain Scheduling easy to implement and understand!
  - Standard configuration
  - Preconfigured faceplate & detailed display
  - Tuning updates at loop execution
- Adaptive Control provides additional benefit and ease of implementation
  - Learns models during normal operation – NO STEP TESTs!
  - More Regions – Up to 5
  - Automatic tuning updates when process model changes
- Improved control performance saves money!

Agenda

- Introduction
- Examples – Best Practices & Lessons Learned
  - Split Range Duty
  - Smith Predictor
  - Gain Scheduling/Adaptive Control
  - Signal Characterization
    - Control Challenge
    - Legacy Solution
    - New Technology Solution
    - Summary
- Summary

Signal Characterization - Control Challenge

- Input Characterization (AI)
  - Combustion control – characterize fuel and air
  - Level – characterize pv (e.g. level to volume)
- Output characterization (AO)
  - modify installed flow characteristics of valves, dampers or other final control elements
  - Use to modify loop gain, e.g. output path to slave loop
- Legacy System Challenges
  - Lack of resolution (few points)
  - External black box characterizers may shift, lose documentation
  - Potential operator/maintenance confusion on signal going to final control element

Signal Characterization – Legacy Solution

- Legacy system using external analog characterizer

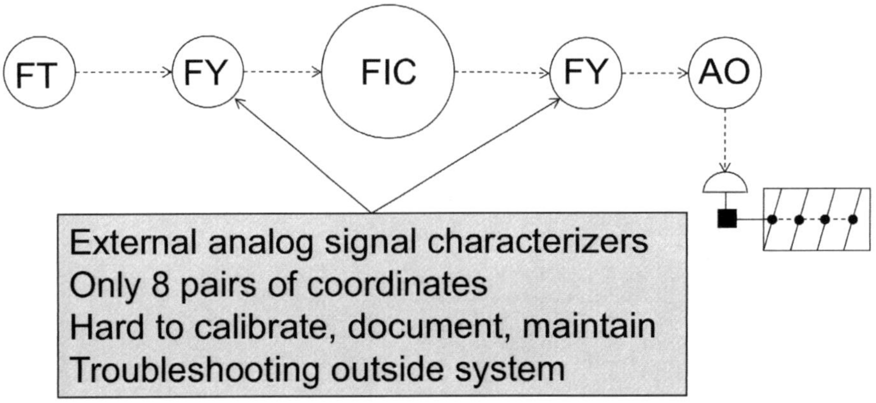

Signal Characterization – New Technology Solution #1

- Standard Module implementation provides 21 point signal characterizer function block (SGCR)

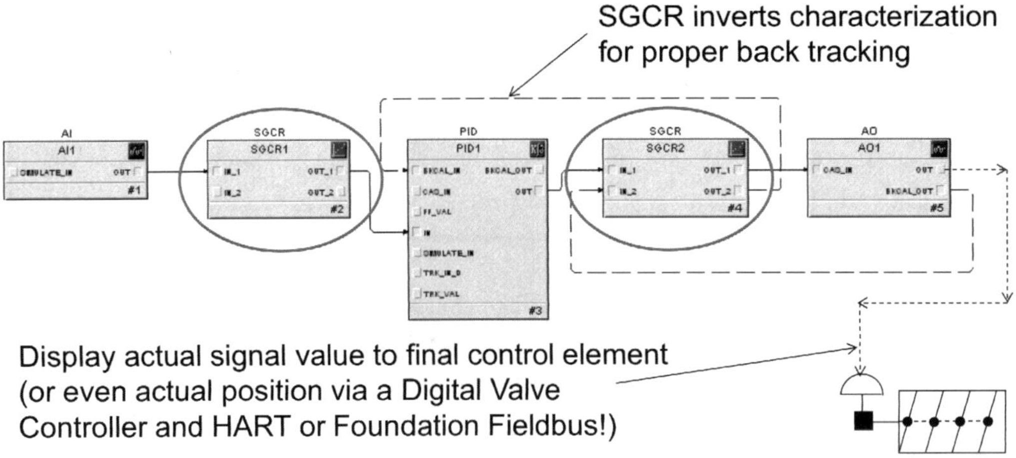

Signal Characterization – New Technology Solution #2

- Adaptive Control
  - Adjust tuning to compensate for final control element characteristics

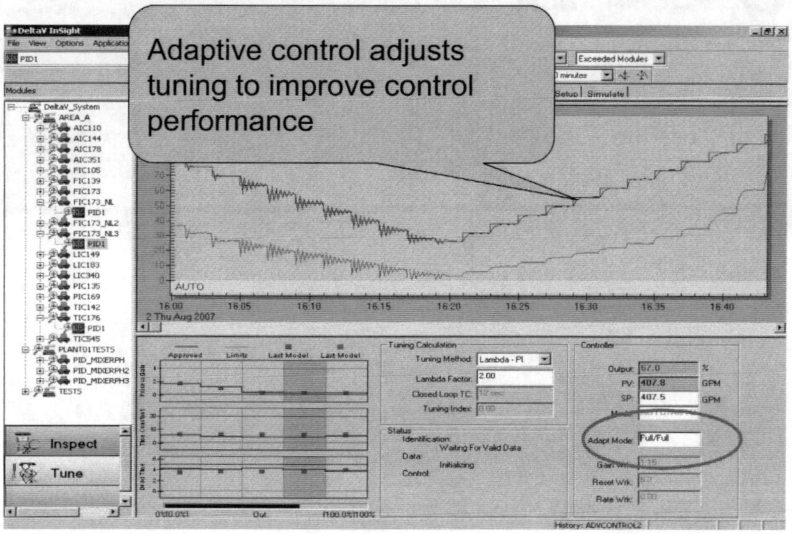

Summary Signal Characterization

- Newer Technology Signal Characterization function block
  - Better resolution (21 pairs of coordinates)
  - Self documenting (exists in control system)
  - No calibration required vs. 'external black box'
  - Provides "inverse" characterization for analog output characterizations
  - Reduce confusion: display signal being sent to final control element
- Applying Adaptive Tuning may provide a superior solution for AO characterization of final control element

Agenda

- Introduction
- Examples – Best Practices & Lessons Learned
  - Split Range Duty
  - Smith Predictor
  - Gain Scheduling/Adaptive Control
  - Controller Characterization
- Summary

Summary

- Practical application of control features that deal with common challenges
- Highlighted best practices & lessons learned
  - Split Range Duty
  - Smith Predictor
  - Gain Scheduling/Adaptive Control
  - Signal Characterization
- Do look at new functionality available when modernizing
  - Allows jump start to process optimization
  - Leads to improved process performance
  - Easier to implement and maintain over the life investment

Contacts & Questions?

James.Beall@Emerson.com
903-235-7935

Laurie.Ben@Emerson.com
512-834-7237

Paralleled Double-Effect Distillation: Simulative Case Studies

*Pei Wang, Leilei Dai, Ming Shi , Fan Wang, Hong Gao, Jinsheng Sun**
School of Chemical Engineering and Technology, Tianjin University, No. 92 Wei Jin Road, Tianjin 300072, P. R. China

Abstract

The increasing scarcity and growing demand for energy have led to an increasingly attention to the energy-efficient methods. Double-effect distillation is a common method to achieve remarkable energy conservation, which has three kinds of typical configurations including the feed splitting (FS), the light split/forward (LSF) and the light split/reverse (LSR). In spite of a good pathway for energy saving, better options are expected to cut down energy expenditure by a bigger margin than double-effect distillation does.

In this paper, case studies of paralleled or combined double-effect distillation schemes are proposed on the basis of current double-effect distillation. In order to verify the feasibility of this method, the improved methanol distillation scheme and organosilicon distillation scheme are optimized under this principle as examples. Exergy losses and the energy consumptions are measured as a tool to compare the two prevalent schemes with their corresponding optimized counterparts, respectively. The results of the total exergy loss and the energy consumption of these schemes show that the optimized schemes can cut the corner of the exergy loss by 22.8% and 40.41% and the energy consumption by 15.61% and 44.28%, respectively, in comparison to their popular cousins.

Keywords: double-effect distillation, paralleled double-effect distillation, methanol distillation, organosilicon monomer distillation, distillation process simulation, energy saving, exergy analysis

1. Introduction

Distillation, in chemical process and related industries, is by far the most widely applied for the separation of liquid mixture. In spite of its well-known benefit and widespread application, the major drawback is its high energy consumption. In classical distillation process, heat is transferred to the reboiler of distillation column. Along with the distillation column to separate the mixture, a large number of heat are directly discharge into the surrounding environment and cooling water. Araujo AB [1] has done some research to achieve that distillation has a relatively low thermodynamic efficiency and it requires the input of high quality energy in the reboiler to push the separation of liquid mixture. Therefore, the focus of energy conservation is to consider how to recover or reuse the discharged heat, or how to reduce the quantity of transferred heat to the column. In the early 20th century, many scholars have devoted to the study of saving energy in distillation, and now most of correlative techniques are widely used in distillation process [2,3]. These energy saving methods involve multi-effect distillation, heat pump distillation, heat integrated distillation and etc [3-8]. But without extra rotating machines by electricity or other power producers, multi-effect distillation has been more favored in research and application [9].

Many researches confirmed multi-effect distillation to be one of the effective methods to reduce the energy consumption [10-12]. The energy saving effect of those schemes used multi-effect distillation is associated with the number of the effects and the characteristics of separation system. Generally speaking, those schemes can offer energy savings around 30% with respect to the original or classical multicomponent distillation scheme [23, 27, 28]. Considering the interaction between investment of plants and energy saving effect, the double-effect distillation has been the most commonly used in not only research but also field application.

The use of the double-effect distillation technology for optimizing the conventional distillation schemes can cut a large amount of the energy requirement in comparison with the correspondingly prevalent schemes [13]. Although the system is very difficult to control in the sense that it is nonlinear, multivariable and interacting [14]. More and more published studies [15-19] have been focused on further questing the way of energy saving on the basis of the double-effect distillation schemes. However, the overhead vapor of one high pressured column only can integrate with the reboiler of one low pressured column in those published studies. In view of this, in this paper, the paralleled or combined double-effect distillation is proposed to broaden the thought for the application of the double-effect distillation, in which the overhead vapor from one of the high pressured column is divided into two or more streams to successively integrate with the reboiler of the other columns [9].

2. Theory

2.1 Double-effect or heat integrated distillation system

A double-effect distillation system is made up of two adjacent distillation columns running in different pressure, which can lead to a temperature difference between the top of the high pressured column and the bottom of the low pressured column. Through the thermodynamic connection, the reboiler of the low pressure column plays a role of the condenser for the high pressure column. According to the direction of the mass flow and the heat integration, the double-effect distillation can fall into three kinds of configuration: the feed splitting (FS), the light split/forward (LSF) and the light split/reverse (LSR), as showed in Figure. 1.

One of the configurations with a high thermodynamic efficiency is the feed splitting (FS), in which the feed is divided roughly equal between the two columns running in different pressure, and the products can be obtained from both of the top and bottom of the two columns, as illustrated in figure. 1 (a). The overhead vapor form the high pressured column is used to heat the reboiler from the low pressured column. There is only energy exchange between the two columns, so the heat duties of the feed splitting (FS) are approximately equal to the heat duty for single column with the same conditions divided by the number of effect.

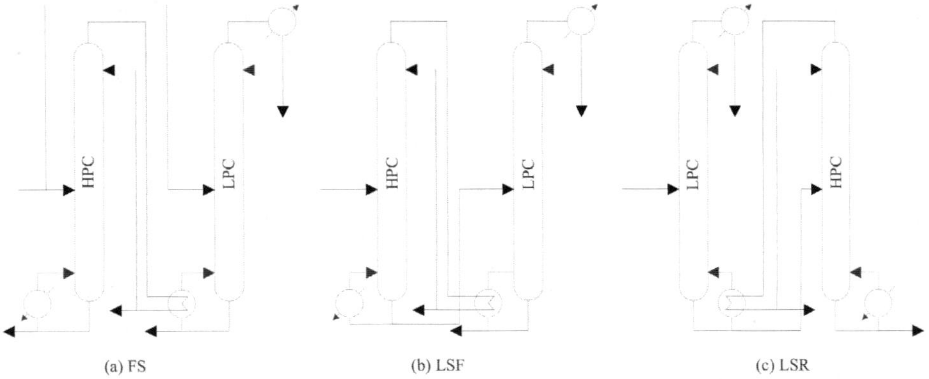

Fig. 1. Three configurations of double-effect distillation

Figure. 1 (b) and (c) respectively shows the configuration of the light split/forward heat integration (LSF) and the light split/reverse heat integration (LSR). Different from the feed splitting (FS), the total feed stream was sent to one column in the other two configurations. In the light split/forward (LSF), the total feed is pumped to the high-pressure column (HPC) and the bottom stream of HPC flows into the low-pressure column (LPC). In this way, all of the heavy components are from LPC and the light components from the top both of the two columns. Figure. 1 (b) indicates the direction of mass flow and heat integration is the same. The last configuration is the light split/reverse (LSR), as shown in figure. 1 (c), all of the feed is sent to the low-pressure column (LPC). It can remove the two kinds of light components from the top both of the two columns and one kind of heavy components from the bottom of HPC. Moreover, the direction of heat integration is in the reverse with the mass flow. These two kinds of configurations have not only energy exchange but also substantial exchange, therefore, when the operating parameters of one column are changed, the other column will also change.

Theoretically, the three kinds of configurations are empirically 30% less energy consumption than the single column distillation. As for choosing what kind of the configurations of the double-effect distillation, it depends on the properties of the multi-component mixture, the pressure of saturated steam, the temperature of cooling water, structure parameters of existing distillation column and so on.

2.2 Exergy analysis of the distillation process

According to the First Law of Thermodynamics, also called the Law of Energetic Balances, no matter what happens, the total amount of energy has not yet changed. However, when the various types of energy generate the mutual conversion, the quality of the energy is always decreased. The separation process in distillation is irreversible. In order to separate the mixture in distillation column, it needs the driving force which must consume a certain amount of external work, by this means, the decline of the quality of energy is inevitable. Therefore, we should choose a reasonable driving force in distillation process to avoid unnecessary loss of the quality of energy. It defined the concept of exergy (E_x) to compare the quality of the different types of energy. Exergy, which is a measure for evaluating the extent of available energy or the quality of energy, is based on the Second Law of Thermodynamics. Exergy analysis reveals the essence of the energy, and can be used as an important tool to evaluate the effect of energy saving. And it usually combines with process optimization studies, and as its guidance, to identify cost saving opportunities and amend the existing scheme.

Szargut [29] has summarized a comprehensive definition of the exergy, that is: 'Exergy is the amount of work obtainable when some matter is brought to a state thermodynamic equilibrium with the common components of the natural surroundings by means of reversible processes, involving interaction only with the above mentioned components of nature'. The state of one stream recorded as 1, according to the definition of exergy, the exergy of the stream can be expressed as:

$$E_x = (H_1 - H_0) - T_0(S_1 - S_0) \quad (2)$$

where, H_1 and S_1 is the enthalpy and the entropy of the stream 1, and T_0 is the temperature of the surrounding and H_0 and S_0 represent the enthalpy and the entropy of the surrounding.

The state of another stream recorded as 2, when the system stream change from state 1 (T_1, P_1) to state 2 (T_2, P_2), the variation of exergy (ΔE_x) is given by

$$\Delta E_x = E_{x,1} - E_{x,2} = \Delta H - T_0 \Delta S \quad (3)$$

In the conventional distillation process, it must input the heat or energy to drive the occurrence of separation process. The bottom of the column input the heat (Q_R) at this temperature (T_R), and the top of the column was sent to the heat (Q_C) at this temperature (T_C). The net work consumption (W) in distillation is

$$W = Q_R(1 - \frac{T_0}{T_R}) - Q_C(1 - \frac{T_0}{T_C}) \tag{4}$$

When the enthalpy difference between the product and the feed is very small, $Q_R=Q_C=Q$ can be approximately gotten, then Equation (4) is rewritten to be:

$$W = QT_0(\frac{1}{T_C} - \frac{1}{T_R}) \tag{5}$$

Then, the total exergy loss ($E_{X\ loss}$) of the distillation column can be calculated by

$$Ex_{Loss} = W - \Delta E_x \tag{6}$$

3. *Prototype case: the prevalent distillation schemes*

3.1 The improved four-column methanol distillation scheme

In the methanol industry, it has gradually appeared two-column, three-column, four-column and five-column distillation scheme [20-23]. With the increase of the number of distillation column, the corresponding distillation scheme can not only increase the efficiency of the process, but also cut the energy consumption. At present, the four-column scheme has a dominant status to produce the purified methanol. However, the improved four-column distillation scheme can cut some more energy requirement with respect to the traditional four-column distillation scheme. The improved four-column distillation scheme is seen as the comparative reference in the paper.

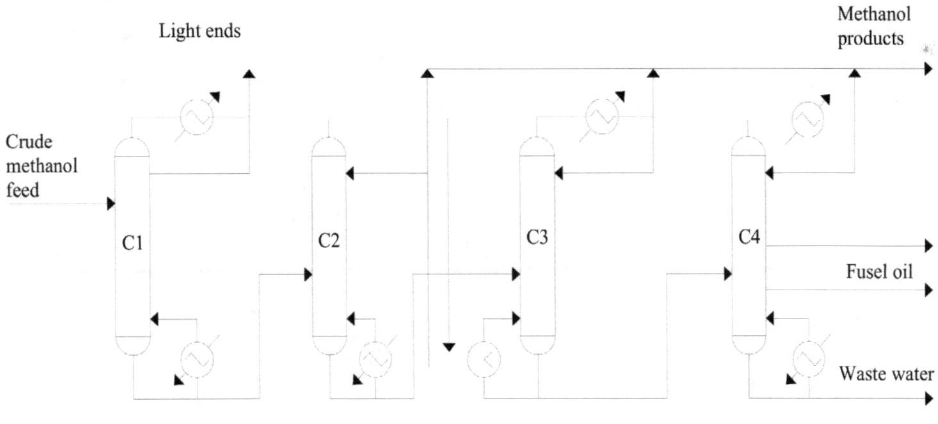

Fig. 2. The improved four-column double-effect methanol distillation scheme (C1, light ends column; C2, pressured column; C3, atmospheric column; C4, water column)

In the improved four-column distillation scheme, as shown in figure. 2 [24], the crude methanol flows into the light ends column (C1) in 87.5 ℃ for removing the light ends and the remnants of dissolved gas from the top of C1. Besides, in order to raise the methanol recovery rates, the light ends column (C1) designs a two-stage condenser system. The crude methanol was distillated successively from the pressured column (C2) and the atmospheric column (C3), and most of the methanol products can be received at the top both of C2 and C3. The bottom stream of C1, C2 and C3 is pumped respectively as the feed into C2, C3 and the water column (C4). Residual methanol, fusel oil and waste water are extracted from the corresponding position of the top, side stream and the bottom of the water column (C4).

In the improved scheme, the operation pressure of C2 is 0.9MPa, and the other three columns run under the atmospheric pressure. Due to the different of operating pressure, the overhead vapor of C2 can undertake thermal coupling connection with the reboiler of C3.

3.2 The current organosilicon monomer distillation scheme

According to the survey, there are several kinds of organosilicon distillation schemes. However, the 9-column organosilicon distillation scheme [9], in China, has an important role in the organosilicon industrial, as illustrated in Figure. 3, it is used as the base case.

Figure. 3 shows that the crude monomer is delivered to the heavies removal column (C1') and the bottom stream of C1' has a lot of the unsteady substances of high boiling points. These unsteady substances of high boiling points are pumped to the high boiling point column (C9') as the feed, and the products from the top of C9' continue to serve as the crude monomer by circumfluence. By this way, it can accomplish the separation between the unsteady substances of high boiling points and the useful substances at a relatively low temperature in the high boiling point column (C9') and avoid the unsteady substances of high boiling points from thermal cracking to generate many unknown substances which would greatly affect the subsequent separation of product. The low removal column (C2') provides trimethylchlorosilane (Me3), methyldichlorosilane (MeH), and other light components for lights removal column (C3'), MeH column (C4'), azeotrope removal column (C5') and Me3 column (C6'), which can finally achieve two products (MeH and Me3) and three wastes. It also provides methyltrichlorosilane (Me1) and dimethylchlorosilane (Me2) and other heavy components for Me1 column (C7') and Me2 column (C8'), which can finally get two products (Me1 and Me2).

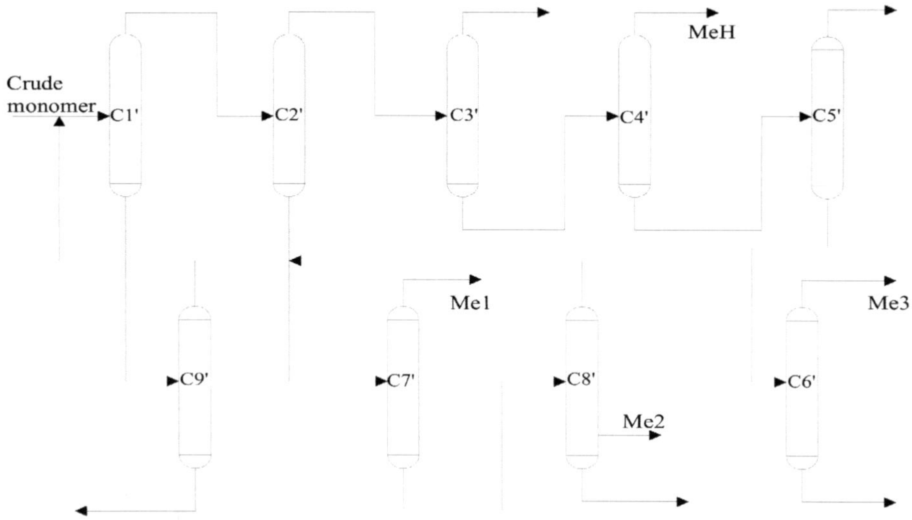

Fig. 3. The prevailing 9-column distillation scheme (C1', heavies removal column; C2', low removal column; C3', lights removal column; C4', MeH column; C5', azeotrope removal column; C6', Me3 column; C7', Me1 column; C8', Me2 column; C9', high boiling point column)

In this scheme, the operating pressure of all nine columns is relatively low, and the highest operating pressure is 260 kPa in the lights removal column (C3'). Although it has a different operating pressure between C3' and the other columns, but the temperature difference is not enough to make them use double-effect or heat integrated distillation.

4. Simulation and analysis both of the prevalent schemes and the optimized schemes

4.1 The novel five-column heat integrated methanol distillation scheme

Although the energy saving of the improved four-column methanol distillation scheme can reach 10% compared with the traditional four-column scheme, it still exist several shortages, which will be illustrated by the results of computer simulation to the improved four-column methanol distillation scheme. The flow rate of the crude methanol is 239230 kg/hr, and the temperature is 87.5 ℃, and the composition of the crude methanol is exhibited in table 1. The inlet and outlet temperature of the cooling water are 30 ℃ and 40 ℃.The methanol content is specified as 99.99 wt% in the methanol products and 10ppm in the wastewater of the water column. Table 2 shows the operating parameters of each column in the improved four-column scheme.

Table 1. Crude methanol feed data

Component	Mass fraction%	Component	Mass fraction%
H_2	0.0100	Pentanol	0.0338
N_2	0.0710	Formic acid	0.0030
Methane	0.0100	Acetic acid	0.0030
CO	0.0360	Methyl formate	0.0092
CO_2	0.6640	Acetone	0.0193
Methanol	95.0000	Methyl ethyl ketone	0.0030
Dimethyl ether	0.0309	Diisopropylether	0.0056
Trimethylamine	0.0010	Water	3.7100
Ethanol	0.0900	i-Pentane	0.0562
i-Propanol	0.0024	n-Pentane	0.0169
Propanol	0.0264	n-Hexane	0.0487
i-Butanol	0.0338	n-Heptane	0.0009
Butanol	0.0185	n-Octane	0.0050
3-Pentanol	0.1407	n-Nonane	0.0050

Table 2. The operating parameters of improved four-column scheme

The operating parameters	C1	C2	C3	C4
Top pressure (kPa)	150	900	110	110
Top temperature (℃)	81.0	132.6	67.7	67.7
Bottom temperature (℃)	83.8	139.1	96.5	104.5
Reflux ratio	0.58	2.44	1.68	11.57
Number of theoretical stage	33	52	58	61

The results of simulation indicate that the hot utility is still in a high level because the reboilers of C1, C2 and (C4) still need the saturated steam to transmit the heat. Table 2 shows the reflux ratio of C4 is 11.57 and this large reflux ratio can cause the increasing of the energy requirement of the water column (C4). Besides, according to the flowsheet of this scheme, the diameter of column is so large that the emergence of many urgent cases.

In view of above these defects in the improved four-column scheme, the novel five-column heat integrated methanol distillation scheme (shown as Fig. 4) is designed in this paper to reduce the impact of these factors as far as possible by using the paralleled or combined double-effect distillation. Juntao Zhang [23] and Jinsheng Sun [25] are also proposed a five-column methanol distillation scheme by using triple-effect distillation and double-effect distillation respectively, and the effect of energy saving of the two schemes is outstanding.

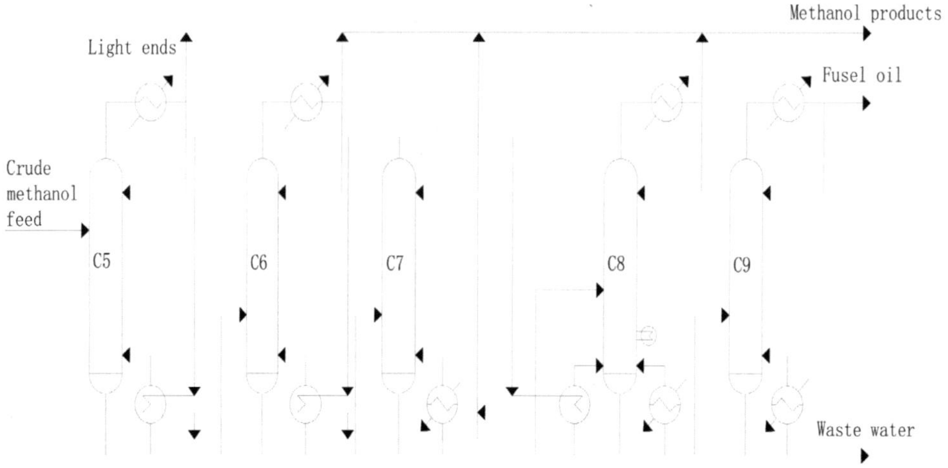

Fig. 4. The five-column heat integrated methanol distillation scheme (C5, light ends column; C6, prepositive atmospheric column; C7, pressured column; C8, postpositive atmospheric column; C9, water column)

According to the principle of the paralleled double-effect distillation, the four columns are unable to constitute the heat integration network in the improved four-column scheme. Through repeated verification, the prepositive atmospheric column (C6) is added after the light ends column (C5), meanwhile, the operating pressure of the pressured column (C7) is reduced to increase its throughput. In this way, the overhead vapor of C7 is split into three streams to respectively integrate with the light ends column (C5), the prepositive atmospheric column (C6) and the postpositive atmospheric column (C8). Due to use the paralleled double-effect distillation, the duties of the hot utility have decreased in a great margin. However, there are still part of heat from the background process is absorbed by the cooling water. So an inter-reboiler with a low temperature level is designed in the postpositive atmospheric column (C8) and it can share the heat with the background process streams, besides, we have proved the feasibility by using the column grand composite curve analysis of columns. Figure. 4 shows that the product of the top of the water column (C9) is fusel oil instead of purified methanol, which can not only reduce the reflux ratio of C9, but also make the diameter of column smaller to reduce the difficulty of operation.

Due to produce the fusel oil instead of purified methanol in the water column (C9) in the novel five-column distillation scheme, the methanol content of the feed for the water column have to limit at 0.016 wt%. In the five-column scheme, the operating pressure and the number of theoretical stage for each column are determined through the existing process, the onsite data and the requirement of the heat coupling. The proportion of the methanol products from the C6, C7 and C8 is determined by their own double-effect or heat integrated distillation system. The simulation results of the novel five-column scheme are listed in table 3.

Table 3. The operating parameters of five-column scheme

Column	C5	C6	C7	C8	C9
Top pressure (kPa)	160	110	400	110	110
Top temperature (℃)	83	66.6	104	66.6	86.5
Bottom temperature (℃)	85.6	84	118.8	104.5	103.2
Reflux ratio	1.46	1.49	2.23	1.87	4
Number of theoretical stage	33	54	54	65	29

4.2 Energy consumption and exergy loss analyses for the two methanol schemes

In the improved four-column scheme, the pressured column and the atmospheric column constitute the distillation system of heat coupling, there are three columns (C1, C2, C4) need the hot utility, and the entire duty of reboiler is 145.4×10^3 kw. As described above, the light ends column (C5) and the prepositive atmospheric column (C6) don't need the saturated steam in the five-column scheme and the total duty of reboiler is reduced to 122.7×10^3 kw. The detailed data of the reboiler duty as shown in table 4.

Table 4. The duty of reboiler in the two schemes

Column	C1	C2	C4	C7	C8	C9
Duty of reboiler (10^4 kw)	3.30	10.04	1.20	8.97	2.86	0.44

Obviously, the novel five-column scheme is much better than the improved four-column scheme in energy saving and the energy consumption of the novel scheme can be reduced by 15.61%. However, the recovery rate of methanol of the novel scheme is approximately equal with the improved four-column scheme.

Only the energy consumption analysis is not revealed the energy saving essence of the novel five-column scheme. Exergy analyses are used to further compare the two schemes, and the exergy flow rates are calculated through using the computer simulation data. The activity coefficient of the liquid is calculated using the Alcohol Package thermodynamic model. The calculation of the net work consumption of distillation column should pay more attention to the temperature of the cooling water. When the cooling water exchanges heat with other streams, the temperature of the cooling water is momentarily changed, so it is replaced the temperature of cooling water with the arithmetic average of inlet and outlet temperature of cooling water. Figure. 5 show the overall exergy loss of each column in the two methanol distillation schemes, in addition to the overall exergy loss of C6. Due to optimize the four-column scheme, the prepositive atmospheric column (C6) is added in the transformed scheme, and the overall exergy loss of C6 is 9157.6 kw. The exergy loss of each column, besides the pressured column (C7), in the novel five-column scheme is obviously less than their counterparts in the improved four-column scheme.

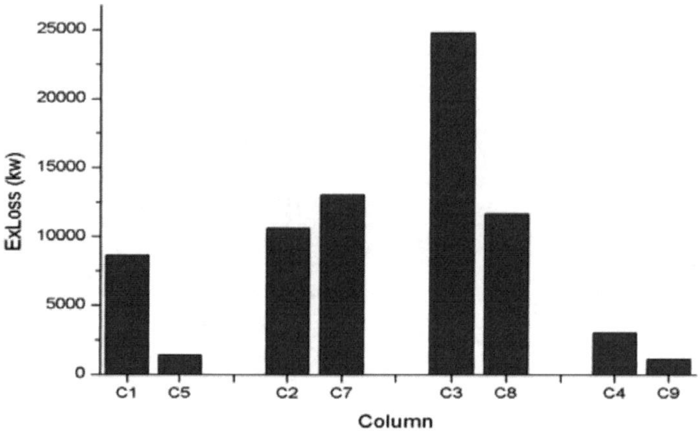

Fig. 5. The overall exergy loss analysis of each column in the two schemes

The total exergy loss of the improved four-column scheme and the transformed scheme is, respectively, 47100.1 kw and 36355.1 kw. The total exergy loss of the optimized scheme is dropped by 22.8%. Through the comparison between the exergy loss and energy consumption of the two schemes, it is indicated that the novel five-column scheme should be applied to produce the purified methanol in industry.

4.3 The novel heat coupling organosilicon distillation scheme

The primary source of the crude organosilicon monomers is the catalytic reaction products between chloromethane and silicon at high temperature and the table 5 displays the normal boiling points and mass percentage of the main components of the crude organosilicon monomers. Methyltrichlorosilane (Me1), dimethylchlorosilane (Me2), trimethylchlorosilane (Me3) and methyldichlorosilane (MeH) are the most valuable components in the crude organosilicon monomers. The purification requirements on organosilicon of these materials are shown in table 6.

Table 5. The crude organosilicon monomer feed data

Component	Normal boiling point at 1 atm/℃	Mass percentage %
LB		1
Methyldichlorosilane	41.8	4
Silicon tetrachloride	57.6	0.5
Trimethylchlorosilane	57.9	2.5
Methyltrichlorosilane	66.4	11
Dimethyldichlorosilane	70.2	80
HB		1
Temperature (℃)	40	
Pressure（kPa）	100	
Flow rate (kg/h)	26200	

Table 6. Purification demand on organosilicon

Monomer	Purification demand
MeH	99.5%
Me3	99.5%
Me1	99.5%
Me2	99.95%

The conventional nine-column scheme is simulated with the aid of existing process and onsite data to achieve the operating parameters of each column, as illustrated in table 7. The bottom temperature of the heavies removal column (C1') and the high boiling point column (C9') are obviously much bigger than that of the other seven columns in table 7. According to the principle of the double-effect distillation, the top temperature of the high pressured column is relatively high in comparison with the bottom temperature of the low pressured column, and the high pressured column can integrate with the low pressured column. Although the top temperature of C1' is less than the bottom temperature of the other seven columns, the overhead vapor of C1' still has a great opportunity to conduct heat with the other seven columns with the increasing of its top temperature.

Table 7. The simulation results of prevailing 9-column

Column	C1'	C2'	C3'	C4'	C5'	C6'	C7'	C8'	C9'
Top temperature (℃)	70.8	65.3	57.3	66.7	72.1	70.6	71.2	74.9	72.7
Bottom temperature (℃)	124.1	89.1	80.4	85.2	74.9	75.4	75.3	75.3	126.5
Reboiler duty (MW)	11.67	5.61	1.13	0.51	1.19	0.46	12.27	16.08	0.02
Condenser duty (MW)	-12.47	-5.53	-1.08	-0.53	-1.20	-0.47	-12.69	-16.12	-0.02
Top pressure (kPa)	110	180	260	220	170	150	110	110	110

According to the way of the paralleled double-effect distillation, the increasing operating pressure in the heavies removal column (C1') can make the top temperature of C1' ascending to integrate with the other seven columns. The idea is still in study to overcome some difficulties. However, we have got another feasible method, in which the operating pressure of C1' and the Me1 column (C7') is raised to 360 kPa and 280 kPa, respectively. In this way, the top temperature of C1' and C7' is increased to ensure they can integrate with the other columns, besides the high boiling point column, and the parallel double-effect or heat integrated organosilicon monomer distillation scheme is designed as shown in figure. 6 [26]. The feasibility of this heat coupling scheme has confirmed through three x-y diagrams including the heavies component and Me1, the heavies component and Me2, Me1 and Me2 [9].

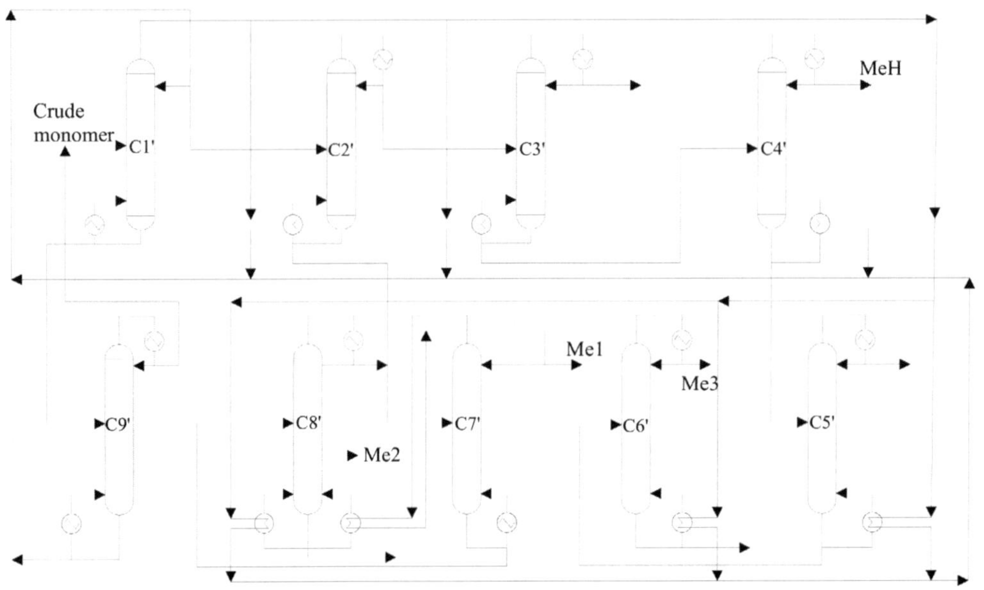

Fig. 6. The heat coupling 9-column scheme (C1', heavies removal column; C2', low removal column; C3', lights removal column; C4', MeH column; C5', azeotrope removal column; C6', Me3 column; C7', Me1 column; C8', Me2 column; C9', high boiling point column)

Figure. 6 shows the heavies removal column (C1') make the heat coupling with the low removal column (C2'), the lights removal column (C3'), the MeH column (C4'), the azeotrope removal column (C5'), the Me3 column (C6') and the Me2 column (C8'), successively. In order to gain a better energy saving effect, C8' is designed with two reboilers, the one integrated with C1' and the other integrated with the Me1 column (C7'). The operating parameters and the duty of reboiler and condenser of each column in the heat integrated 9-column are enumerated in table 8.

Table 8. The simulation results of heat integrated 9-column

Column	C1'	C2'	C3'	C4'	C5'	C6'	C7'	C8'	C9'
Top temperature (℃)	114.3	65.3	57.3	66.7	72.1	70.6	105.1	74.9	72.7
Bottom temperature (℃)	124.9	89.9	78.8	84.5	75.4	75.4	110.9	76.6	121.4
Reboiler duty (MW)	13.06	5.62	1.13	0.51	1.19	0.46	13.83	14.12	0.38
Condenser duty (MW)	-13.72	-5.54	-1.08	-0.53	-1.20	-0.47	-14.02	-14.43	-0.41
Top pressure (kPa)	360	180	260	220	170	150	280	110	110

4.4 Energy consumption and exergy loss analyses for the two organosilicon schemes

Figure. 4 indicate that each column of the prevalent nine-column scheme is not integrated with the other columns and the reboilers of all of these columns are required the saturated steam (600 kPa). According to the data from the table 7, the total duty of reboiler is 48.94 MW for the prevailing 9-column scheme. In the new nine-column scheme, C1' integrate with C2', C3', C4', C5', C6' and C8' successively and C7' integrate with C8', so C2', C3', C4', C5', C6' and C8' don't consume the saturated steam. The total duty of reboiler is 27.27 MW. Obviously, the heat integrated 9-column scheme has an absolute advantage over the prevailing 9-column scheme in energy saving, and the energy savings can reach 44.28%.

As the same with the methanol distillation scheme, the inlet and outlet temperature of cooling water are 30 ℃ and 40 ℃, and the total exergy loss is calculated through the results of computer simulation of the two schemes. The overall exergy loss comparison of each column in the two schemes is illustrated in figure. 7. In the normal scheme, the throughput of C9' is so small that the overall exergy loss of C9' is very near to zero, so it only have the exergy loss of C9' of the new scheme in figure. 7. The overall exergy loss of each column in the optimized scheme is declined in comparison with that of the corresponding column in the normal scheme except the high boiling point column (C9'). Because the crude organosilicon monomers contains some unstable components, in order to avoid these components happening thermal cracking in the heavies removal column (C1') in the novel scheme, more liquid from the bottom of C1' is pumped to C9' as the feed. That can cause the ascent of the throughput and the overall exergy loss of C9'.

The total exergy loss is 14035.5 kw and 8363.2 kw, correspondingly, in the conventional scheme and the novel scheme, and the exergy loss of the new scheme can be reduced by 40.41%. Through the above analysis and comparison, the heat integrated nine-column organosilicon distillation scheme should be a substitute for the prevalent nine-column scheme.

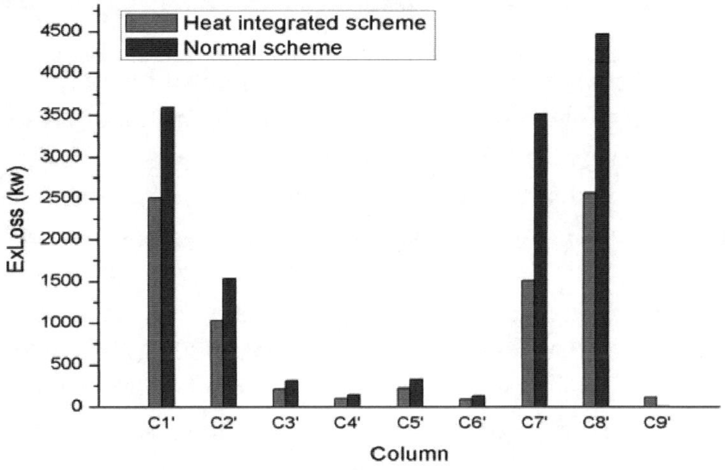

Fig. 7. The exergy loss comparison between the heat integrated scheme and the normal scheme

5. Conclusions

Paralleled double-effect or heat integrated distillation for energy saving is proposed, exampled by heat integrated five-column methanol distillation scheme and the heat coupling 9-column organosilicon monomer distillation scheme, with the aid of computer simulation. Besides, the optimized methanol distillation scheme and the transformed organosilicon scheme are comprehensively analyzed through the energy consumption and the exergy loss in comparison with the correspondingly prevalent scheme.

Adding a prepositive atmospheric column and decreasing the operating pressure of the pressured column in the novel five-column scheme can make the pressured column to simultaneously integrate with the light ends column, the prepositive atmospheric column and the postpositive atmospheric column. Besides, an inter-reboiler is added to integrate with the background process streams. The energy consumption and the exergy loss of the new scheme are 15.61% and 22.8% lower than the improved scheme.

Increasing the operating pressure of the heavies removal column and the Me1 column can make them to integrate with the other columns, in addition to the high boiling point column, to constitute a paralleled double-effect distillation system in the optimized nine-column scheme. The energy consumption and the exergy loss of the novel scheme are 44.28% and 40.41% less than the normal scheme.

The two cases expound that the key of paralleled double-effect distillation process lies in the reasonable adjustment of the energy distribution, the operating parameters and the configuration of some columns. It not only reduces the quantity of energy consumption, but also avoids the decrease of the quality of energy. For factories with a large numbers of distillation schemes of high energy requirement, the paralleled schemes lead to improve economic benefit.

References

[1] Araujo AB, Brito RP, Vasconcelos LS, "Exergetic analysis of distillation processes-a case study," Energy 2007;32:1185-93.

[2] Michele Mascia, Francesca Ferrara, Annalisa Vacca, Giuseppe Tola, Massimiliano Errico, "Design of heat integrated distillation systems for a light ends separation plant," Applied Thermal Engineering 2007;27:1205-1211.

[3] Amiya K. Jana, "Heat integrated distillation operation," Applied Energy 2010;87:1477-1494.

[4] Hilde K. Engelien, Sigurd Skogestad, "Multi-effect distillation applied to an industrial case study," Chemical Engineering and Processing 2005;44:819-826.

[5] James G. Gebbie, Michael K. Jensen, Piotr A. Domanski, "Experimental transient performance of a heat pump equipped with a distillation column," International Journal of Refrigeration 2007;30:499-505.

[6] Massimiliano Errico,Giuseppe Tola, Michele Mascia, "Energy saving in a crude distillation unit by a preflash implementation," Applied Thermal Engineering2009;29:1642-1647.

[7] Massimiliano Erricoa, Giuseppe Tola, Ben-Guang Rong, Daniele Demurtas, Ilkka Turunen, "Energy saving and capital cost evaluation in distillation column sequences with a divided wall column," chemical engineering research and design 2009;87:1649-1657.

[8] Ivar J. Halvorsen, Sigurd Skogestad, "Energy efficient distillation," Journal of Natural Gas Science and Engineering 2011;3:571-580.

[9] Jinsheng Sun, Fan Wang, Tingting Ma, Hong Gao, Yanzhen Liu, Fang Cai, "Exergy analysis of a parallel double-effect organosilicon monomer distillation scheme," Energy 2012;47:498-504.

[10] Hilde K. Engelien, Sigurd Skogestad, "Multi-effect distillation applied to an industrial case study," Chemical Engineering and Processing 2005;44:819-826.

[11] Kazuo Matsuda, Kenichi Kawazuishi, Yasuki Kansha, Chihiro Fushimi, Masaki Nagao, Hiroshi Kunikiyo, Fusao Masuda, Atsushi Tsutsumi, "Advanced energy saving in distillation process with self-heat recuperation technology," Energy 2011;36:4640-4645.

[12] M.A. Sharaf, A.S. Nafey, Lourdes García-Rodríguez, "Exergy and thermo-economic analyses of a combined solar organic cycle with multi-effect distillation (MED) desalination process," Desalination2011;272:135-147.

[13] Weiwen Wang, Wenjian Wang, "Energy Saving in Methylchlorosilane Distillation," International Conference on Computer Distributed Control and Intelligent Enviromental Monitoring 2012:290-293.

[14] Myungwan Han, Sunwon Park, "Multivariable control of double-effect distillation

configurations," J. Proc. Cont. 1996:247-253.

[15] Hilde K. Engelien, Sigurd Skogestad, "Multi-effect distillation applied to an industrial case study," Chemical Engineering and Processing2005;44:819-826.

[16] Debadrita Maiti, Amiya K. Jana, Amar Nath Samanta, "A novel heat integrated batch distillation scheme," Applied Energy2011;88:5221-5225.

[17] Anton A. Kiss, Servando J. Flores Landaeta, Carlos A. Infante Ferreira, "Towards energy efficient distillation technologies-Making the right choice," Energy2012;47:531-542.

[18] Diego C. Alarcón-Padilla, Lourdes García-Rodríguez, Julián Blanco-Gálvez, "Experimental assessment of connection of an absorption heat pump to a multi-effect distillation unit," Desalination2010;250:500-505.

[19] Diego C. Alarcón-Padilla, Lourdes García-Rodríguez, Julián Blanco-Gálvez, "Design recommendations for a multi-effect distillation plant connected to a double-effect absorption heat pump: A solar desalination case study," Desalination2010;262:11-14.

[20] Fiedler E, Grossman G. Methanol, "Ullmann's Encyclopedia of Industrial Chemistry," Weinheim : Wiley-VCH Verlag/Gmbh & Co; 2002.

[21] Meyers RA, "Handbook of synfuels Technology," New York : McGraw-Hill; 1984.

[22] Chu YZ, Qin LP, "Optimum design of methanol distillation process and columns," Chem Ind Eng Prog 2008;(27):1659-62.

[23] Juntao Zhang, Shengrong Liang, Xiao Feng, "A novel multi-effect methanol distillation process," Chem Eng Pro 2010;(49):1031-37.

[24] Jinsheng Sun, Fan Wang, Tingting Ma, Hong Gao, Peng Wu, Lili Liu, "Energy and exergy analysis of a five-column methanol distillation scheme," Energy2012;45:696-703.

[25] J. S. Sun, P. Wu, Y. H. Wang, H. Gao, L.L. Liu.CN Patent102039058 A. 2011.

[26] J. S. Sun, P. Wu, Y. H. Wang, Y. X. Su, H. Gao, J. Li, D. Yan. CN Patent 101798322 A. 2010.

[27] Michele Mascia, Francesca Ferrara, Annalisa Vacca, Giuseppe Tola, Massimiliano Errico, "Design of heat integrated distillation systems for a light ends separation plant," Applied Thermal Engineering 2007;27:1205-1211.

[28] Olga A. Flores, J. Carlos Ca´rdenas, Salvador Herna´ndez, Vicente Rico-Ramı´rez, "Thermodynamic Analysis of Thermally Coupled Distillation Sequences," Ind. Eng. Chem. Res. 2003;42:5940-5945.

[29] Szargut J, "Exergy Analysis of Thermal, Chemical and Metallurgical Processes," New York: Hemisphere Publishing Corporation;1988:55-152.

(45n) In Situ Transesterification of Jatropha Curcas Seed Particles with Alkaline Phase Transfer Catalyst

Subbarao Duvvuri[1], Shuhaimi Mahadzir[2] and Sintayehu Hailegiorgis[2], (1)Department of Chemical Engineering, Universiti Teknologi PETRONAS, Tronoh, Malaysia, (2)Chemical Engineering, Universiti Teknologi PETRONAS, Tronoh, Malaysia

Economic viability of biodiesel technology is at stake due to the high cost of edible oil feedstock and slow reaction rates due to limited solubility of oil in methanol. In situ transesterification of non edible oil seed particles can minimize the cost of feedstock. Microwave pretreatment of seeds can make oil molecules more reactive. Use of phase transfer catalysts can enhance the reaction rates. In the present work, these three concepts are utilized together to investigate alkaline in situ transesterification of microwave pretreated jatropha curcas seed particles in presence of benzyltrimethylammonium hydroxide (BTMAOH) as a phase transfer catalyst (PTC). It was observed that addition of PTC to alkaline catalyst has improved conversion and reduced reaction time. Microwave pretreatment of jatropha seed particles reduced reaction time drastically. The quality of biodiesel produced was within the limits of the international standard requirements of ATME D6751 and EN-14214. Thus, combined use of microwave pretreatment of seed particles, phase transfer catalysis in the presence of alkaline catalyst can increase the rate of in situ transesterification and product yield.

(45o) Catalyzed-Assisted Manufacture of Olefins (CAMOL): Updated for Use in Naphtha Service

Maximilian Walter, Petrochemicals, BASF SE, Ludwigshafen, Germany and **Suzanne Rech**, BASF Qtech, BASF Corporation, Iselin, NJ

Since the commercialization of the olefins plant, operators have searched for ways to be more competitive. One area always targeted includes keeping their furnaces in operation longer between decoking steps. Over the years, many anti-coking methods/technologies have become commercially available using different approaches to prevent coils from coking. One of the latest available technologies is the novel CAMOL^{TM} (Catalyzed-assisted Manufacture of Olefins) Technology from BASF Qtech. As reported in the introduction of the technology to the EPC in 2008 & 2010, the CAMOL^{TM}technology differs from other available anti-coking technologies in that the technology not only eliminates the catalytic (or filamentous) coke make through a physical barrier, but also gasifies the thermal (or amorphous) coke for an even greater coking resistance. The coating process has generated a coating with almost identical thermo-mechanical properties as the base metal substrate (alloy). This allows bonding to the base metal similar to being welded with a minimum risk of spallation when going through thermal cycling. Commercially, the CAMOL Technology has been available since 2006 and is now in operation in furnaces of various designs.

With CAMOL^{TM}, operators have experienced extended run lengths, reduced energy requirement and GHG emissions, lower average TMT operating temperatures, reduced DMDS consumption, and less carburization. In addition, the technology has exhibited high operating temperature stability (>1130°C/ 2065°F).

At BASF SE in Ludwigshafen, Germany, the technology has been in operation since early 2010 in mostly naphtha service. After overcoming some initial challenges associated with a catalytic surface technology, such as reactivating the catalytic species after decoking, the coil is now successfully operating with run lengths significantly higher than uncoated coils. This paper will report on the results and learnings from the experiences gathered over the past three years along with the advancements made since the last report in 2010.

(45p) Extending Furnace Run Length and Coil Life by Optimizing Decoking

Vinod Mishra, Ingenero, Mumbai, India and **James Brigman**, Ingenero, Houston, TX

Abstract:

In the Ethylene manufacturing process, coke deposited in the radiant coils and Transfer Line Exchangers (TLEs) during normal operation is removed periodically by decoking the furnaces using a mixture of steam and air. There is a significant variation in operating parameters adopted by different plants for decoking. These are often either too conservative or aggressive, leading to production loss from increased off-stream time, TLE limitation and deterioration in coil life. This paper describes how the decoking procedure was modified based on furnace and TLE operating conditions and the decoking steps changed to obtain the best possible Ethylene yield, without adversely impacting coil life and TLE performance. This optimization has now been done in several major ethylene facilities and furnace run length has more than doubled and production has significantly increased.

SOLVING OPERATIONAL PROBLEMS IN A NAPHTHA SPLITTER

Flávio Weissheimer, GE Water & Process Technologies do Brasil Ltda, São Paulo/Brasil
Giorgio Franceschetti, Braskem, Santo André/Brasil
Rafael Henrique Camparin, GE Water & Process Technologies do Brasil Ltda, São Paulo/Brasil
Patrick Lucas, GE Water & Process Technologies, The Woodlands/US

Abstract

This paper sets out how the polymerization process of Reboilers of the (A-100) Crude Naphtha Fractionation Unit of Braskem UNIB 3 ABC can be controlled by applying polymerization inhibitors developed by GE Water & Process Technologies Ltda and Braskem´s engineering solutions.

These heat exchangers have a high degree of polymerization because of the content of olefinic compounds present in the feed composition which, when in contact with high temperatures in the heat exchange areas of reboilers, tend to polymerize. The joint solution of GE technology and Braskem engineering solutions applied to this system ensures regular operation of such equipment for periods of time greater than those hitherto recorded without manifesting loss of heat transfer efficiency.

In the 2012 shutdown while the opening of these reboilers was being inspected, it was observed that pipes had not been clogged nor was surface fouling present, which is a very positive state when compared to the historical records of previous campaigns.

The strategy adopted by GE/Braskem to ensure a longer campaign for these pieces of equipment was based on the quality control of feeding by adjusting dosages and monitoring performance using statistical analysis tools.

The use of statistical models based on Multivariable Regression Analysis (MRA) to evaluate performance, which is validated by inspecting reboilers internally, has been an efficient way to determine the degree to which the system has been fouled and also to schedule shut-downs for cleaning the mechanical parts.

Description of the Process

The (A-100) Naphta Fractionation Unit includes two distillation columns, the objective of which is to separate the light (C_6), medium (C_6 - C_8) and heavy (C_{9+}) cuts, included in the feed supply of crude naphtha.

The light cut and a portion of the heavy cut are sent to pyrolysis furnaces so as to produce ethylene. The medium cut is sent to the Hydrotreating and Catalytic Reforming Unit, with the objective of increasing the aromaticity of the system so as to produce BTX.

The A-100 processes 100 tons/h of crude naphtha, and approximately 35% of this current represents the average cut. Figure 1 below shows the schematic flow diagram of this fractionation process.

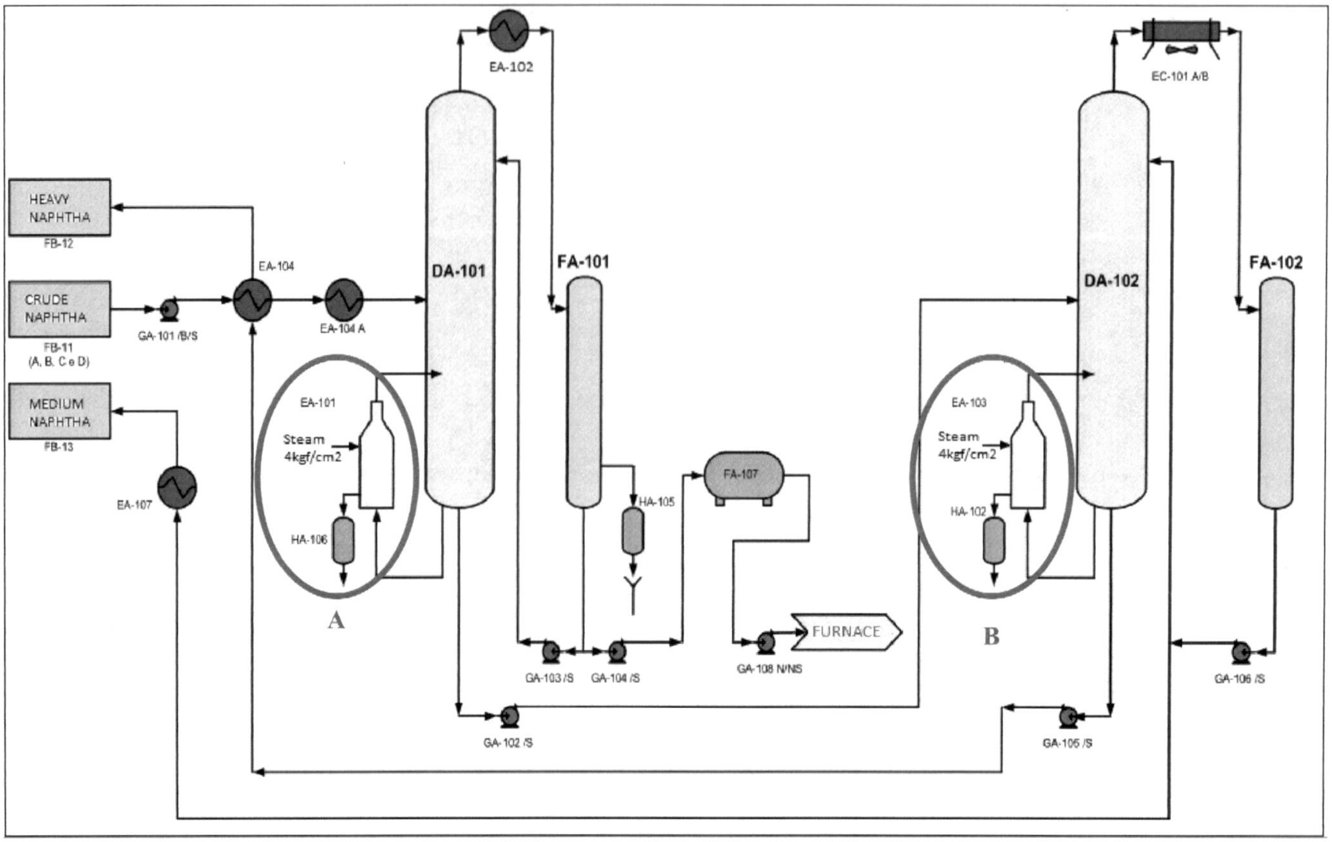

Figure 1. Process Flow Diagram (A-100)

The Reboilers, indicated by the letters A and B in Figure 1 above, operate with an injection of low steam (4kgf/cm2) to promote vaporization of the bottom stream of the A-100 distillation columns. These devices operate over a temperature range of about 150°C, which is sufficient to promote the polymerization of certain olefinic compounds present in the feed of crude naphtha.

Poor performance of these Reboilers can lead to production stoppages for maintenance and cleaning, which very often affects up to five days of production. The possibility of raising the times of campaigns and predicting the degree of fouling in downstream units helps to minimize production losses due to unscheduled stoppages.

This study will discuss the results of Reboiler A, which is the most critical equipment, with a view to synthesizing data, but gains in equipment B were also significant.

Evaluation System

Historical Record of the Campaign

Reboiler A was installed in 1972 but examination of the historical records did not reveal that the sheaf of tubes had ever been replaced. This exchanger has currently 11 tubes plugged into it.

The Campaign times of this equipment are shown in Table 1, with the last campaign being highlighted in blue so as to set it apart from the other ones. Table 1 gives a summary in the order of the timings of the campaigns, the average load processed in the period, the total load processed, the reason for operational stoppages etc.

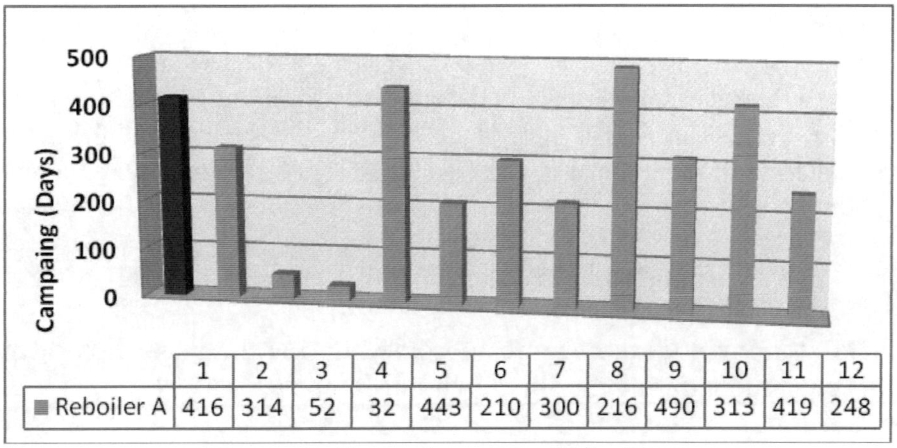

Figure 2. Campaign of the Reboiler A

Table 1. Historical Record of Reboiler A Campaigns

Date	Campaign *(days)*	Capacity *(ton/h)*	Capacity *(ton/campaign)*	Occurrence	Tubes Clogged *(%)*
13/06/12	416.0			Programmed Shutdown	0
17/03/11	314.0	93.1	701601.6	Heat exchange performance	30
03/05/10	52.0	90.8	113318.4	Stuck tube	-
11/03/10	32.0	71.1	54604.8	Heat exchange performance	50
09/02/10	443.0	72.0	765716.6	Heat exchange performance	20
16/11/08	210.0	98.1	494575.2	Heat exchange performance	-
16/04/08	300.0	104.0	748800.0	Programmed Shutdown	-
16/06/07	216.0	92.8	481075.2	Programmed Shutdown	-
10/11/06	490.0	-	-	Programmed Shutdown	-
30/06/05	313.0	-	-	Stuck tube	-
17/08/04	419.0	-	-	-	-
18/06/03	248.0	-	-	-	-
10/10/02	-	-	-	-	-

Conditions of Opening

In this review, a comparison is presented between two operational shut-downs of A-100, which occurred in March 2011 and May 2012, in order to observe the improvement in performance from the last campaign associated with the change in the GE Treatment strategy.

Campaign I: Production shutdown on 03/2011

Table 2 below summarizes the features observed during the inspection of the state of the inlet of this piece of this equipment during the shutdown of March 2011:

Table 2. Internal inspection of Reboiler A in 03/2011

Bottom Cover	Surface residue
Bottom Mirror	No deposits or internal blockage of the tubes were observed.
Top Mirror	Approximately 30% of the tubes were clogged, especially in the region of the vapor inlet, which has higher temperatures. Magnetic and non-magnetic deposits.

May/2011

Figure 3. The mirror of Reboiler A before it was cleaned.

May/2011

Figure 4. The mirror of Reboiler A at the vapor inlet before it was cleaned.

Note: It is difficult to carry out internal cleansing by Hydro jetting (pump pressure during cleaning is 20,000 psi). This reboiler has 11 tubes plugged into it, but none of them are plugged in during this shutdown.

Campaign II: Production shutdown on 03/2011

Table 3 below summarizes the features observed during the inspection of the state of the inlet of this piece of equipment during the shutdown in May 2012:

Table 3. Internal inspection of Reboiler A in 05/2012

Bottom Cover	Surface residue
Bottom Mirror	No deposits or internal blockage of the tubes were observed.
Top Mirror	Clogged tubes were not identified, not even in the region of the inlet vapor. There was only a thin layer of deposit that was easy to remove and which was evenly distributed over the upper surface of the mirror.
Spool	A thin deposit on the inner surface of the caisson.

May/2012

Figure 5. The mirror of Reboiler A before it was cleaned.

May/2012

Figure 6. The mirror of Reboiler A before it was cleaned.

Note: Hydro jetting was done quickly, without any obstructions due to deposits having been identified. This equipment has 11 tubes plugged into it, but none of them were during this shutdown.

Analysis of Deposits

The characterization of the deposits found in Reboiler A showed an 86.0% loss on ignition, thus indicating the strong presence of organic compounds in their composition, with the functional groups of carboxylic acid and ester standing out. The inorganic analysis of the sample indicated a high concentration of iron oxides and also sulfur oxides in the form of sulfides on the cover and spool of this reboiler. This is summarized in Table 4 below:

Table 4. Analysis of the Deposits of Reboiler A

Organic Compounds - LOI (%)	86.0
Water of hydration	Smaller
Sulfur compounds	Smaller
Carboxylic Acid and Ester	Greater
Dichloromethane soluble (%)	2.0
Elemental Sulfur	Greater
Dichloromethane insoluble (%)	98.0
Water of hydration	Smaller
Inorganics	Smaller
Carboxylic Acid and Ester	Greater
Composition of Dichloromethane insolubles (%)	
Carbon	56.0
Hydrogen	5.0
Nitrogen	3.0
Sulfur	8.0
Inorganic Compounds (%)	
Iron, $Fe_2O_3 + Fe_3O_4$	82.0
Sulfide, S^{2-}	16.0
Phosphate, P_2O_5	1.0
Sodium, Na_2O	1.0

Evaluation of Performance

The evaluation of performance of Reboiler A was prepared using the MRA tool, which is a GE good practice to predict fouling in critical systems. This statistical tool enables a dependent variable to be determined, in this case the percentage of opening the steam valve, by means of the numerical regression of the other operational variables of the column. This method generates a mathematical equation to predict the percentage of inlet steam when the equipment is clean, given that the equation obtained by regression includes data of this condition.

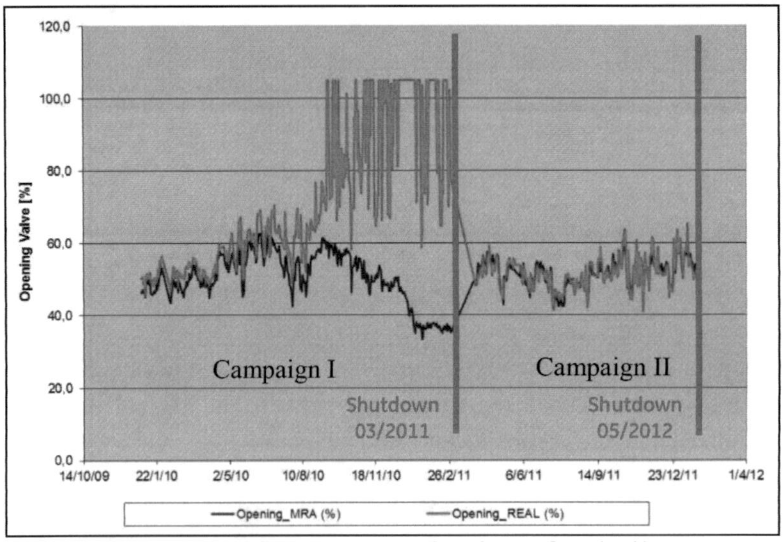

Figure 7. Performance Evaluation of Reboiler A

In Figure 7 the dark blue line represents the percentage of inlet steam when the equipment is clean, as per the equation defined by the MRA. The pink line represents the actual percentage of inlet steam. The sharp upward shift evidenced in the period between November 2010 and March 2011 shows the fouling of the system during Campaign I (prior to the 2011 operational shutdown).

After the 2011 shutdown, it can be seen that the two curves follow the same trend, without a shift, indicating that the unit has the equivalent performance of a clean piece of equipment. The change in the GE treatment strategy, taking the control of the dosages of the product Petroflo 20Y25 into account, was seen to have been effective in increasing the campaign time of this equipment.

Conclusion

The change of strategy in applying dispersant and controlling the dosages after the March 2011 shutdown, presented positive results, given the states of the inlet seen during the inspection of reboilers A and B in May 2012..

The low degree of fouling of these reboilers evidenced in the 2012 shutdown led to the production process being considered reliable, as the A-100 reboilers are no longer the pieces of equipment responsible for the loading limit of the unit. The downtime during the cleaning of the mechanical parts of the reboilers was about 10 times less than that observed in the previous shutdown, thus reducing Braskem´s maintenance costs.

In addition, the MRA statistical tool was able to produce a reliable representation of the degree of fouling of the system. This tool can be effective for scheduling shutdowns more than 03 months in advance.

References

1. Setal Lummus, "Manual de Operação da Unidade de Processamento de Nafta Bruta (Área 100)", Santo André, Brazil.
2. Ponce de Leon, Carolina; Weissheimer, Flavio and Camparin, Rafael H. (2011), "Proposta Técnica de Fornecimento de Especialidades Químicas, Equipamentos e Serviços para Tratamento Químico para a Unidade de Processamento de Nafta Bruta (Área 100)", Santo André, Brazil.
3. Kleinbaum, David G., Kupper, Lawrence L., Muller, Keith E. (2008) "Applied Regression Analysis and Other Multivariable Methods", Belmont, USA.

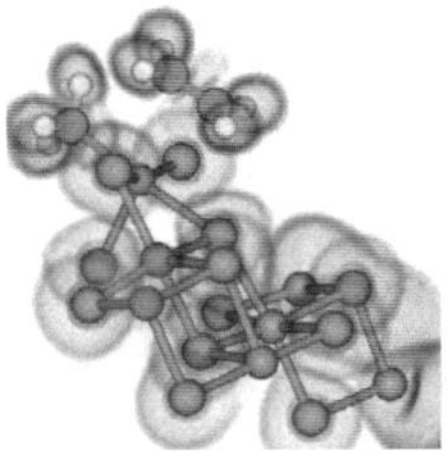

13 AIChE
Spring Meeting, San Antonio, TX

SmartProcess® Application Strategy Achieves Faster and More Efficient Control of an Ethylene Complex

Patrick Truesdale
Principal Consultant
Industry Solution Group
Emerson Process Management

April 29, 2013
San Antonio, TX

EMERSON
Process Management

Speaker Introduction

- Principal Consultant
  Emerson Process Systems & Solutions
- BS Electrical Eng, PE
- 40+ Years experience with Refinery & Petrochem Process Control
- Joined Emerson in 2007
- Previous employers: duPont, Bonner & Moore, SETPOINT, AspenTech, Comcept Technology

Patrick Truesdale

E-mail: patrick.truesdale@emerson.com

13 AIChE
Spring Meeting, San Antonio, TX

EMERSON
Process Management

Agenda

- Performance Package Review
- SmartProcess Ethylene Overview
  - Furnaces
  - Distillation
- Benefits

13 AIChE Spring Meeting, San Antonio, TX

EMERSON. Process Management

SmartProcess Ethylene Packages

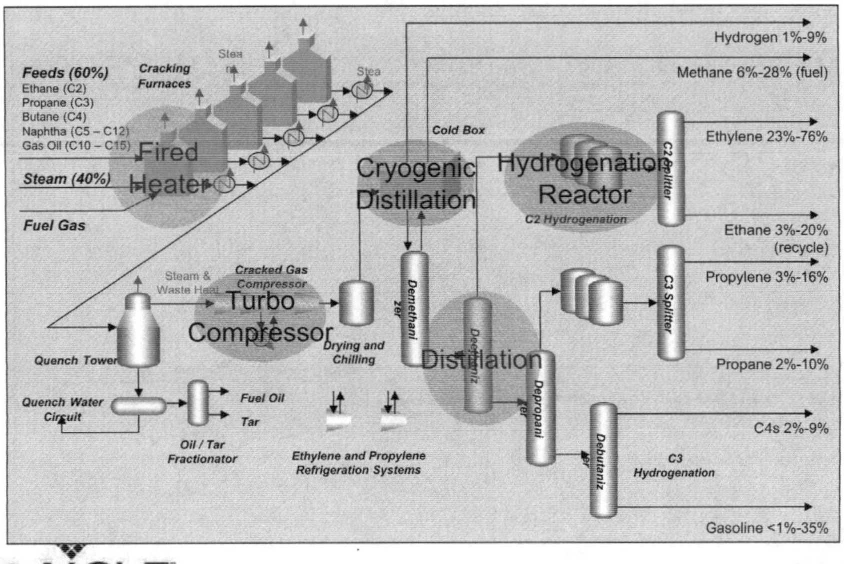

13 AIChE Spring Meeting, San Antonio, TX

EMERSON. Process Management

SmartProcess Scope

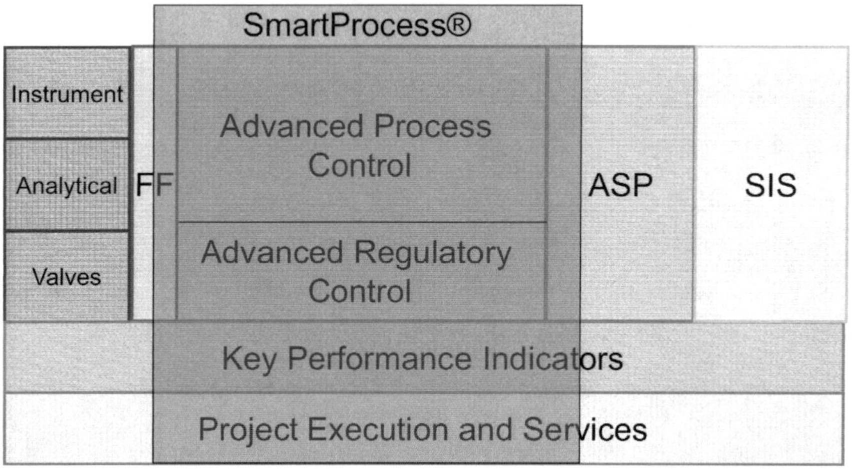

13 AIChE Spring Meeting, San Antonio, TX

EMERSON. Process Management

Basic Ethylene Furnace Variables

Key Operating Objectives:

- **Constant severity**
- **Maintain HC/Stm ratios**
- **Min excess air**
- **Maintain within constraints**
- **Max charge (if needed)**

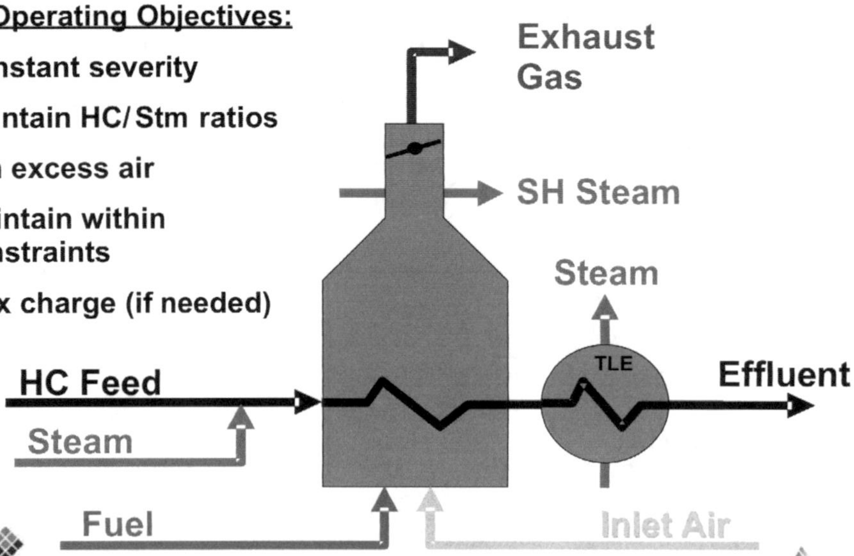

13 AIChE Spring Meeting, San Antonio, TX

EMERSON. Process Management

SmartProcess Furnace Module

- Design allows for different furnace configurations
  - Process Zones
  - Firing Zones
  - Multiple TLE analyzers
- Advanced Regulatory
  - Charge Control
  - Steam Control
  - Severity Calculation
  - Combustion Control
  - KPI's
- Advanced
  - Pass Balancing
  - Heat Balancing
  - Severity Control
  - O2 Control

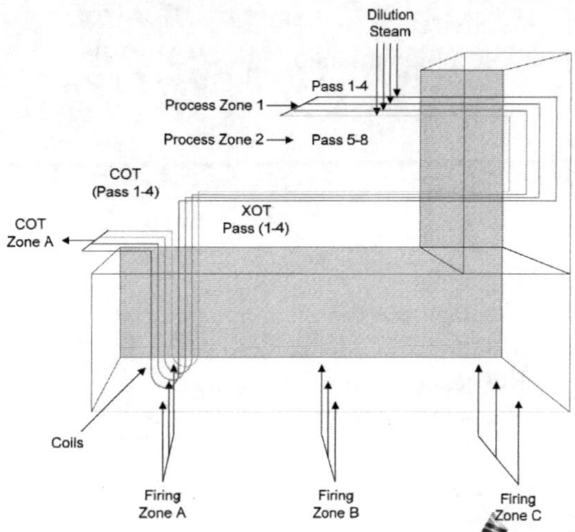

13 AIChE
Spring Meeting, San Antonio, TX

EMERSON. Process Management

Standard Furnace Module Design

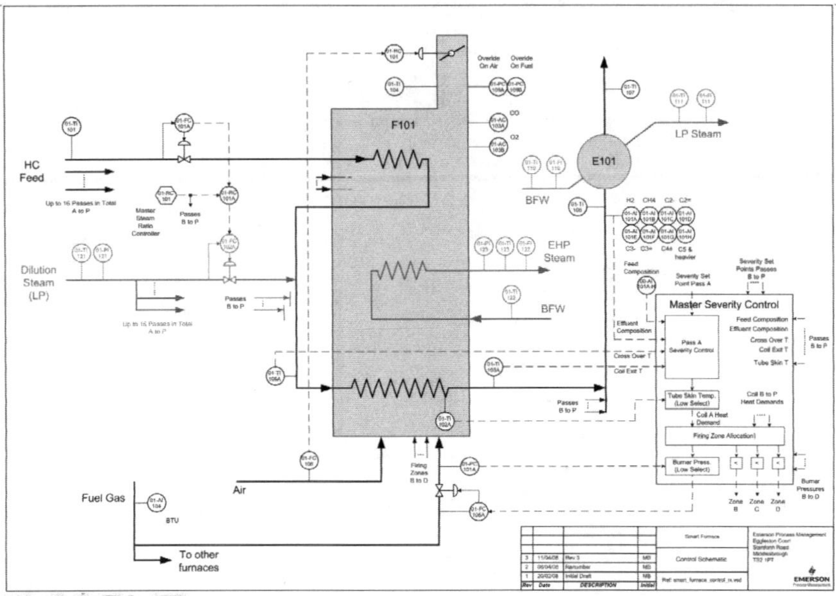

13 AIChE
Spring Meeting, San Antonio, TX

EMERSON. Process Management

Charge Rate Control

Functions:
- Pass Balancing
- Constraints
- Feed Maximization

Functions:
- Ramping
- Limits & Alarms
- Multiple Feed Types
- Furnace Status
- Pass Offsets

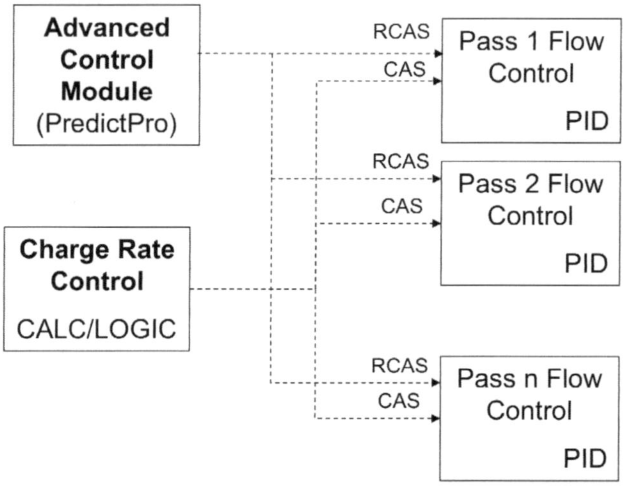

13 AIChE Spring Meeting, San Antonio, TX

EMERSON Process Management

Steam Ratio Control

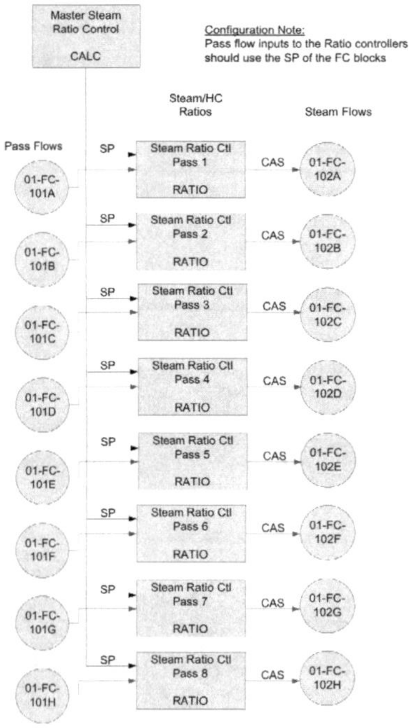

13 AIChE Spring Meeting, San Antonio, TX

EMERSON Process Management

Severity Control

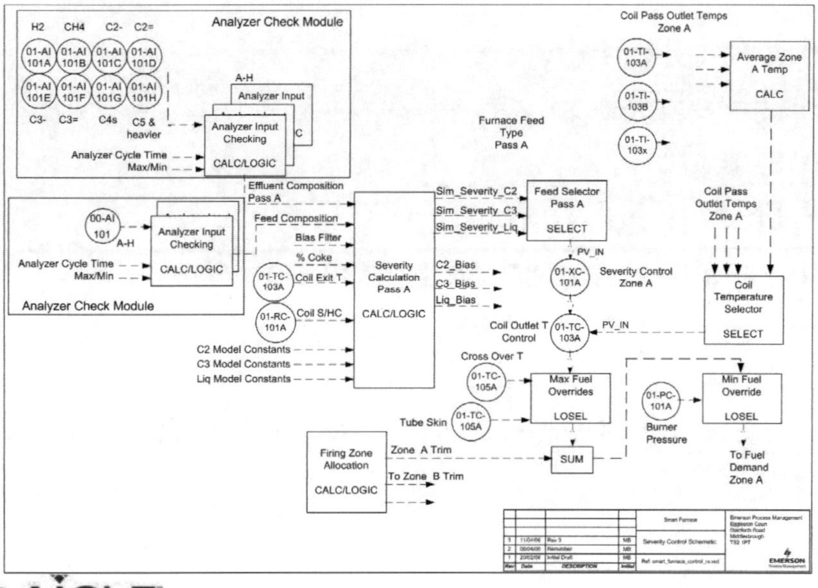

Automated Furnace Sequences

- Standard module templates to be used for Furnace Startup, Shutdown and Decoking sequences.
- Utilizes preconfigured Sequence Function Charts (SFC) running in DeltaV controllers or App Station
- Standard sequence functions can be customized to match site requirements
- Manual operator steps with confirmation can be mixed with automated steps
- Sequence execution control graphics and faceplate templates

Key Performance Calculations

- Furnace efficiency
  - Heat loss method
- Furnace Duty
- TLE Duty, Fouling factor
- Stack gas estimated composition
- Stack gas daily total tons exhaust for CO, CO2, S, N, etc.
- Stack gas NOx emissions (optional – based on CEMS)

13 AIChE
Spring Meeting, San Antonio, TX

EMERSON. Process Management

Distillation - Classic Multivariable Problem

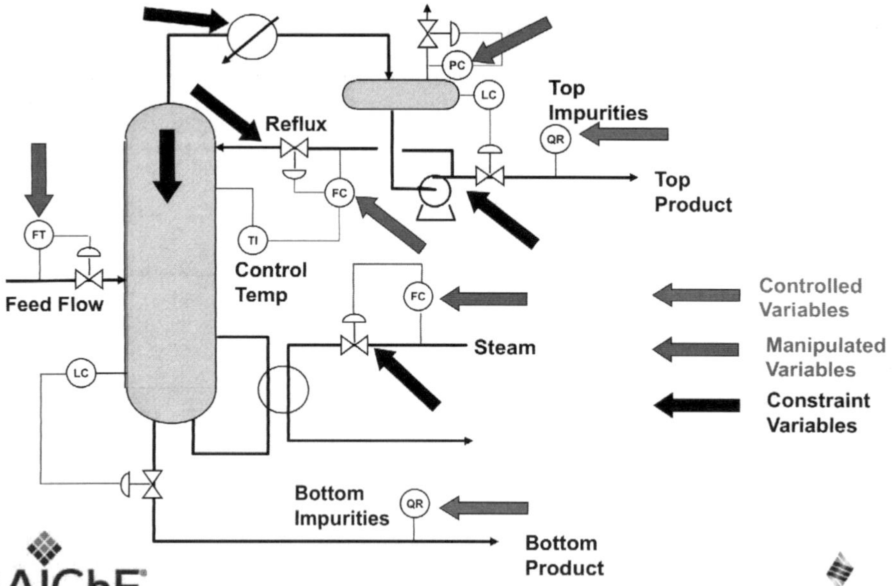

13 AIChE
Spring Meeting, San Antonio, TX

EMERSON. Process Management

Distillation Control Module

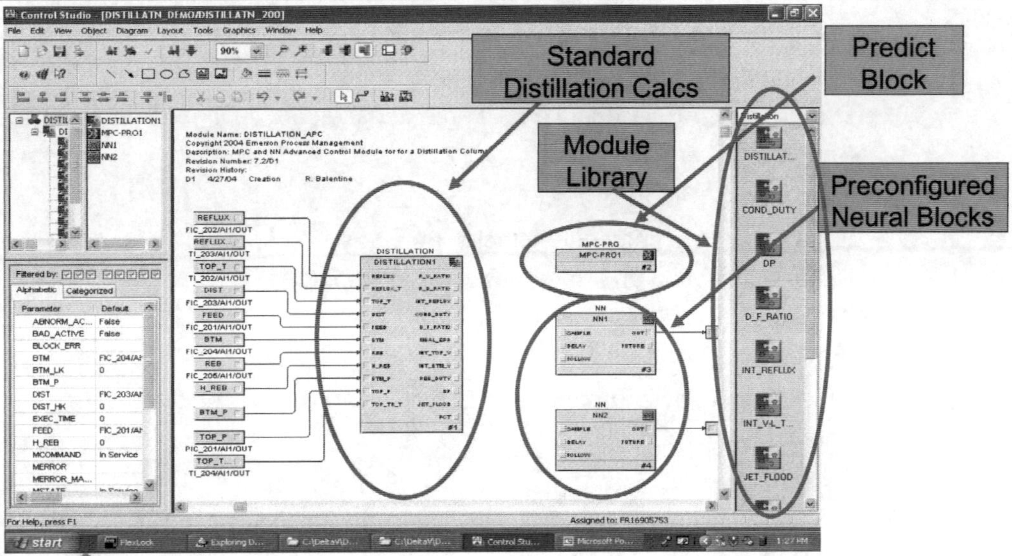

SmartProcess Distillation

- DeltaV PredictPro with up to 4 MV's
- DeltaV Neural for up to 2 product qualities
- Standard set of calculations & constraints
  - Internal reflux
  - Pressure compensated temperatures
  - Jet flooding
  - Reboiler & condenser duty
- Engineering Services to design, implement and commission

SmartProcess Applications

- Licensed industry applications for common processes that are:
  - Pre-engineered, tested, documented
  - Designed by APC Consultants with extensive practical experience
  - Supported & enhanced
- Bundled combination of:
  - DeltaV libraries: composite blocks, modules & templates
  - DeltaV simulation and demonstration example
  - APC tools (e.g. Predict, Neural)
  - Application code (e.g. VB, .Net)
  - Other databases and apps (e.g. SQL*Server, Excel, TankMaster)
  - Sample DeltaV Graphics
  - Documentation
- Templates to be used to configure for specific site requirements
- May be implemented by Emerson or Customer

Benefits

Ethylene Furnace High Temperature Imaging Cameras to Monitor Burner Flames ,Tube Temperature and Tube Failure Detection

Thomas M. Canty, P.E.
President
J.M. Canty Inc
Buffalo, NY USA

Abstract

The development of explosion proof, high temperature imaging cameras which operate in ethylene furnaces without cooling air now allows monitoring of ethylene furnaces to improve energy/operating efficiency along with providing important safety benefits. Imaging cameras work off the visual wavelength which provides 2 dimensional view and temperature measurement of the tube temperature which is more accurate than thermocouples and IR probes. In addition visual images of the furnace burners lets you detect flame detachment, smoking burners and flame shape in order to increase burner efficiency and save energy. The safety aspects of the Camera imaging systems is significant as it allows visual detection of pin hole leaks and tube failure in the furnaces.

Introduction

Ethylene furnaces, or any type of cracking unit, rely on thermal management in order to make product efficiently and to operate in a safe manner. Having 'eyes' in the furnace can help to identify problems in their early stages of development. High temperature camera technology is able to survive the environment and give a wide angle view of the tubes from which temperature and tube blockage may be determined. Achievable benefits from a camera system include:

1. Temperature analysis capacity for all areas of the image, not just single point. Multiple zone capability for user customization of analysis.
2. Flame detection and size measurement
3. Detection of tube movement and ability for software to automatically move measurement zones as tubes distorts.

Imaging Fundamentals

Basics of an imaging system include a high resolution sensor, proper illumination, appropriate hardware for the process and a robust software analysis package. For this particular application illumination is created by the process and need only be managed optimally for best results. A high density pixel array yields resolution of the image and measurement as well. The hardware and software are the more challenging aspects of the application and will be discussed in some detail.

Hardware

The design of the camera system is unique in that it's internally exposed lens assemblies are **designed to survive without cooling air**. When installed properly, the combination of system materials and heat transfer coefficients at the camera tip limit heat transfer to the system and allow it to run hot without chance of damage. This feature, when system is installed in a clean, gas burning environment, does not disrupt the temperature or air balances of the furnace because no cooling air is being introduced to the furnace. In addition, the design includes a disposable, quarts shield which protects the process exposed lenses from direct contact with the internal atmosphere. Should any particulate contact the system, the quartz shield is there to protect the optical lenses. Replacement of the protective quartz shield is inexpensive. Replacement of damaged lenses is very expensive and so the Canty design guards against that possibility.

Internally, the system is a two piece assembly consisting of the internal insertion lens and the external camera package. As mentioned earlier in this document, Canty employs a fused glass window as part of the electronic housing package. This has a dual purpose; Firstly, the camera housing assembly is a separate entity which carries WP and EXP ratings. Secondly, the camera housing bolts to the internal lens system and transmits the optical rays through the fused glass window to the CCD. This mounting also creates a hermetic barrier between the process and the electronic housing thereby preventing migration of hot, process gasses outside the pressure boundary of the system.

Figure 1. High Temp Camera Assembly

The camera electronics come with a full array of controls that allow it to be optimized to the software function to achieve accurate, consistent data output.

As previously mentioned, the installation is an important feature of the design. When installed correctly, the camera sits inside the refractory (see Figure 2) where its exposure to the process environment is minimized thus minimizing heat transfer into the internal lens.

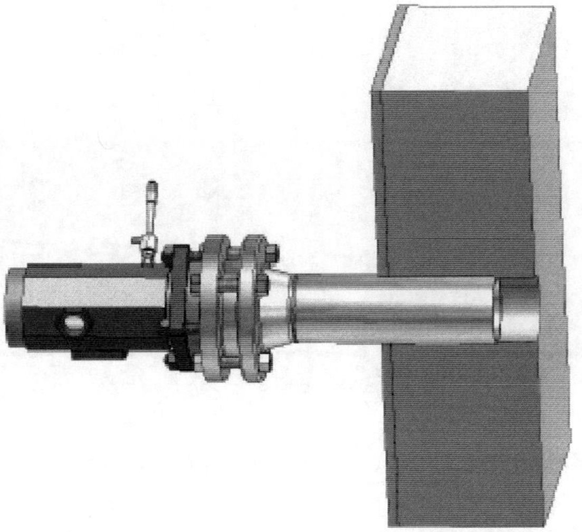

Figure 2. Camera Installation Schematic

Software
With respect to software design, the Canty software is fully flexible in analyzing any portion of the captured image. Multiple analysis zones can be drawn to monitor discreet areas of the view and each zone can run its own particular algorithm depending on need. The benefit here is that multiple tubes can be continually surveyed for changes and trends. The analysis data is also not single point, but is comprised of all the data points in the analysis zone. By choosing proper software filters various information can be obtained from the full array of data such as hot or cold spots. This data can also be archived in total or as snap shots at determined increments of time. The on screen data output can also be updated in the short or near term and can thus be used to judge what is happening in the furnace over time with just a glance at the screen. For instance, using the histogram feature with a 1x daily output a long term trend can be presented on screen in real time.

The basic theory of temperature measurement is based on the radiative emissions of a body at temperature – as the body temperature increases, the radiation curve shifts towards the shorter wavelengths and that is why as a body heats up the eye cannot detect it until it reaches 1000F which is the approximate visible point where emitted radiation is dark red in color. As the temperature increases, the color graduates to bright red, then orange then toward white. A pyrometer functions by measuring a narrow band of wavelength intensity and comparing it to the intensity of a black body to determine the temperature. From the graph below it can be seen that if a single wavelength is monitored the intensity of that measurement changes as the body temperature changes (curve shifts) which then corresponds to a different temperature on the calibration scale.

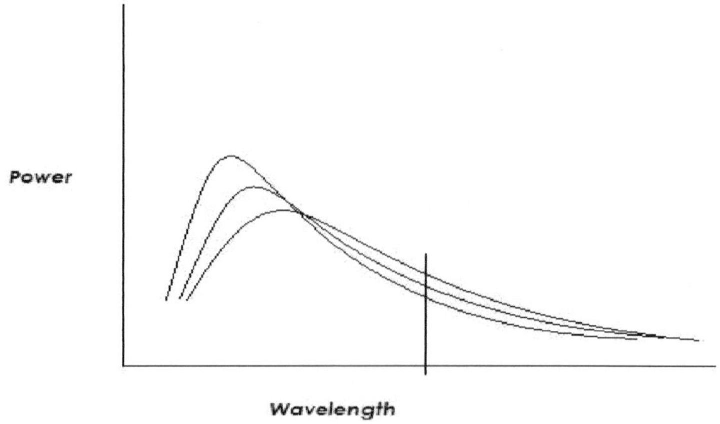

Figure 3. Power Curve at Varying Temperatures

The same is true of a camera system measuring the increasing intensity of an object as it radiates from approx 1000 F (dim, dark red color) on up the temperature scale to 3000+ F for instance which would be a bright white. From the graph above one can see that as color goes from red toward white the peak intensity is always increasing and it is the measure of this intensity which indicates the object temperature. Using an RGB camera the measurement can also be made by monitoring the individual color components of the signal impinging on the chip. For industrial uses, the measurement of intensity usually suffices.

Standard instrumentation used in industry include thermocouple type devices (contact) and thermal imaging and pyrometer systems (non contact). In severe temperatures and atmospheres, the contact devices deteriorate quickly and provide gradually less reliable information leaving the operator running blind. They also create a discontinuity on the surface they are welded to which, in the case of a furnace tube, can lead to premature tube failure.

Non contact systems have difficulties with calibration drift and emissivity adjustments to account for the fact that many items being measured are not black body emitters. In addition, these technologies are single point measurement systems and are best used as hand held devices which can be difficult to use and provide subjective measurement results based on operator preferences for targeting.

Vision technology offers several benefits these other technologies do not. The visual aspect of the system allows the operator to view the process and confirm the readings received. It also allows flexibility to set up measurement zones on screen at any time to assess any or all areas of the process in view should a change be required. Since vision systems are generally fixed installations, an added benefit is that the information is taken from the same process points and features within a given measurement zone at all times which means there is no error in sighting of the instrument.

The vision system is also unique in that it does not suffer calibration drift as almost all instruments do. Once the calibration is set the readings repeat themselves within a couple degrees F. The calibration is based on thermocouple or pyrometer readings initially, however after that calibration is finished it does not have to be redone in the future as long as the system remains in its location.

Process history and performance can also be best documented by a vision system. First off, the camera can analyze any area of process that is within the systems view. This affords the operator the capability of viewing many points of interest instead of just single point temperature readings. As an example, if the system is monitoring a tube bank it can assess the temperature of the tubes for the full length of view. If there is a local flow problem in a tube it will become overheated in one section. The camera can locate that section and indicate its temperature rise over the rest of the tube or adjacent tubes.
This information allows the tube location to be identified as well as the position in the tube that has encountered the upset condition. Over time this historical data, recorded in video or still image format by the control software, may be significant in improving the process. In addition to the ability to monitor the full area of the view, the full area of view can also be segmented into multiple zones of measurement, i.e. each tube having its own measurement zone for automated process control capabilities. This capability allows the tube to be operated at optimum efficiency since the addition of excess heat can be avoided thereby reducing energy draw to operate the furnace.

Thermal Analysis

Tube Temperature
The following figures illustrate the temperature analysis capability for tubes in a live test situation. Measurement zones are placed over the tubes (figure 4) as the furnace is heated and cooled. The temperature plots are shown in figures 5 and 6. The results are as expected and uniform from tube to tube, although a small number are included in this demonstration.

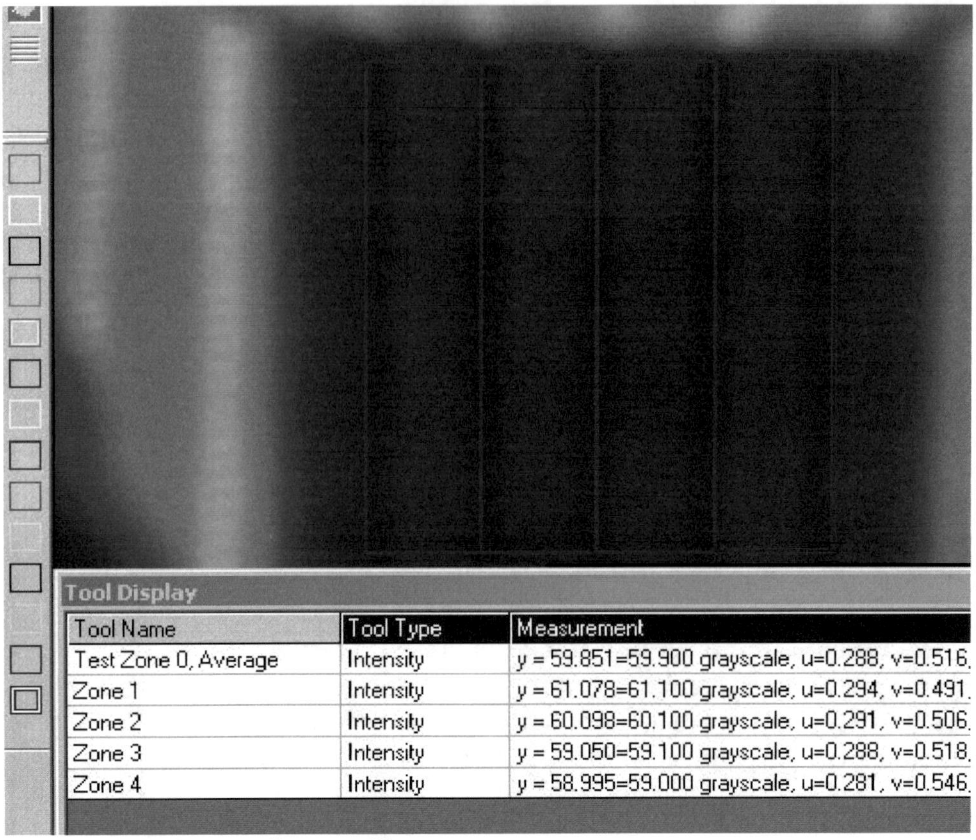

Figure 4. Software Screen Shot

As thermal loads are fluctuated up and down the following graphs track those changes. The cooling cycles are left to natural convection and thus are not linear in nature.

For the following curves, the vertical scale is gray scale intensity which is measured 0 – 255 (pitch black to bright white). The hotter the tube, the higher, or more intense, the gray scale reading. This value in process would be calibrated to the temperature. The horizontal scale is the image number, or time.

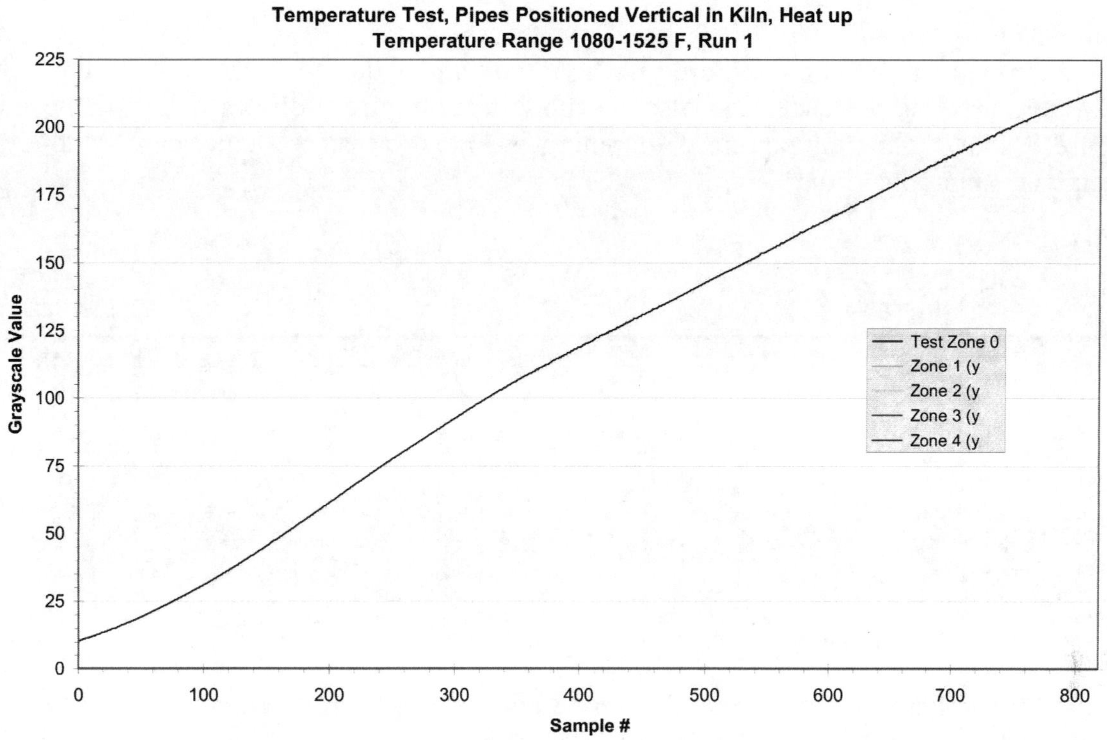

Figure 5. Heating Cycle

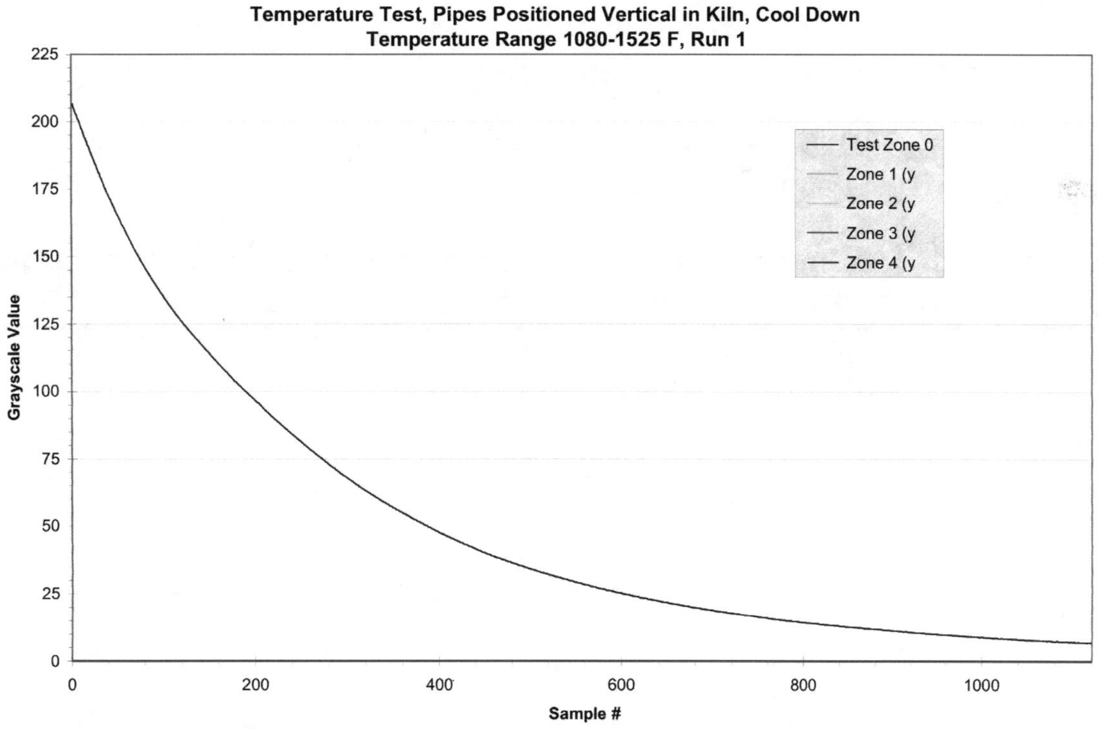

Figure 6. Cooling Cycle

Flame Measurement

Flame temperature can be monitored in the same way as described previously, however the flame size, or presence, can also be monitored. Figure 7 shows a live flame and Figure 8 shows a digital capture. Computing the flame area, or shape, can be done to track its changes over time.

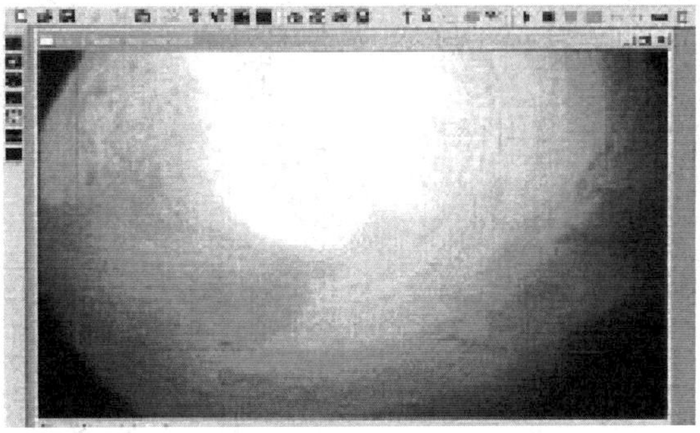

Figure 7. Kiln Flame (Live)

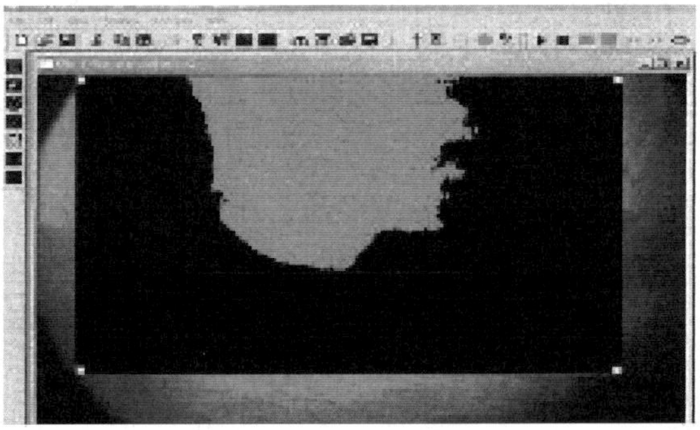

Figure 8. Kiln Flame (Digital)

Conclusion

Vision Technology presents an opportunity to increase effectiveness of furnace monitoring through hardware and software capabilities. The thermal design at the core of the hardware enables the system to provide the operator a view into the furnace and feed back on its thermal performance over time. This data can assist in detecting building upset conditions, but also provide a record for study to assist in improving the operational method of the furnace.

(45v) Straighten up! Non-Linear Compensation Using Feed Forward Control

Barbara Hamilton, Process Systems and Solutions, Emerson Process Management, Hampden , ME

Non Linear processes occur throughout the petrochemical plant. While the PID loop is surprisingly robust for most situations, there are cases where compensation of these non-linearity's is critical for stable control. Similarly, disturbances can cause major disruptions in control stability.

A method of feed forward control, using data based models, can provide robust control throughout the process range. Historical data are used to build models to drive the output toward the solution with PID control providing the final drive to zero error.

The model is also used to develop dynamic tuning. The result is a simple to configure, robust solution to manage non-linear loop and/or disturbance variations.

(45w) Adsorption of CO2 by Alkaline Activated Ca-Al2O3-SiO2 Material

Naim Faqir, King Fahd University of Petroleum & Minerals, Dhahran, Saudi Arabia and Reyad Shawabkeh, Chemical Engineering, King Fahd University of Petroleum & Minerals, Dhahran, Saudi Arabia

Natural gas contains a slight level of carbon dioxide which leads to several industrial and domestic heating problems. In addition, it is one of main greenhouse gases that contribute for global warming, where its pollution is a wide-spread problem arising from a wide range of industries, leading to complex contaminant in atmosphere. Given the increasing pressures worldwide to minimize the emission of greenhouse gases into atmosphere, CO_2 must be remediated efficiently and cost effectively.

In this short abstract a chemically treated limestone and alumino-silicate material were utilized for removal of CO_2 from natural gas using a combined adsorption-chemical reaction processes. This functionalized Ca-Al_2O_3-SiO_2 material was characterized according to its crystalline phase, surface area, pore size distribution and surface morphology. Then it was tested against different concentrations of CO_2 in CO_2-N_2 gas streams.

Pressure Gauges: Hidden Danger or Visible Safety

by Jeff Placek, WIKA Instrument, LP, Lawrenceville, GA

Introduction

In recent years, refining plants, petrochemical plants and chemical plants have been required to increase capacity, but have been provided with very few upgrades to existing equipment to accommodate the increased capacity requirements. This has caused plants to have increased stress and strain on their systems and operating units. For example, higher pulsating pressures and heavy mechanically-induced vibrations are translating to all areas of the processing equipment, especially pressure instrumentation.

Pressure gauges are one of the most important, yet most overlooked or ignored keys to ensuring the continued safety, productivity and profitability of process plants. Process pressure gauges are used to read pressure on transfer lines, discharge pumps, and other places where it is important to be able to read pressures.

Problems Associated with Misapplied, Faulty or Ignored Gauges

Incorrect, misapplied, damaged, faulty or ignored gauges can lead to a number of issues, including safety problems, downtime, and costly regulatory agency fines and other penalties.

Process Safety:

One group of problems relates to process safety issues, such as loss of containment that leads to fugitive emissions, fires and explosions. Bourdon tubes in malfunctioning gauges can rupture, causing the release of media, such as gasoline, oil or gas. Because this media is often volatile and corrosive, it can cause fires, explosions, or other serious safety issues, such as a toxic gas releases or acids spraying onto workers.

Media releases are common results of misapplied, faulty or ignored gauges. Fires and explosions are less common but they do occur. When reviewing plant instrumentation, findings have revealed instances where the wrong gauges have been installed in locations on pumps. In one instance, a gauge vibrated so severely when the pump came out of alignment that it caused the Bourdon tube in the gauge to crack. The media coming out of the tube leaked onto the pump, which was excessively hot because of age and misalignment, causing a fire.

Expensive Downtime:

A second group of problems relates to expensive downtime, which can result from the breakdown of process safety just mentioned, or simply from having to shut down a process in order to replace incorrect, misapplied, broken or otherwise damaged gauges.

1 - Safety-Related Downtime:

As a result of the problems identified above (emissions, fires, explosions, etc.), there can be any number of causes of subsequent downtime, including fire suppression activities, released media containment, line shutdown, personnel evacuation, and then eventually having to rebuild or repair the lines, replace equipment, etc. In addition, if there is a media release with no injuries or fatalities, the plant is required to notify the EPA. However, if there is a media release that does cause injuries or deaths, the plant is required to notify OSHA. These can lead to very long process shutdowns, anywhere from a day to a month or more. (More on OSHA and the EPA below.)

2 - Non-Safety-Related Downtime:

When a line fails, causing a gauge to fail, even if there is no release or safety incident, downtime can still occur as a result of the process needing to be shut down in order to replace the gauge. Gauge failure rarely causes other parts of the process to fail, although gauge failure is usually an alert that there is a problem. In most cases, a plant worker will simply replace the pressure gauge and move on. When another failure occurs, the worker will do the same, until the root cause of the failure, such as a misaligned pump, causes the plant to shut down the line, resulting in more costly, unplanned downtime and a major maintenance time.

Regulatory Agency Fines and Penalties:

A third group of problems relates to the potential for significant fines and other penalties from federal regulatory agencies (primarily OSHA and the EPA), as well as from state and local jurisdictions, that result from compromised process safety.

1 - OSHA:

OSHA addresses many process-related safety issues in 1910 Subpart H, "Hazardous Materials." More specific is 1910.119, "Process safety management of highly hazardous chemicals." In its Process Safety Management section, OSHA notes that, "Unexpected releases of toxic, reactive, or flammable liquids and gases in processes involving highly hazardous chemicals have been reported for many years in various industries that use chemicals with such properties. Regardless of the industry that uses these hazardous chemicals, there is a potential for an accidental release at times they are not properly controlled, creating the possibility of disaster."

OSHA goes on to note that: "To help ensure safe and healthful workforces, OSHA has issued the Process Safety Management of Highly Hazardous Chemicals standard (29 CFR 1910.119), which contains requirements for the management of hazards associated with processes using highly hazardous chemicals."

OSHA's standard emphasizes the management of hazards associated with highly hazardous chemicals and establishes a comprehensive management program that integrates technologies, procedures, and management practices.

While OSHA does not monitor gauges in specifically, they will enforce guidelines for worker safety. OSHA investigates injuries and deaths, and may levy fines or other penalties.

2 - EPA:

According to Greg Neneman, director of business development solutions for Valin Corporation (San Jose, Calif.), fugitive emission regulations from the Federal level have tightened up over the past several years. In many states, such as California, Oregon and Washington, the regulations are even tighter.

"One challenge companies face is ensuring compliance with emission regulations when they upgrade or expand their processes," points out Neneman. "They are not allowed to produce additional emissions beyond the original limit set by regulations."

From the standpoint of federal EPA regulations, each company that produces or emits volatile organic compounds (VOCs) is required to come up with a leak detection and repair program. "In some areas, county and local requirements are even tighter than federal requirements," continues Neneman.

In certain areas, companies that produce or emit VOCs are required by regulation to do audits on valves, flanges, gauges, and all other process connections on a quarterly basis. "And in some jurisdictions, the regulatory agencies can give plants 24 hours notice before they descend on that plant to check all of their records and conduct other investigations," adds Neneman.

In May 2012, the U.S. Environmental Protection Agency (EPA) levied a $1 million civil penalty on a plastics company for alleged violations of the Clean Air Act at two of its chemical manufacturing plants. In addition to paying the penalty, the company agreed to implement a comprehensive program to reduce fugitive emissions of hazardous air pollutants (HAPs) from leaking equipment such as valves and pumps. The settlement also required the company to implement enhanced work practices, including more frequent leak monitoring, better repair practices, and new efforts designed to prevent leaks. After being fined by the EPA, the company realized that instrumentation was one of the areas where they having fugitive emissions. Until that time, workers had been replacing gauges left and right, because they weren't aware of the problems that could occur. The gauges weren't a big expense to them, so they considered the replacement activity to be more of a nuisance than anything. It's when they had the rupture that they realized there was a real problem.

An audit of the plant, which involved reviewing at all of the instrumentation, found that in many applications, they were using the wrong instrumentation. To correct the issues, the right instrumentation was specified for them, and the correct gauges were ordered. The plant now has its instrumentation under control.

While conforming to EPA regulations has always been a challenge for process facilities, it is going to become even more challenging in the future as a result of EPA's recent introduction of new requirements related to fugitive emissions and leak detection and repair (LDAR) strategies. On April 17, 2012, the EPA Administrator signed a suite of final air quality regulations that affect both upstream and midstream oil and gas sectors - the production/processing and transmission/storage sectors of the oil and natural gas industry.

For the first time, EPA will regulate air emissions from natural gas wells that are hydraulically fractured, as well as other emission sources associated with exploration, production, processing, and transportation of oil and natural gas.

Promulgation of the final rules will result in new control equipment requirements, as well as increased requirements for monitoring, recordkeeping, and reporting for owners, operators, contractors and vendors.

The final rules require the use of emission controls, as well as work practices, through two different air regulatory programs. One is the New Source Performance Standards (NSPS). The other is the National Emissions Standards for Hazardous Air Pollutants (NESHAP).

EPA finalized a new NSPS at 40 CFR Part 60, Subpart OOOO, which regulates emissions of volatile organic compounds (VOC) and sulfur dioxide (SO2) from a variety of oil and gas exploration, production, processing, and transportation facilities, including natural gas wells, certain compressors and pneumatic controllers, storage vessels, and sweetening units at onshore natural gas processing plants. Subpart OOOO also includes startup, shutdown and malfunction provisions.

The EPA has also made it clear that oil and gas production and processing sources will be the focus of its National Enforcement Initiative (NEI) for fiscal years 2012 and 2013. With the new NSPS and revised NESHAPs in place, EPA now has the tools it needs to enforce in a comprehensive manner.

As a result of these new regulations, plants should look at everything more closely, including gauges.

How Problems Can Occur With Gauges

Technical problems can occur with process pressure gauges in a number of ways, which can be conveniently categorized into seven broad categories with specialties or subsets. The most common reason for gauge problems is that workers often install gauges without knowing how they should be installed or accessorized. As a result, the gauges can fail because they were never designed to function in those specific situations. They can also fail because they were installed incorrectly, or because they were not properly maintained.

One way to eliminate or minimize these problems is to consider specific pressure gauge accessories. These can prove to be the extra step needed to ensure a successful, proactive preventive maintenance program, and for extending the life of the gauges. Accessories can provide additional protection for the gauges during their lifetimes.

1 - Mechanical Vibration.

This is oscillating, reciprocating or other periodic motion of a rigid or elastic body or medium forced from a position or state of equilibrium. Typically, vibration is the main cause of pressure gauge failure in process plants. Pump applications produce most of this vibration.

Gauges that aren't designed to withstand vibration may be installed in applications where vibration occurs, causing the gauges to come out of accuracy. Even more serious, a gauge can vibrate so much that the Bourdon tube inside it can crack, releasing media."

One way to prevent vibration-related problems is to install liquid-filled gauges, which help to dampen the vibration and also protect the internal components. The liquid case filling also lubricates the internals and prevents corrosion.

2 - Pulsation.

When media rapidly cycles through the pressure system of the gauge, pulsation occurs. If the pointer pulsation increments are greater than 5 percent of the full scale value, you must intervene to prevent damage to the gauge.

A restrictor or throttle screw can solve most pulsation problems. It will slow down the media by reducing the size of the orifice and averaging out pressure fluctuations. For extreme pulsation, such as spiking and surging, a snubber works best. It allows more control over the amount of snubbing and the flow of media into the pressure system.

3 - Temperature:

If a gauge is exposed to a media with very high temperature, the metal alloy in the gauge may not be able to handle the temperature, causing the Bourdon tube to fail and release media.

Long pipes that separate the pressure gauge from the process, or small cooling elements with radiator fins, protect the pressure instrument from high process temperatures by allowing ambient air to flow over the surface area. In environments with steam, using a siphon with the pressure gauge cools the steam down and drives it back to its liquid state before it gets to the gauge. With very high-temperature media, a diaphragm seal with a cooling element will isolate and prevent the hot media from reaching the gauge.

4 - Spikes or Overpressure.

In some cases, process media may be under extreme pressure. Pumps also can cause spikes and over-pressure, both of which shorten gauge life. Gauges can fail as a result of these pressure spikes. When workers turn pumps on or off, or open or close valves, a surge of media or oil goes through the pipe and hits the pressure gauge, which causes a huge spike. For example, if a line is pumping 100 psi of oil, someone may install a 100 psi gauge, not realizing that this leaves no allowance for any overpressure, surging, or spiking. When pressure does increase, it causes overpressure on the gauge, sometimes to the point of where the Bourdon tube ruptures and leaks media.

One way to prevent this problem is to use a different type of gauge, such as 200 psi gauge for a 100 psi line. Another option is to accessorize the gauge with an over-pressure protector, which would shut off the pressure if there were a surge or spike. Overpressure protectors are designed to avert spikes and the resulting overpressure damage. These protectors provide protection for pressure gauges against the effects of pressures exceeding their maximum pressure ratings.

5 - Corrosion.

Corrosion problems can occur with gauges if the process media is not compatible with the pressure gauges' wetted parts material. Media can sometimes be so corrosive that it corrodes through the Bourdon tube and leaks onto the ground.

When the media is not compatible with the wetted parts material of the pressure gauge, a diaphragm seal is required to isolate the gauge from the media. The benefit of a diaphragm seal is that it can be made of special alloys that are designed to withstand corrosion. It is essential to select a diaphragm material that is compatible to the process media.

6 - Clogging.

In some processes, the media may eventually crystallize or may otherwise harden. If a standard gauge is used in such an application, it will eventually get clogged and will not be able to read the pressure accurately. For example, if a gauge gets clogged and ends up reading zero pressure, when in fact it is under tremendous pressure, there will be an unexpected and dangerous release of media if and when the piping system is opened. Diaphragm seals protect the pressure gauge from solids, particles and viscous media by creating a barrier between the process and the pressure inlet of the gauge. There is also an option to equip the diaphragm seal with flushing ports to constantly flush material out from the diaphragm surface.

7 - Mishandling and Misapplication.

Gauges can also fail prematurely if they are misapplied. For example, a gauge may have been ideal for the application when it was installed. However, if there are equipment modifications or more components subsequently added, this can change process parameters, resulting in the gauge not longer being right for that application. Misapplication can lead to serious problems associated with failure and subsequent media release.

It is also possible to accidentally damage gauges during installation. Gauges should be installed using the proper sized wrench on the flats. Gauges should never be installed by turning the case. After installation, gauges can also fail if they are mishandled or used improperly.

Why Problems Can Occur With Gauges

In a typical process plant, about 25 percent of pressure gauges have failed or are about to fail.[1]
A lack of management focus and worker education and experience are key reasons plants experience issues with gauges.

Management Focus.

Gauges are often viewed as inexpensive commodities, with management failing to realize that they are very critical parts of the processes. Often, plants will replace failed gauges with the same replacement gauges, which will also fail, only to have to be replaced again. Eventually, their luck may run out and a serious failure may occur because the incorrect gauge has been installed.

Why do gauges tend to receive such little attention from management? Management tends to focus on the expensive equipment - $100,000 or million dollars items. They often don't understand the importance of focusing attention on $100 gauges. This is not the fault of management. The problem is that no one has explained to them just how important gauges are.

As noted earlier, though, ignoring instrumentation can cause serious safety and financial problems. Plant process gauges can be thought of in the same way as vehicle gauges. If the temperature gauge on an automobile is defective, it is possible to destroy the engine as a result of overheating that is not registered by the defective temperature gauge. Similarly, if the oil pressure gauge on an automobile is defective, it is again possible to destroy the engine as a result of seized components resulting from loss of oil that is not indicated by the defective oil pressure gauge.

Worker Education and Experience.

A lack of worker education and experience also contributes to gauge issues. In many instances, when maintenance technicians look at gauges, they don't pay much attention to them unless they leaking or look as if they are going to explode. They don't realize all of the information that that gauge can provide. In addition, technicians may not realize that a gauge has failed or is about to fail. Again, as long as it is not leaking media, they will often just keep walking. This is rarely the fault of the technicians. Many have not been properly trained to understand process pressure gauge performance.

Along the same lines, the industry has been experiencing a "brain drain" for several years, with large numbers of experienced technical workers and engineers retiring and being replaced by younger people who lack the education and experience to understand what gauges are telling them, or, worse, not being replaced at all.

According to research conducted by Cambridge Energy Research Associates in 2007, more than half of all oil-field professionals will reach retirement age during the next decade. The Interstate Oil & Gas Compact Commission found that the average age of production and service company employees is 46 to 49 years, and the average retirement age in the oil and gas industry is 55 years. [2]

Historically, plants had specialists whose jobs focused exclusively on instrumentation. They spec'd the gauges, maintained them, and made sure they were operating correctly. However, as plants got leaner in recent years, they started letting these people go. Plants first shifted gauge responsibility to the pipefitter whose main responsibility was to make sure the pipes were put together correctly and not leaking.

[1]WIKA data aggregated from more than 250 instrument audits.
[2]Peter Peek, Jocelyn Fenard, Patrick Gantes, Christine Theiler. December 2008, "Skills Shortage in the Global Oil and Gas Industry: How to close the gap" Part I. Pages 14-15.

Then, as plants began letting pipefitters go, they shifted the responsibility to the unit operators or technicians, who now install the gauges. These people usually don't have any experience or knowledge of how to install instrumentation. They usually just replace old gauges with similar gauges. The problem is that the gauges are usually the wrong ones to begin with. The result is that, these days, few plants have people who have much if any education and experience with instrumentation.

Taking Stock of Process Pressure Gauges

When it comes to improving the safety, productivity and profitability of process plants through an effective gauge management program, companies can take a couple of steps. First, an understanding of the vital importance of gauges must be established. Second, workers must understand what kind of gauges will be effective.

Focus on Gauges:

World-class process plants maintain their assets proactively. In such plants, pressure gauges receive as much attention as pumps and control valves. Management understands the importance of proactively maintaining their assets, and they realize that one of the best ways to maintain their most expensive assets is to properly source and maintain the gauges that protect these assets. That is, they understand the value of spending $100 for a gauge that can subsequently prevent damage or destruction to a $100,000 or $500,000 piece of equipment and the resulting negative safety, downtime and regulatory implications.

There is an interest in predictive maintenance these days in process plants, which is a good thing. A lot of this, though, involves expensive software. In addition, when something is out of kilter, the system may be set up such that technicians may have to wait for it to get far enough out of the normal range before it will generate an alert on the computer. However, properly-maintained pressure gauges, studied by people who know what to look for, can be very inexpensive but reliable and useful mechanical predictive maintenance tools.

Gauge Procurement Considerations:

Over time, as noted earlier, the status of pressure gauges has unfortunately diminished to that of commodities in many process plants. In order to ensure process plant safety, productivity and profitability, it is essential that plant professionals understand the importance of selecting the right gauges for applications.

It is also important that these be high-quality gauges, not "commodity" gauges. That is, not all process gauges are created equal. The so-called "lowest price" gauges appear attractive at first, but are rarely the "lowest cost" in the long run, because they usually have to be replaced more often, leading to excessive downtime and inconvenience. Worse, they may fail unexpectedly, leading to catastrophic results.

Consider these 10 steps to select the right gauges that will help to ensure safe, productive and profitable plant processes.

1 - Know what is in the pipe. This way, you can select a gauge alloy that will be compatible with the process media.

2 - Next, know the temperature of the media. If it is very hot, for example, it will be unsafe to use a standard pressure gauge. A long pipe, siphon or diaphragm seal assembly may be needed to cool the media sufficiently before it reaches the gauge internals.

3 – Third, know the ambient conditions surrounding the gauge. For example, if it is corrosive around the gauge, you need to make sure that the external components of the gauge such as the case and window can handle the environmental conditions.

4 - Next, know the pressure in the piping that the gauge is meant to measure. A best practice is using a gauge that is designed for two times the normal operating pressure. For example, if normal pressure is 100 psi, use a 200 psi gauge.

5 – Know the conditions that affect gauge wear. These could include mechanical vibration, pulsation, pressure spikes, etc. For example, if vibration exists, a gauge with a liquid-filled case is recommended. If there is pulsation, use a throttle plug or restrictor accessory.

6 - Select gauges that are designed for a particular service. Process type gauges are designed with strong Bourdon tubes and other heavy duty components. If a gauge is misapplied, the internals of the gauge may not be rugged enough to resist failure and subsequent media release in these applications.

7 - The gauge case must be rugged and have the correct safety features. In some case designs, if there is a rupture of the Bourdon tube, media will be released out the front of the gauge, which can spray workers. Select a gauge where media would release out from the back of the gauge.

8 - Accuracy is important. In most process plants, +/-0.5% accuracy is the norm, and +/-1% is acceptable.

9 - Size is also important. The gauge dial needs to be large enough that operators can read it from a distance – four inches or larger.

10 - Finally, the gauge should be mounted properly. A gauge is usually mounted by its stem, but it can also be mounted in a panel or against a flat surface.

Looking Outside for Assistance

As noted earlier, it can be difficult for plant management to devote enough time and attention to an effective process pressure gauge program. Similarly, it can be difficult for technical personnel to get up-to-speed in terms of education and experience related to process pressure gauges.

There has been a growing trend over the last 30 years in virtually all industries for companies to outsource services that are not considered "core competencies." That is, companies have come to realize that it makes sense to focus their attention on their missions and what they do best, and leave secondary and peripheral activities to third parties that specialize in these other activities and services. Becoming experts in pressure gauges is certainly not a core competency for refining plants, petrochemical, and chemical plants, so this is an area that can easily be outsourced to an experienced third-party audit team.

If a plant elects not to gain the experience that it needs on its own, free assistance is available. To ensure that gauges are working properly and are able to protect the valuable assets that they were designed to protect, a five-phase instrument review is recommended. A best practices review identifies gauges with unwanted conditions and recommends corrective actions to ensure optimum performance. The review identifies issues that affect reliability, longevity and safety. Such a review can uncover evidence of pulsation, vibration, pressure spikes, excess process or ambient temperature, installation damage, clogging, chemical incompatibility, and other problems that can lead to serious problems.

Phase 1 - Interview. This involves discussions between the third-party audit team and plant personnel. During the interview, plant personnel learn the purpose of the instrument audit, how it will be conducted, and how they will benefit from the audit. The interview also helps determine the best way to match the audit to the needs of that specific plant.

Phase 2 - Pre-Meeting. This involves determining the units or sections of the plant to audit, as well as outline roles and responsibilities. It also identifies prohibited and exempt areas of the plant. The third-party team reviews problematic instrument installations and discusses safety procedures with plant personnel.

Phase 3 - Instrument Application Audit. This involves having key plant personnel lead the third-party team on an actual plant walk-through. The audit team reviews the target gauge installations, identifies gauges that demonstrate unwanted conditions, as well as those that demonstrate installation best practices.

Phase 4 - Post-Meeting. This involves explaining the instrument audit results to key plant personnel, specifically noting gauge installations that pose immediate safety concerns. Together, the audit team and plant personnel determine which installations to investigate immediately and which to block out. The audit team and plant personnel then set schedules for the follow-up meeting.

Phase 5 - Follow-up Meeting and Presentation. The third-party team presents its general findings and the executive summary resulting from the instrument audit. The team also presents a spreadsheet that lists details about the gauges from the instrument review, color-coded by severity. The team also provides recommendations for improvements based on experience and best practices. Together, the audit team and plant personnel develop a strategy for moving forward.

The plant should also consider leveraging the audit team in other ways. For example, over time a plant's storeroom inventory can get out of hand. A qualified third-party audit team can help plant personnel rationalize, consolidate and manage their inventory of pressure gauges. For example, if a plant stocks spares for 10 or more ranges, the audit team may be able to help reduce that number to five or six ranges.

Essentially, when used properly, gauges play a significant role in plant safety, productivity and profitability. They act like silent alarms, indicating when something is seriously wrong or helping to diagnose deeper issues within systems and processes that can eventually lead to serious safety issues. To ensure safety and reliability at your plant, make certain the proper gauges are in place by reviewing your processes, gauges and your inventory. Everything is at stake.

REFERENCES

1. WIKA data aggregated from more than 250 instrument audits.

2. Peter Peek, Jocelyn Fenard, Patrick Gantes, Christine Theiler. December 2008, "Skills Shortage in the Global Oil and Gas Industry: How to close the gap" Part I. Pages 14-15.

13 AIChE
Spring Meeting, San Antonio, TX

Design & Revamp of Cooling Water Networks

E. K. Tamakloe, G. T. Polley & M. Picón Núñez

University of Guanajuato,

Mexico

Appear Complex

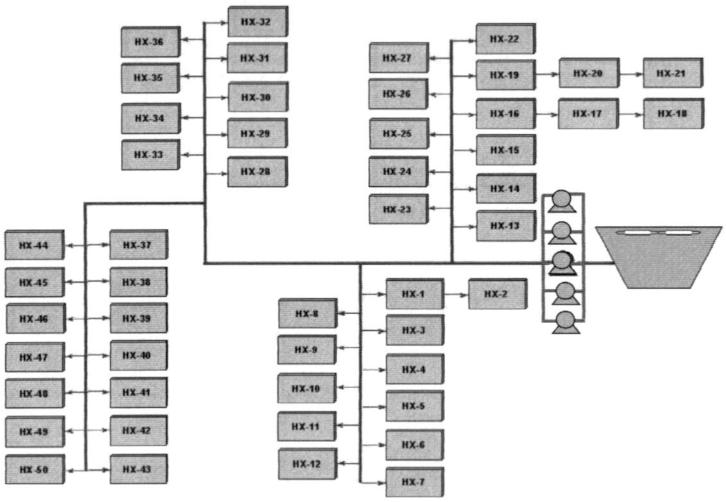

13 AIChE
Spring Meeting, San Antonio, TX

Reality

- Simulations can be set up quickly
- Analysis quite simple
- Significant Benefits

13 AIChE Spring Meeting, San Antonio, TX

Design of New Networks

13 AIChE Spring Meeting, San Antonio, TX

Topics

- Design Principles
- Flow Simulation
- Fouling Analysis
- Blowdown & Make-up

13 AIChE
Spring Meeting, San Antonio, TX

Design Principles

13 AIChE
Spring Meeting, San Antonio, TX

Cooling Towers

- Individual Plants – Plant Regions – Centralized System

- Packaged Units – Specialist Suppliers

- Selected with future expansion in mind

13AIChE
Spring Meeting, San Antonio, TX

Cooling Towers

- Minimum Wetting Rates apply: so, system may have flow set by cooling tower rather than process plant

- Maximum water return temperatures apply

  (function of water chemistry)

- Thermal Behaviour : Effectivess-Ntu Model

  Tested, Reliable

13AIChE
Spring Meeting, San Antonio, TX

Piping

- Dominates Capital Cost
- Set with allowance for plant expansion
- Equations for economic pipe size
- Capital Cost : related to diameter to power 3

  (exception being very long straight runs)
- Installation: again diameter to power 3

13 AIChE Spring Meeting, San Antonio, TX

Piping

- Function of Plant Layout – not use of temperature driving force

13 AIChE Spring Meeting, San Antonio, TX

Heat Exchanger Sizing

- Must be undertaken in "system context"
- Cannot arbitrarily set pressure drop or cooling water temperature rise

13AIChE Spring Meeting, San Antonio, TX

Flow Simulation

13AIChE Spring Meeting, San Antonio, TX

Components

- Gravitational head
- Frictional losses
- Momentum changes

13 AIChE

Special Feature

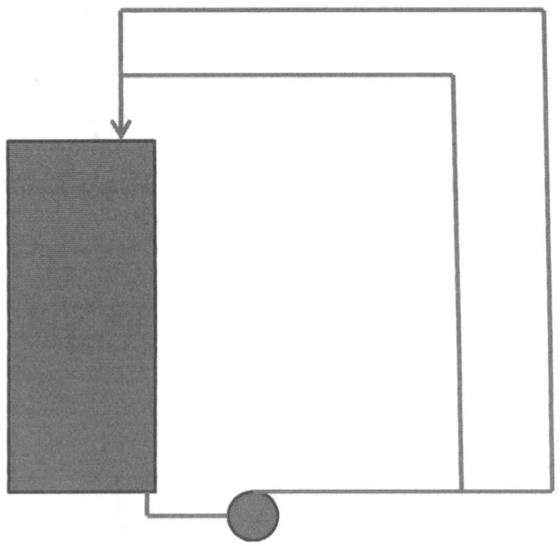

CLOSED SYSTEM

13 AIChE

IMPLICATIONS

- MOMENTUM EFFECTS INDEPENDENT OF FLOW PATH
- GRAVITATIONAL EFFECTS INDEPENDENT OF FLOW PATH (Provide flow initiated)

13 AIChE
Spring Meeting, San Antonio, TX

Gravitation

- Initiation of flow may require provision of back pressure – restriction at cooling tower return
- Vacuum can be pulled in heat exchangers – checks for vapour blocking are required

13 AIChE
Spring Meeting, San Antonio, TX

Friction Equation

- Great advantage in using equation based on Volumetric Flow Rate rather than velocity

- Velocity can vary along specific flow path – but volume flow remains constant

13 AIChE Spring Meeting, San Antonio, TX

Pipe

$$\Delta P_f = 2f\left(\frac{l}{d}\right)\left(\frac{\rho u^2}{g_c}\right)$$

13 AIChE Spring Meeting, San Antonio, TX

Valves & Fittings

$$K_{valve} = \left(\frac{8\rho}{\pi^2 d_{pip}^4} \right) k_{valve}$$

13 AIChE

Exchanger Pressure Drop

Function of Square of volumetric throughput

$$\Delta P = f(V^2)$$

13 AIChE

Components are Additive

Pipe + Valve + Heat Exchanger + Valve

$K_{leg} = 2K_{valve} + K_{exchanger} + K_{pipe}$

13 AIChE

Branch

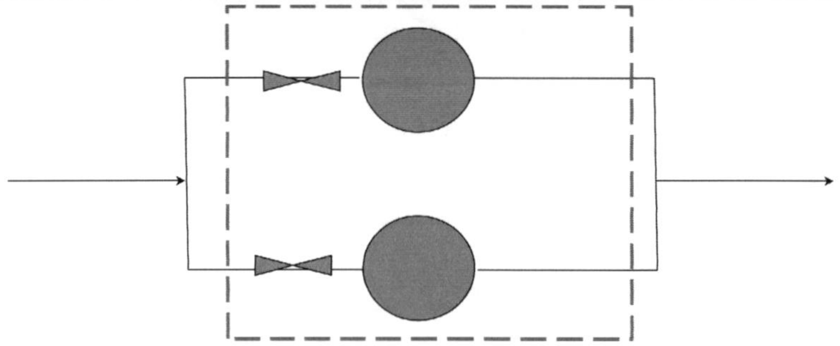

$$K_{branch} = \frac{1}{\left(\left(\frac{1}{K_{branch\,1}}\right)^{0.5} + \left(\frac{1}{K_{branch\,2}}\right)^{0.5}\right)^{2}}$$

13 AIChE

K for Whole System

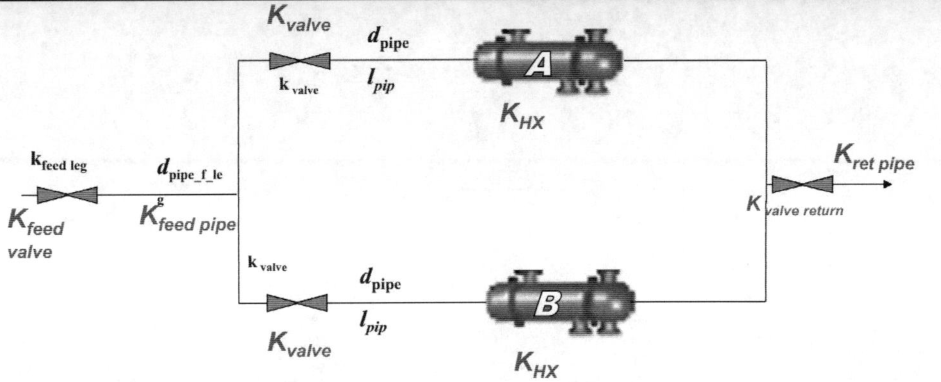

$$K_{sys} = K_{feed-valve} + K_{feed-pipe} + K_{branch} + K_{ret-valve} + K_{ret-pipe}$$

13 AIChE Spring Meeting, San Antonio, TX

Flow Through Network

$$V_T = \left(\frac{\Delta P_T}{K_{sys}} \right)^{0.5}$$

Solved simultaneously with Pump Curve Equation

13 AIChE Spring Meeting, San Antonio, TX

Distribution Through Network

$$\frac{v_j}{V_T} = \frac{\sqrt{K_j}}{\sum_k \sqrt{K_k}}$$

13 AIChE
Spring Meeting, San Antonio, TX

Heat Exchanger Sizing

- Relationship between heat ransfer coefficient and pressure drop

$$\Delta P = \frac{bA}{V}\alpha^m = KV^2$$

V = Volumetric flowrate, A = Area

b = constant (physical properties, characteristic dimension – e.g. Tube i.d.

m = exponent (tubes = 3.5) based on exchanger type (available for tubes, shells, plate-and-frame)

- Solve together with basic exchanger design equation

13 AIChE
Spring Meeting, San Antonio, TX

Fouling

13 AIChE
Spring Meeting, San Antonio, TX

Calcium Sulphate

- Najibi Kraus (1997) developed a model for fouling due to crystallization of calcium sulphate.
- The rate of fouling is defined by two consecutive processes:
  - First, transport of dissolved salts to the convective heat transfer surface as defined by the mass transfer coefficient and concentration gradient between the flow and the surface
  - Second, the reaction and precipitation of the salt ions in the flow to the heat transfer surface to form deposits. This reaction is defined by the Arrhenius relationship.

13 AIChE
Spring Meeting, San Antonio, TX

Predictions of Model

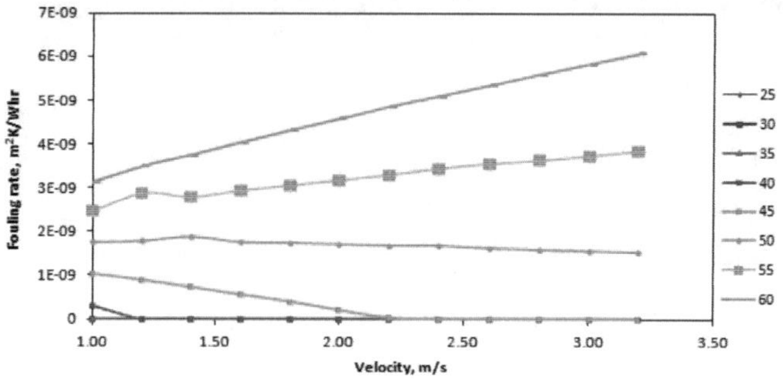

High water temperatures: fouling rate increases as velocity increases

Moderate temperatures: fouling rate decreases as velocity increases

A fouling "threshold" can be observed

13 AIChE

Fouling Thresholds

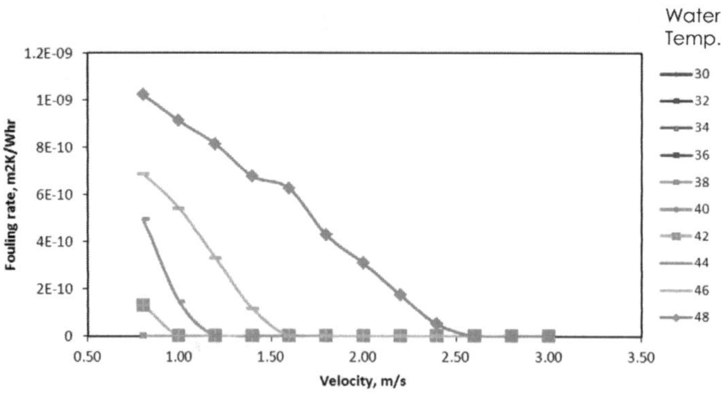

Since water temperature (rather than "film temperature" has been used, plots are function of process side h.t.c.

13 AIChE

Blowdown & Make-up

13 AIChE
Spring Meeting, San Antonio, TX

Governing Equations

- Concentration of calcium sulphate in make-up water is known.
- Concentration in recycled water needs to be controlled.

Water balance:

Make-up = Evaporation + Blowdown

Salt balance:

Make-up(feed conc) = Blowdown (system conc)

Approximate Heat balance:

Evaporation = Tower Load/Latent Heat of water

13 AIChE
Spring Meeting, San Antonio, TX

Typical Results

Blowdown: kg/s

		Feed (ppm)			
		10	50	100	200
System (ppm)	1000	0.03	0.14	0.3	0.68
	1200	0.02	0.12	0.25	0.55
	1500	0.02	0.09	0.19	0.42
	1800	0.02	0.08	0.16	0.34
	2000	0.01	0.07	0.14	0.3

Tower Heat Load: 6 MW

Vaporisation: 2.73 kg/s

13AIChE

Revamping Networks

13AIChE

Two cases considered

1. Reduction of Load on Tower

2. Installation of New Cooler

Other cases exist: e.g. elimination of pumps

Result of Improved Heat Recovery

- Load on coolers is reduced
- **BUT** Process streams enter and leave coolers at lower temperatures

Case study

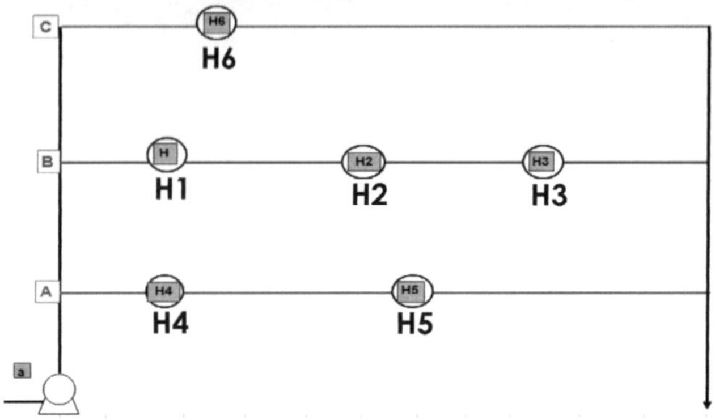

- 3 Branches in parrallel
- 6 Heat exchangers

13AIChE

Current Performance

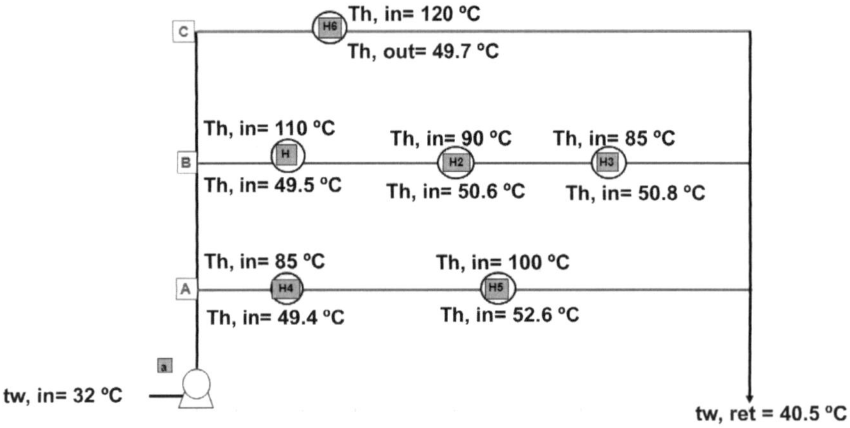

13AIChE

Coolers

Coolers		
Unit	Tin	Tout
H1	110	49.59
H2	90	50.6
H3	85	50.8
H4	85	49.49
H5	100	52.6
H6	120	49.7
Load on CT = 9.58 MW		
T_{Wret} = 40.5° C		

13 AIChE Spring Meeting, San Antonio, TX

Result of improved Heat recovery

- Reduced entrance temperatures
- "Leads to reduced outlet temperatures"

13 AIChE Spring Meeting, San Antonio, TX

After energy savings

After energy savings			
Unit	Tin	***Tout***	Tout (current)
H1	70	***40.57***	49.59
H2	65	***41.99***	50.6
H3	65	***42.49***	50.8
H4	70	***44.54***	49.49
H5	75	***45.29***	52.6
H6	75	***40.68***	49.7
Load on CT = 5.27 MW T_{Wret} = 36.7° C			

13 AIChE

Observations

- Streams go to storage at lower temperatures
- "Over cooled"
- Load on cooling tower higher than necessary

13 AIChE

Options for revamp of coolers

1. Reduce cooling water flow
   - Throttling
   - Bypass on cold side
2. Bypass on hot side
3. Reduce area – plug tubes

13 AIChE

Effects of bypass, plugged fraction and flow reduction

- Hot bypass fraction:

  $h_h \downarrow$

- Flow reduction (Cold bypass fraction):

  $F\downarrow \quad h_c \downarrow$

- Plugged fraction area (plugged tubes):

  $A \downarrow \; Q \downarrow$

13 AIChE

Restore Cold Outlet Temperatures

One option

Unit	Bypass, cold	Plug. Tubes	Bypass, hot	Tout
H1	0	0.4768	0	50
H2	0	0.5299	0	50
H3	0	0.6	0	49.97
H4	0	0.2943	0	50
H5	0	0.2611	0	49.99
H6	0	0.423	0	49.99
Load on CT = 3.81MW T_{Wret} = 35.5° C				

13 AIChE

Summary

- Current operation 9.58 MW
- After improved Heat recovery 5.27 MW

  $\Delta Q = 4.31$ MW

- After revamp of coolers 3.81 MW

$\Delta Q_{saved} = 5.77$ MW **75% through heat recovery additional 25% through cooler modification**

13 AIChE

Identifying ALL Options

13 AIChE
Spring Meeting, San Antonio, TX

Procedure Producing Matrix – Each Unit

Cold Bypass

Hot Bypass

	0	0.02051952	0.041039	0.061559	0.082078	0.1026	0.12312	0.164156142	0.16415614	0.1846757
0	0.476744938	0.47454218	0.4722194	0.469767	0.467173	0.46443	0.46151	0.458415376	0.45511867	0.4516022
0.04737	0.457881777	0.45558658	0.4531654	0.450608	0.447902	0.44504	0.44199	0.438757875	0.43531245	0.4316355
0.09474	0.435785526	0.43337403	0.4308291	0.42814	0.425293	0.42228	0.41907	0.415663636	0.41203165	0.4081534
0.14211	0.409378219	0.40681567	0.4041101	0.401249	0.39822	0.39501	0.39159	0.387961156	0.38408721	0.3799479
0.18947	0.376985097	0.37421817	0.3712952	0.368203	0.364927	0.36145	0.35775	0.353815384	0.34961382	0.3451209
0.23684	0.335816061	0.3327568	0.3295228	0.326099	0.322469	0.31861	0.31451	0.310135588	0.30546364	0.3004627
0.28421	0.280744013	0.27723108	0.2735139	0.269575	0.265393	0.26095	0.25621	0.251153062	0.24574479	0.2399472
0.37895	0.200827021	0.19650992	0.1919345	0.187077	0.181912	0.17641	0.17053	0.164249423	0.15751146	0.1502701
0.37895	0.065656457	0.05948539	0.0529206	0.045924	0.038451	0.03045	0.02187	0.012644932	0.00269392	-0.0080679
0.42632	-0.291636584	-0.3079721	-0.325769	-0.345244	-0.366662	-0.39035	-0.4167	-0.44628658	-0.4797441	-0.5180001

Assumes correct inlet condition
Evaluates exchanger behaviour

Fraction of plugged tubes

13 AIChE
Spring Meeting, San Antonio, TX

Adding a Cooler

13 AIChE

Case Study 2

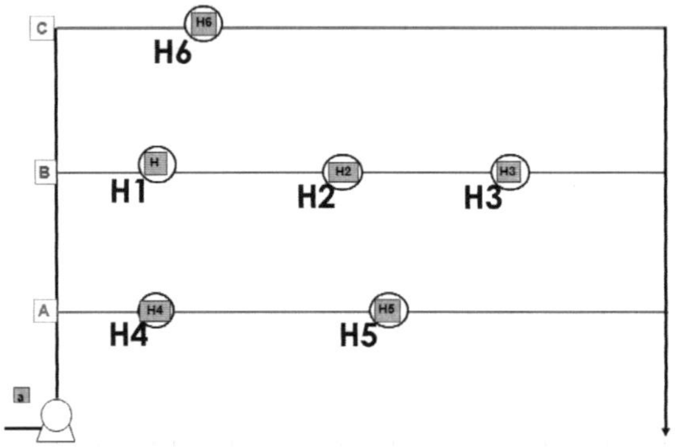

New Cooler to be added. Geographical location close to branch A.

13 AIChE

Options

- New parallel branch
- New branch plus pump revamp/replacement
- In series – after exchanger H5
- In parallel with exchanger H5

Parameters for New Unit

- Thermal behaviour: Load of New Unit
- Hydraulic behaviour: K value

  (variable to be explored)
- Fouling behaviour: model used for new systems

Design of New Unit

- Simulation gives flow and temperature field
- Procedure described for new design applies

13 AIChE
Spring Meeting, San Antonio, TX

Option 1: New Parallel Branch

Unit	Change in Load MW	Change in Hot Outlet Temp.
1	-0.105	+2.3
2	-0.081	+3.2
3	-0.203	+3.8
4	-0.047	+1.2
5	-0.105	+2.6
6	-0.044	+2.1

13 AIChE
Spring Meeting, San Antonio, TX

Change in Fouling

Unit	Original		Revised	
	Inlet	Outlet	Inlet	Outlet
1	0	0.0008	0	0.00145
2	0.0002	0.0023	0.0014	0.00307
3	0.0013	0.0034	0.001	0.0035
4	0	0	0	0
5	0	0.0009	0	0.00154
6	0	0	0	0

13 AIChE Spring Meeting, San Antonio, TX

Option 2: Parallel Branch + Pump Modified

Unit	Change in Load MW	Change in Hot Outlet Temp.
1	+ 0.04	- 0.9
2	+ 0.014	- 0.6
3	+ 0.013	- 0.3
4	+ 0.038	- 0.9
5	+ 0.025	- 0.65
6	+ 0.04	- 0.55

13 AIChE Spring Meeting, San Antonio, TX

Change in Fouling

Unit	Original		Revised	
	Inlet	Outlet	Inlet	Outlet
1	0	0.0008	0	0.0006
2	0.0002	0.0023	0	0.002
3	0.0013	0.0034	0.0012	0.0033
4	0	0	0	0
5	0	0.0009	0	0.0007
6	0	0	0	0

13 AIChE Spring Meeting, San Antonio, TX

Option 3: New Unit in Series with H5

Unit	Change in Load MW	Change in Hot Outlet Temp.
1	-0.006	+0.1
2	+0.003	-0.2
3	-0.048	-0.2
4	-0.045	+1.2
5	-0.067	+1.7
6	-0.013	+0.3

13 AIChE Spring Meeting, San Antonio, TX

Option 4: New Unit in Parallel with H5

Unit	Change in Load MW	Change in Hot Outlet Temp.
1	-0.03	+0.6
2	-0.018	+0.7
3	-0.008	+0.7
4	+0.007	-0.2
5	-0.092	+2.3
6	-0.02	+0.36

13 AIChE Spring Meeting, San Antonio, TX

Conclusion

BETTER OPTIONS ARE:

NEW PARALLEL BRANCH PLUS PUMP MODIFICATION

NEW UNIT PLACED IN BRANCH A IN SERIES WITH H5

13 AIChE Spring Meeting, San Antonio, TX

Overall Conclusions

- Thermo-hydraulic Simulation of Cooling Water Networks is not difficult
- Simulation includes consideration of tower behaviour and pump characteristics
- Fouling models can easily be incorporated into the analysis
- Design principles for new systems have been established

13 AIChE

Overall Conclusions

- In revamping the effect that improved heat recovery has on the performance of the existing system has to be evaluated if true reduction in tower load is to be achieved
- Existing exchangers can be modified in order to achieve reduction in tower load – full range of possible options can be identified
- Care must be exercised when installing new coolers – options easily evaluated
- All coolers need to be designed in "systems context" – procedures established

13 AIChE

(61b) Water / Wastewater Optimization Methods for Industrial Systems

Joseph W Guida, Process Technology, Fluor, Sugar Land, TX

Efficient management of water and wastewater systems in large industrial complexes is important for minimizing costs and optimizing natural resources. This paper provides an overview of methods used to analyze water use efficiency such as water pinch analysis and water cascade analysis as well as other visual techniques. Discussion will also include techniques for water recovery and recycle such as zero liquid discharge and an explanation of when recycling makes sense and when it doesn't.

(61c) Balancing Act - Simple Techniques for Steam Header Controls

Barbara Hamilton, Process Systems and Solutions, Emerson Process Management, Hampden , ME

An industrial steam distribution system with multiple headers, turbines, boilers, and reducing valves is managed with header control strategies that mitigate upsets, increase power generation, and reduce clean condensate makeup. Effective use of override, anti-windup, and TG governor interface strategies insure header stability. Upsets are managed without cascading into multiple equipment failure. An overall coordination scheme is built from optimum use of pressure reducing valves that is easily grasped by operating personnel.

Coordinating interconnected steam headers is impossible without automatic control. A major upset on one header can cascade throughout the system causing multiple equipment failures and eventual plant outage. A control strategy that coordinates turbines, boilers, and reducing valves can balance a multiple header system during normal operation and prevent a catastrophic failure during a major upset.

Coordinated control of the pressure reducing valves (PRVs) is the most effective too for stable header control. Proper configuration of override controls and limiters keep the valves closed until needed. Interfaces with the Woodward TG governor systems coordinate the PRVs with TG extraction to allow PRVs to take over when the TGs are unavailable.

13AIChE
Spring Meeting, San Antonio, TX

Thermal Fluids: the Key to Overall Site Integration

E. K. Tamakloe, G. T. Polley & M. Picón Núñez

University of Guanajuato,

Mexico

Topics

- Key Message
- Advantages of Thermal Fluids
- Integration between adjacent plant sections – example an FCC Plant
- Total Site Analysis
- Linkage with Fired Heaters

13AIChE
Spring Meeting, San Antonio, TX

Key Message

Large number of factors MUST be considered simultaneously when undertaking systems design:

- Energy efficiency
- Heat exchanger behaviour
- Capital costs
- Control
- Flexibility

13 AIChE

Advantages of Thermal Fluids

- Major capital cost advantage – piping costs – dominated by valves and fittings – vary with diameter raised to a power of around 3.
- Pipes carrying steam very much larger than those transporting liquid
- Steam temperature set by mains pressure – becomes superheated when pressure is let down
- Heat can be extracted from thermal fluids over significant temperature range

13 AIChE

Integration between Adjacent Plant Sections – Example
Revamp of an FCC Plant

13 AIChE
Spring Meeting, San Antonio, TX

Process Flow Diagram

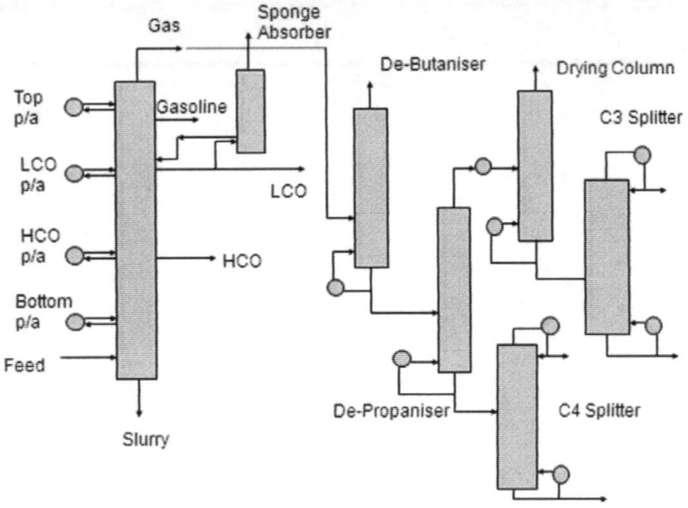

13 AIChE
Spring Meeting, San Antonio, TX

Geographical arrangement

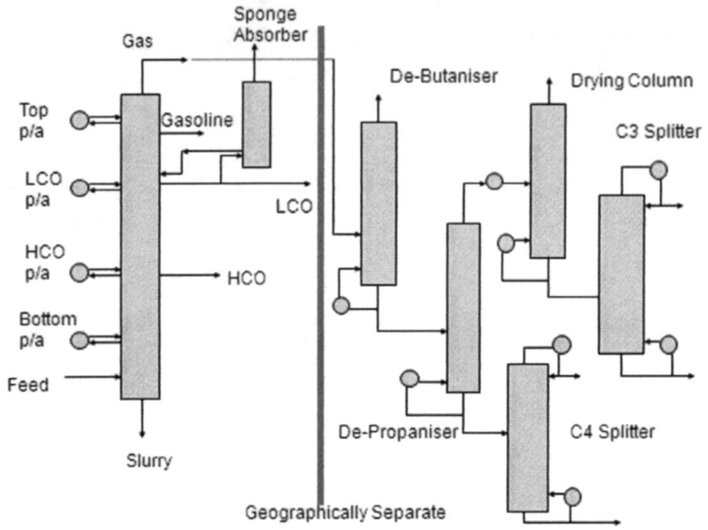

Energy Analysis

- Full Problem Energy 36 MW

  T.A.C. 6.04 M$/annum

- Column Energy 3.2 MW

  T.A.C. 2.5 M$/annum

- Gas Processing Energy 74.1 MW

  T.A.C. 7.5 M$/annum

Energy Efficiency Requires Transfer of Heat from Column Region to Gas Processing Region

Problem: Re-boilers in Gas Processing Region operate at low temperature

Column Zone

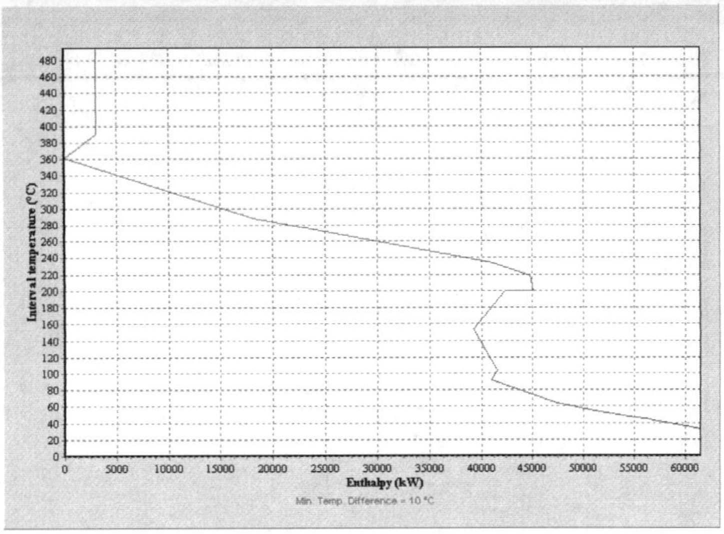

Gas Processing zone

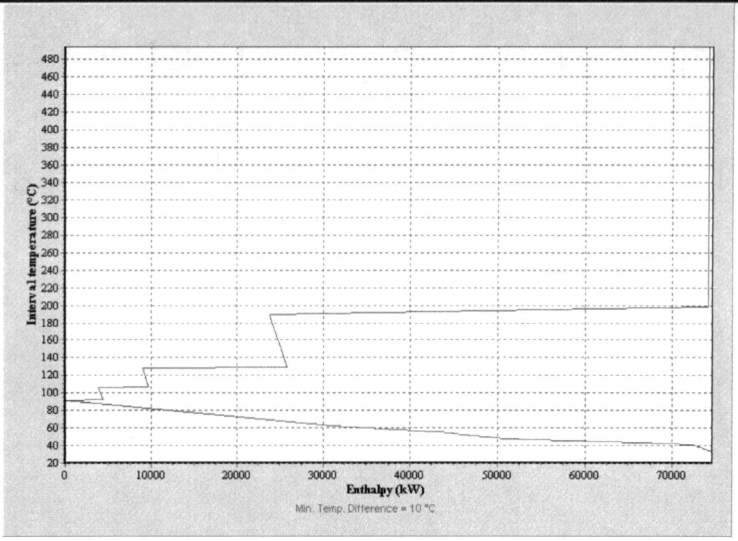

13 AIChE
Spring Meeting, San Antonio, TX

Column Zone

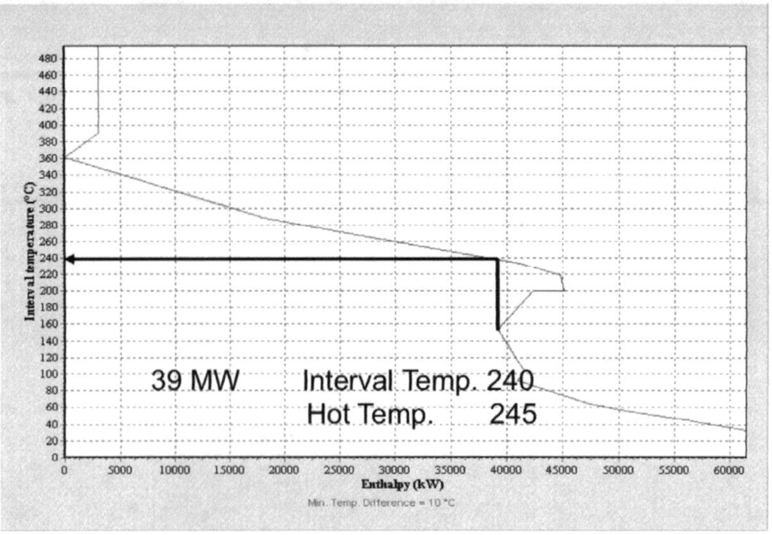

13 AIChE
Spring Meeting, San Antonio, TX

Gas Processing Zone: Option 1

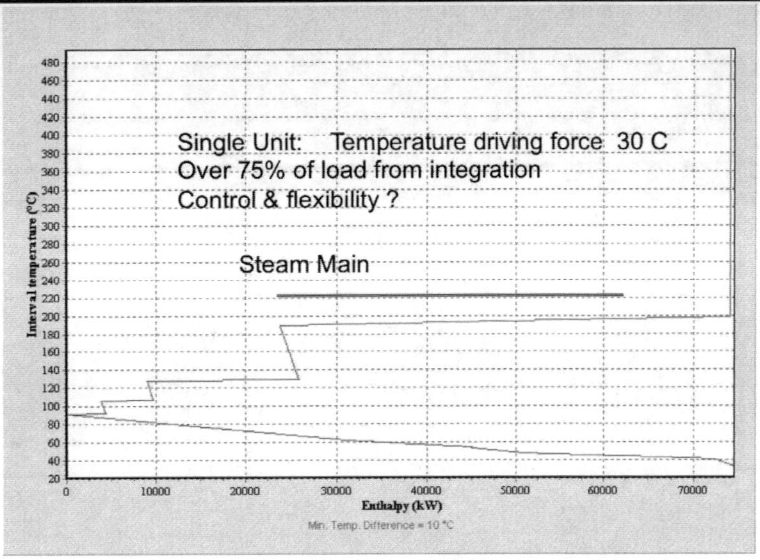

13 AIChE

Gas Processing Zone: Option 2

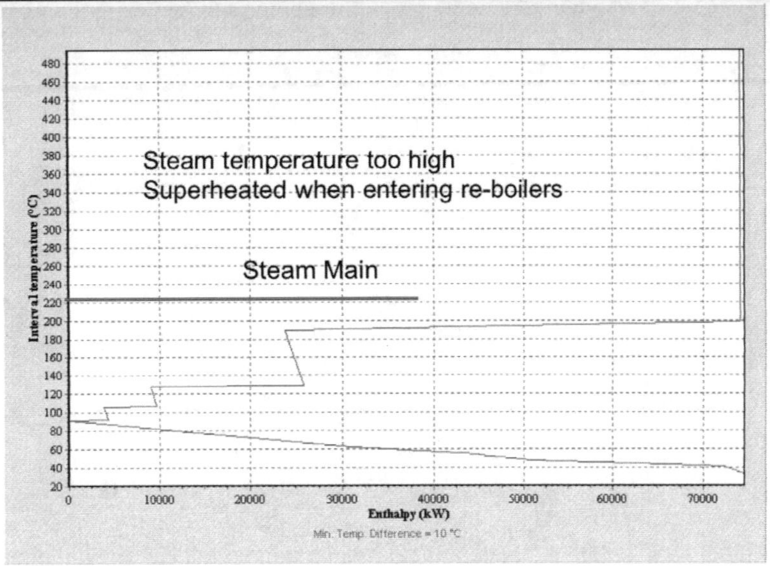

13 AIChE

Column Zone: Used to Heat Thermal Fluid

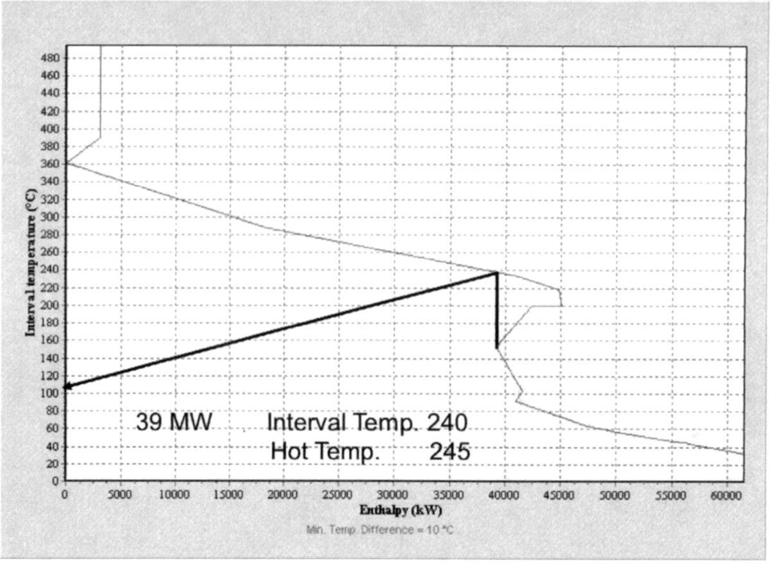

13AIChE

Liquid Re-circulation: Determined from Other Zone

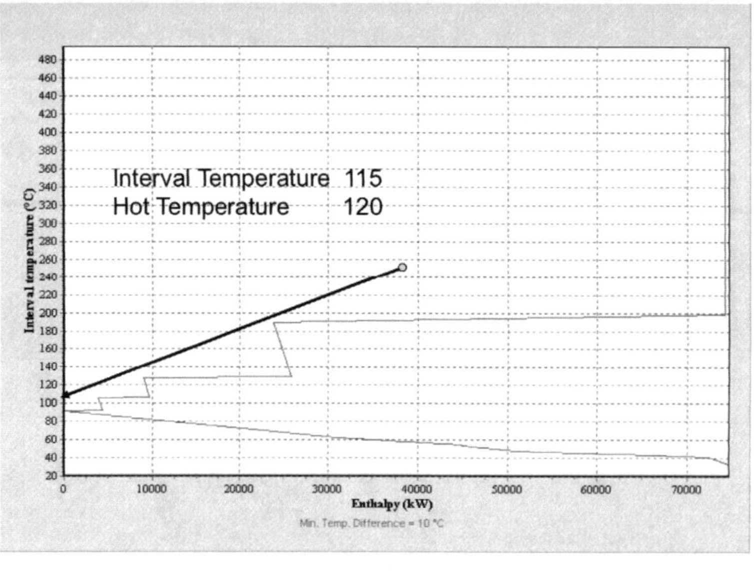

13AIChE

Observations

- Hot oil used to drive 4 individual re-boilers
- Ring-main can be employed
- Piping costs relatively low.
- Temperature at which oil is used consistent with re-boiler duty
- Control very easy – rate at which oil is extracted from ring main
- Supplementary oil heater provides easy start up and good flexibility

13 AIChE

Total Site Analysis

13 AIChE

Overall Production Site

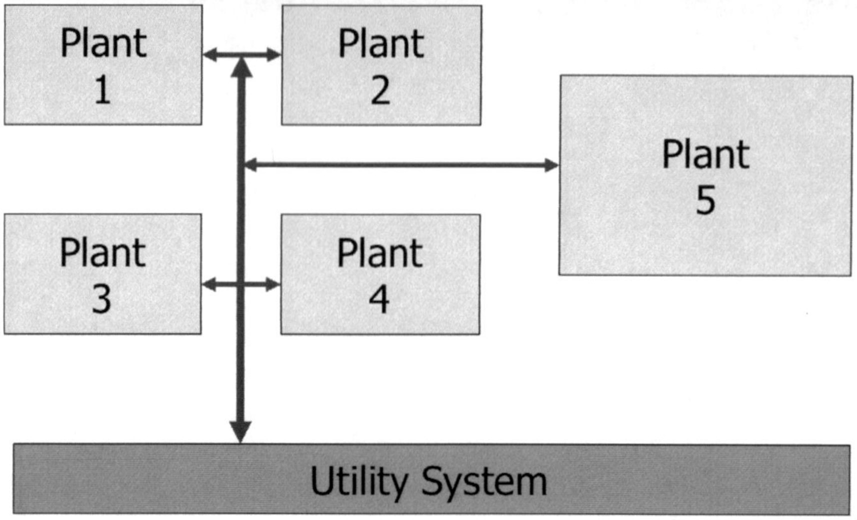

13 AIChE Spring Meeting, San Antonio, TX

Individual GCC's

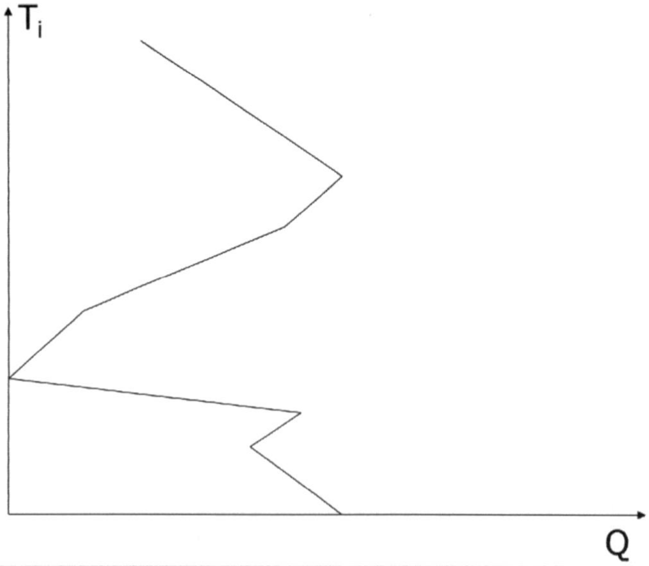

13 AIChE Spring Meeting, San Antonio, TX

Demands

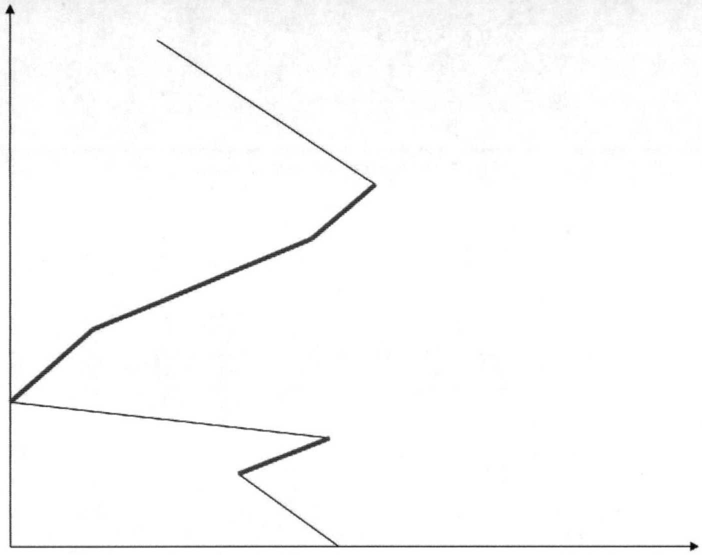

Supplies

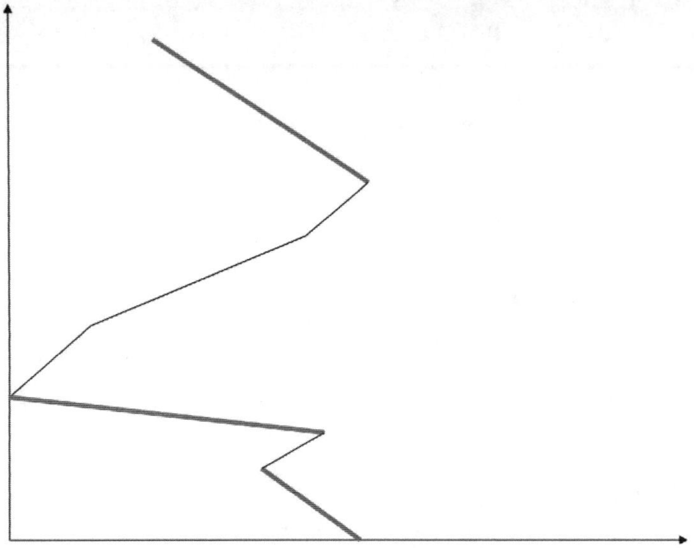

Site Composites

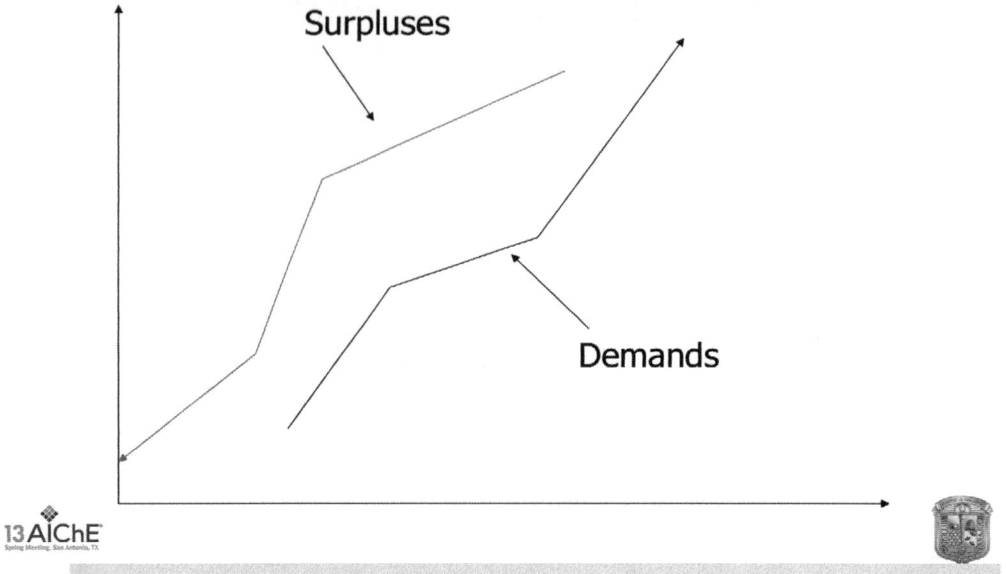

Site Composites

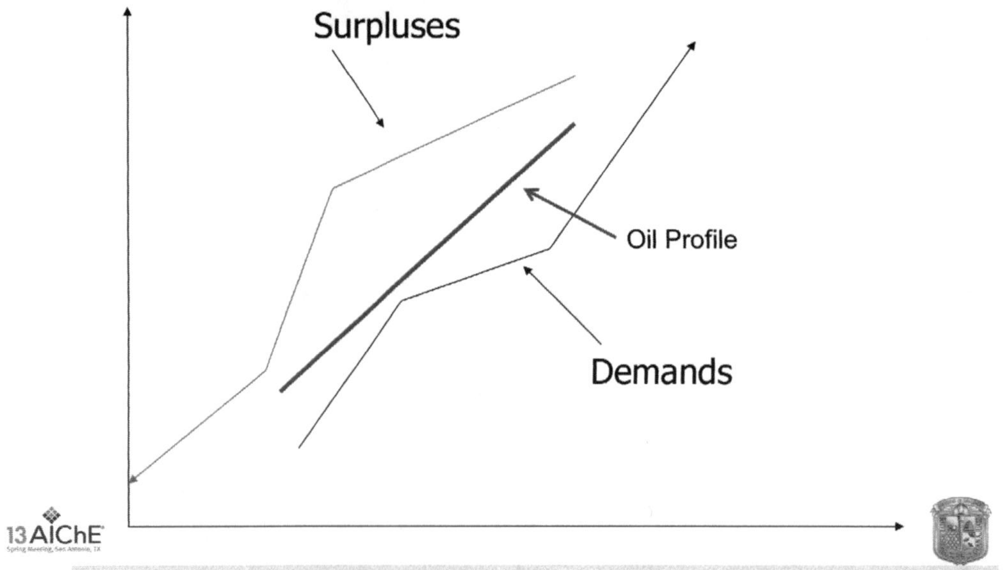

Linkage with Fired Heaters

13 AIChE
Spring Meeting, San Antonio, TX

Typical Furnace

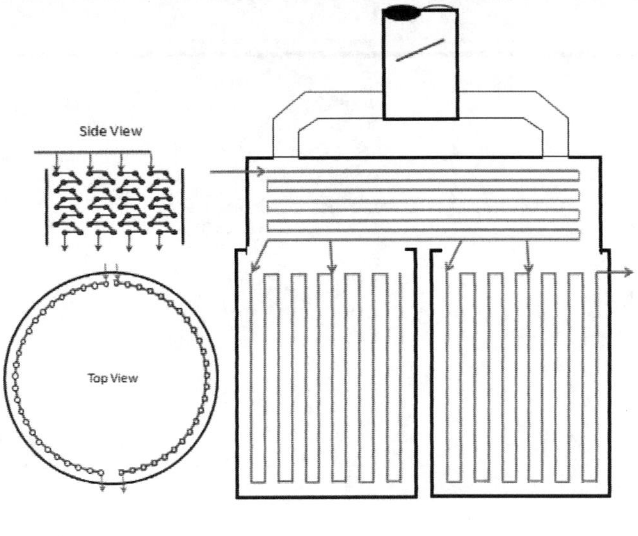

13 AIChE
Spring Meeting, San Antonio, TX

Reducing Fouling Rates in Fired Heaters

- Fouling rates in convective section of a fired heater are affected by the flow conditions on the crude side
- If crude enters as a single phase liquid fouling rates are high

13 AIChE

Improved Efficiency

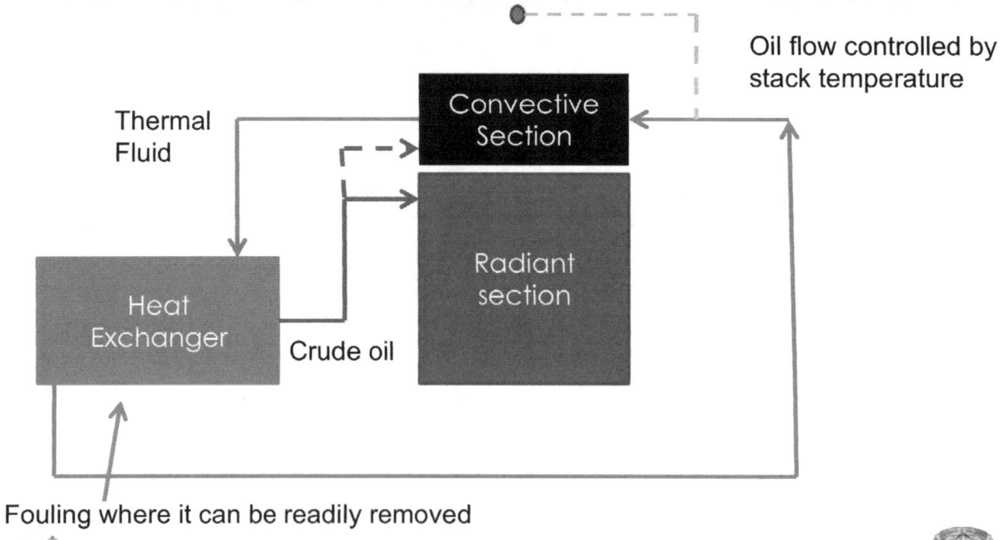

13 AIChE

Overview of P.E. Licensing Process in United States

William R. Parrish, P.E. and Cory D. Jensen, P.E.

Professional engineering licensure was established in the United States to protect the health, safety and welfare of the public from unscrupulous or incompetent decisions of engineers. Licensure provides a proven and accepted assessment of experience, education and ethical standards.

This presentation gives an overview of the benefits of being licensed, the key players involved in becoming licensed and the path to licensure. The overview presents the background information to help put the other session presentations in perspective. More details about licensure are available in the March issue of *Chemical Engineering Progress (CEP).*

Value of being licensed

Being licensed shows other that you have a strong commitment to high technical and ethical standards. It can provide you with better career opportunities and provide a competitive edge when seeking employment. Many employers state that they would choose a licensed engineer over an unlicensed engineer, given both are otherwise equally qualified. Also, companies often offer incentives to encourage their employees to become licensed.

Some engineers think that licensure is important only in consulting or expert testimony. However, licensure is often required in firms doing design, construction and startup of processes. State laws also may include those teaching "advanced" engineering subjects be licensed.

Some states allow an "industrial exemption" for large companies who have the financial resources to cover liabilities arising from employee decisions. Most companies encourage licensure but many oppose eliminating the industrial exemption if they have operations in multiple states because it requires their engineers to be licensed in multiple states. There is a push to eliminate the industrial exemption. This important topic is discussed in another presentation.

No one can predict the exact future course of their career. However, being licensed may provide unexpected rewards and opportunities.

Key players in licensure

Becoming a P.E. involves education, experience and examination. Besides your coworkers, who can help guide your job experience, there are two important players in becoming licensed. The first is your local engineering board that establishes the standards for education and experience and ability to take the examinations. You will work with the board throughout the licensure process.

The second important player is the National Council of Examiners for Engineering and Surveying (NCEES) (www.ncees.org) . This is a nonprofit organization whose members are the various engineering licensing boards. The NCEES was created to provide some uniformity of standards of education, experience and examination. The membership establishes Model Law to offer uniformity. However, Model Law is a set of suggested best practices and no board follows Model Law exactly. All exams are administered and graded by the NCEES.

The licensure process

Obtaining a P.E. involves the following steps:

1. Having the appropriate formal education
2. Passing the Fundamentals of Engineering (FE) exam
3. Obtaining the appropriate experience (engineering internship)
4. Providing required documentation to board
5. Passing the Chemical Professional Engineering (PE) exam

While the general requirements are uniform across all jurisdictions, exact details vary. It is critical to work with your local board so that you understand and meet the exact requirements. A brief synopsis of each step is given below. More details are provided in the *CEP* article and in other presentations in this session.

1. Formal education: All jurisdictions require BS engineering degree, or equivalent, from an EAC/ABET-accredited four-year engineering program. Foreign degrees and other degrees are considered but must be approved by the board.
   Current Model Law states that as of 2020, engineers taking the PE exam must have a Masters or equivalent (MOE). This requirement is strongly opposed by AIChE and many other organizations. More details are presented in another presentation in this session and in the *CEP* article.

2. FE Exam: This exam tests your fundamental knowledge of engineering. It is easiest to pass when you are first eligible to take it (usually senior year or shortly after graduation). Another presentation and the *CEP* article provide more details of the exam and the upcoming changes in the exam.

3. Practical experience: Your work experience must be appropriate engineering practice. Typically four years are required although some boards permit fewer. In some jurisdictions advanced degrees can reduce the number of years required.

4. Documentation: The exact format of documentation and what is required varies between jurisdictions so it critical to work with your local board on this. However, all will require proof of education, passing the FE exam and work experience. The work experience will be supported by several references, who will also vouch for your ethical behavior.

5. PE Exam: This is the final step in becoming a licensed professional engineer. The exam is designed to be passed by a "minimally competent engineer with four years of practical experience". This exam tests the practical application of your engineering knowledge. Both a session presentation and the *CEP* article provide more details on the exam.

Summary

The AIChE recommends that its members become licensed to help ensure a competent and ethical chemical engineering workforce. While the process may seem daunting at first, working with your fellow employees and your local board will make the effort much easier. Each year about 400 chemical engineers become P.E's. They have made a significant investment in themselves and possibly in their future career.

(78b) Licensure--An Academic Perspective

S. Ranil Wickramasinghe, Department of Chemical Engineering, University of Arkansas, Fayetteville, AR

Most Chemical Engineering faculty members are not licensed. Why? Is it important for faculty to be licensed? Most undergraduates take the Fundamentals of Engineering exam but do not take the Professional Engineering exam. Why? Should faculty be licensed? Is licensure important for faculty who are conducting research? Should we encourage faculty to obtain their PE license? In this presentation these questions will be discussed.

TESTING FOR COMPETENCY - THE PROFESSIONAL ENGINEERING EXAMINATION

Anne Bertelsmann, PhD, PE, Galveston Bay Refinery, Marathon Petroleum Company LP, Texas City, TX
And
Denise Chastain-Knight, PE, CFSE, CH2M HILL, Augusta, GA

Introduction

The decision to pursue engineering licensure is a personal one, and the road includes many hurdles to overcome. Engineers take this road for a variety of reasons, with different motivations and experiences along the way, and finish the trek with the PE Exam. The Principals and Practice of Engineering Examination (PE Exam) is a nationwide exam used to ensure that licensed PE's are "minimally competent". The exam must include test content applicable to the competency being measured, effectively differentiate between competent and not competent candidates, and provide a consistent competency measure from administration to administration. This paper will define minimal competency, discuss the purpose of the Professional Engineering Exam, exam format using the Chemical Engineering Exam as the example, development methodology, item quality metrics and offer suggestions for test taking preparation. It will also describe the function of the NCEES Exam Development Committee and their role in developing a fair, effective exam, and explain why it takes weeks to get a result for a computer scored exam.

Path to Licensure

Whether an engineer undertakes the path to licensure early or late in their career, each must successfully negotiate the same hurdles. Each step in the process has a specific goal to either develop or to measure the engineer's capabilities toward professional competency. Figure 1 on the next page illustrates the path to licensure.

The quest to licensure begins with obtaining a Chemical Engineering degree. The next step is to pass an exam that measures competency on engineering basics. This first exam is the Fundamentals of Engineering Exam (FE) – also sometimes called the EIT (Engineer in Training) Exam. The FE exam is typically taken either during the senior year in college or shortly after graduation. The types of items resemble "textbook" questions and are designed to test understanding of the engineering fundamentals as taught in a typical engineering degree program. Currently, the FE examination consists of a general morning session (common to all disciplines) and a choice of seven discipline specific afternoon modules. Starting in the beginning of 2014, the FE examination will convert to a computer based test, which is discipline specific (i.e. no general module). For more information on the FE exam content and the conversion to computer based testing, refer to the NCEES website (http://ncees.org).

Figure 1: Path to PE Licensure

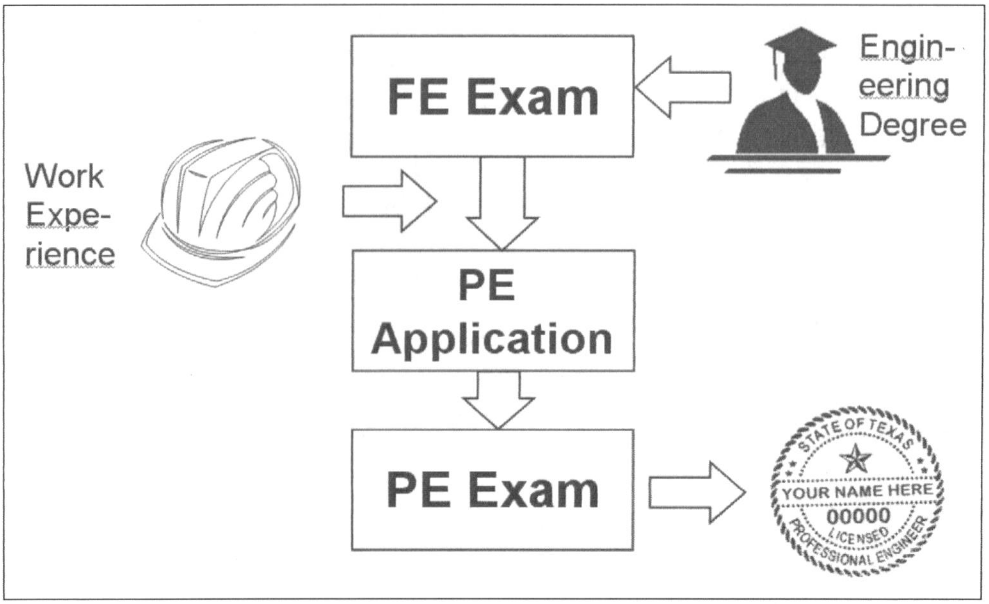

Following graduation, engineers enter the practice phase of their careers where fundamental knowledge is applied to real world application. In this phase of their careers, engineers use the theoretical principals learned from their education to solve problems that ultimately deliver physical products to market. Most states require a minimum of 4 years engineering related work experience before allowing candidates to sit for the PE exam. In order to become licensed, the candidate needs to submit an application to the state licensing board, describing education and experience, documenting passing of FE exam, and providing character and skills performance references. Format of the application and requirements for references, transcripts etc. vary from state to state. Once the application is accepted by the board, the candidate can register for the PE exam. After passing the PE examination, the engineer is licensed and can obtain the PE stamp. It is through this process that the state licensing boards ensure that only qualified candidates become licensed.

PE Exam Purpose

The PE exam is the final step in the road to licensure and its purpose is to ensure that only competent engineers become licensed. This helps to uphold the high standards of licensure and protects the health and welfare of the public. Since the exam is the same nationwide, it ensures consistency between states. Thus, one of the main requirements for a PE examination is that it can distinguish between competence and incompetence and that it can demonstrate this reliably and consistently. It is important to remember that "competent" should not be equated with "expert". The purpose of licensure is to identify individuals who have at least the minimum knowledge at which a person can practice professional engineering in such a manner that will safeguard life, health and property and promote the public welfare. That is why the term "minimally competent" is used.

A minimally competent engineer:

- should be able to perform identified tasks
- should meet minimum education and experience requirements
- is beginning to specialize in a sub-discipline
- may safely practice in another sub-discipline even though he/she does not practice in this area on a daily basis

The definition of "minimally competent" has implications on how the questions on the exam are written. The PE exam differs from the FE exam in that it addresses both the principles and the practice of engineering. So, while the FE questions are typically textbook related, questions on the PE exam are predominantly practice related. The questions build upon the experience that an engineer has gained during his or her first few years of working and need to recognize that chemical engineering experience can be gained in a variety of areas. The difficulty level of the questions is geared towards an engineer with about four years of experience.

PE Exam Content

The content of the examination is determined through a survey of practicing chemical engineers (PAKS study = Professional Activities, Knowledge and Skills), which is repeated about every 6-8 years to ensure that the examination stays current. The PAKS process includes conducting a technical content survey with input from practicing professionals. This survey is conducted over a period of weeks to ensure that the respondent pool size is adequate and is sufficiently diverse to be representative of the overall profession. Survey results are analyzed by a team of licensed Chemical Engineers, most with no prior exposure to PE exam development work, under the direction of a psychometrician who ensures that proper procedures are followed. The PAKS team will determine both the domain (subject area) content and item weight of the subdomains. PE exam content is established based on the PAKS study, and is published in the format of the PE exam specification available from the NCEES website (http://ncees.org/exams/pe-exam/) and is summarized in Table 1 on the next page.

The ChE exam specification was recently updated and became effective for the April 2011 exam administration. It includes six major domains and fifteen sub-domains that investigate theoretical and practical knowledge. Exam content is distributed within the major and subdomains according to the percentages shown in Table 1 to ensure a consistent content basis for every exam. The specification also includes guidelines for topical content within each sub-domain. For example the exam will include 4% of the items from sub-domain VI. C. Operation. Figure 2 includes example content from the 2011 exam specification.

Table 1: April 2011 ChE PE Exam Specification Summary

Knowledge Area	Approximate Percentage of Examination
I. Mass/Energy Balances and Thermodynamics	**23%**
A. Mass Balances	10%
B. Energy Balances and Thermodynamics	13%
II. Heat Transfer	**16%**
A. Mechanisms	9%
B. Applications	7%
III. Kinetics	**11%**
A. Reaction Parameters	6%
B. Reactors	5%
IV. Fluids	**16%**
A. Mechanical-Energy Balance	12%
B. Flow and Pressure Measurement Techniques	4%
V. Mass Transfer	**14%**
A. Phase Equilibria	5%
B. Continuous Vapor-Liquid Contactors	8%
C. Miscellaneous Mass Transfer Processes	1%
VI. Plant Design and Operation	**20%**
A. Economic Considerations	1%
B. Design	10%
C. Operation	4%
D. Safety, Health, and Environment	5%

Figure 2: Sample Exam Specification Sub-domain Content

C. Operation 4%

1. Process and equipment reliability (e.g., testing, preventive maintenance, startup/shutdown procedures, robustness)
2. Process improvement and troubleshooting (e.g., debottlenecking, experimental design and evaluation, optimization)

PE Exam Format:

The PE examination is given twice a year - once in April and once in October. It is an eight hour exam consisting of a four hour morning session and a four hour afternoon session with a lunch break in between. There are a total of 80 multiple choice questions (40 in the morning, 40 in the afternoon). Each question has four answer choices of which one and only one is the correct answer. Each question is stand-alone, i.e. it is independent of the other questions on the exam and can be solved with the information given in the problem, sometimes with the help of standard references. Some questions involve calculations while others are qualitative. All questions count equally and there is no penalty for guessing.

Each of the 80 multiple choice questions on the exam is of the same general format. In exam development, questions are referred to as "items" rather than questions because not every item has a question mark (like the one in the example below). Figure 3 illustrates the anatomy of an item, i.e. its general format and identifies the different parts of the item.

Figure 3: Anatomy of an Item

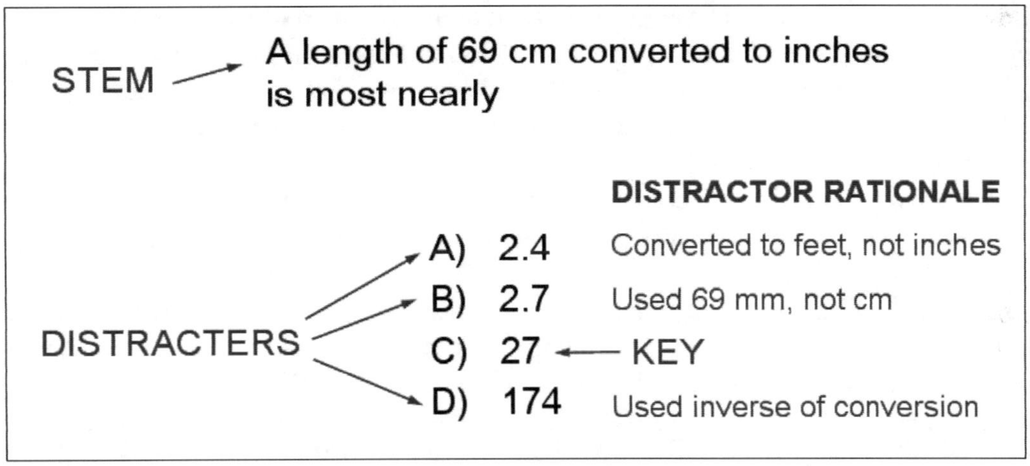

Each item consists of a stem and four answer choices. The stem (or problem statement) identifies what calculations are needed. It includes the information that is provided to solve the problem (sometimes even in table format) and states what is required for an answer. Beneath the stem are the four answer choices. Of these four answer choices, one (and only one) is the correct answer – the key. The other three are so-called distracters. They are incorrect answers – but they are not random numbers.

Each of the wrong choices represents a typical error that a less than minimally competent engineer (or someone not familiar with this specific topic) may make. Each distracter has to have a "rationale", i.e. an explanation which error was made. So, when a candidate is taking the exam and calculating a number that is exactly one of the given answer choices it does not necessarily mean that the problem was solved correctly. Maybe the candidate just made one of the "expected likely errors"…

The distracters can have a significant impact on the statistics of an item. If one of the distracters is obviously wrong (such as a negative absolute pressure), then even less than competent candidates will disregard it. That reduces the feasible choices, improves the "guessing" performance, and increases the chance that a not minimally competent candidate gets the question right. The Exam Development Committee strives to develop items that include the key, and three equally likely errors as distracters.

Exam Development

The PE Examination (as well as the FE, and the Surveying Exams) is a service that NCEES provides to the licensing boards. As such, NCEES provides the framework and the oversight for the exams, but the responsibility for the content rests with each exam committee. The committee writes the items, assembles the exams and reviews the statistics. NCEES has established rules and procedures that the exam committees follow to ensure consistency. Committees report back to NCEES with their exam quality measures. NCEES also provides the logistical support (psychometric consultant, publishing and printing, administrative support, maintaining records and the item bank, administering the exam, scoring, etc) and provides all security functions. For the larger exams, NCEES also pays and arranges for volunteer travel. For the smaller exams, this is the responsibility of the sponsoring society.

The Chemical Exam Committee is the group that actually writes and reviews the Chemical PE Exam. To be on the committee, one must be a licensed PE and express interest through the NCEES website or through a committee member. The committee has about 50 active members and meets twice a year in Clemson, South Carolina at the NCEES headquarters. Not everyone comes to every meeting, of course, so typically 30 – 35 committee members present at any given meeting. During the meeting items are written and reviewed by committee members who work with one another to critique and improve the items. Task teams also work to assemble, review and pretest exams.

Before an exam is administered, each item will have no less than six documented formal reviews and likely have been assessed by more than 20 people. Committee members are asked to complete detail oriented and often challenging engineering work and can earn PDH (professional development hours) for this activity. Developing the PE exam helps committee members to keep their engineering skills current and to give back to the profession. Without the work of this committee, the PE exam and with it the licensing wouldn't happen.

Preparing an exam that effectively and consistently measures minimal competency is a challenging task that succeeds or fails based on the ability of the committee to predict item performance. Each committee relies on a structured process and employees a variety of quality tools and techniques to improve the opportunity for success. Figure 4 is a workflow of the exam development activities. Items are written by a PE and submitted for review. Many items are prepared by the committee members at each meeting, while others are submitted by PEs who do not attend the meetings. The individuals who write and review items are referred to as Subject Matter Experts (SME). The author and two reviewers will independently work the item and develop the rationale for each distracter. Through this process the item is often revised and edited to improve its presentation and quality. This step is considered one of the most critical activities of the committee because items are the building blocks of the exam and each must meet high quality standards. Each committee member is responsible for item quality and is granted authority to modify draft items as needed to meet quality objectives, or reject items that fail criteria and cannot be edited to usable items. Finally the item is given a final check by the committee chair (or authorized designee) to verify that documentation is complete, both reviewers agree to the item content, and all quality checks have been completed, and approves the item for acceptance into the item bank. The item bank is a repository for items that are ready to be placed on exams.

Figure 4: Exam Development Process

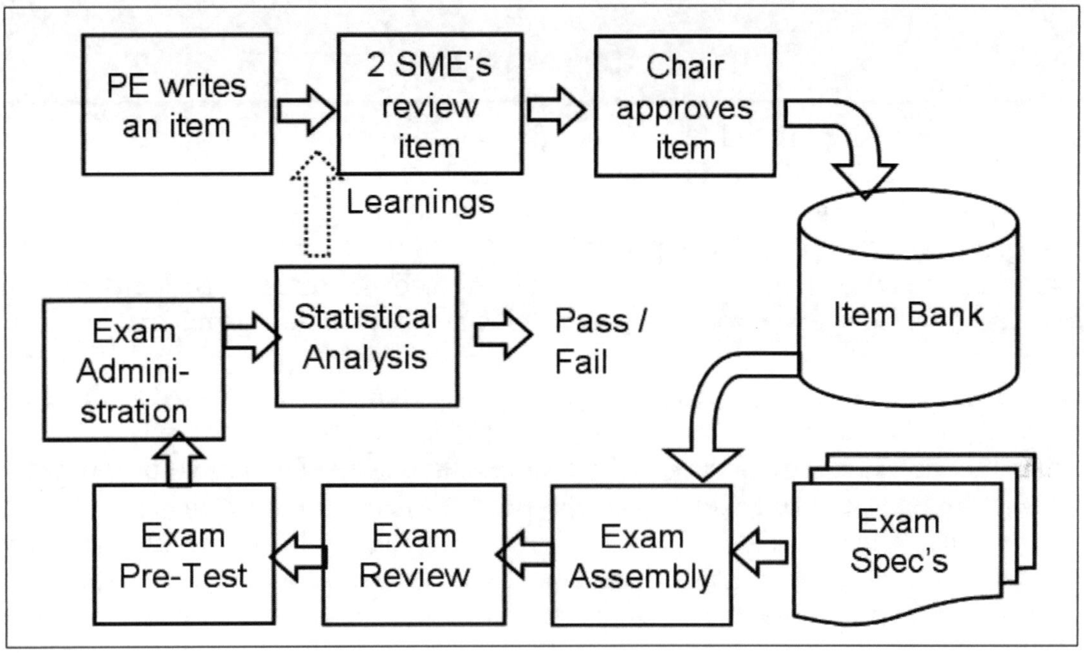

Exams are built, reviewed, revised, reviewed, quality tested and reviewed over a series of meetings. Each exam will have three independent team reviews, a pretest and a quality assessment. There are specific checks required at each workflow gate with task teams assigned to maximize the number of eyes that evaluate any given exam. An exam is compiled using items from the bank following the content distribution of the specification and sequenced in the specification order. The build team will then conduct a complete review of the exam, looking at each exam item for individual quality and checking for overall content flow and replacing items if necessary. The exam is then reviewed in two more meetings by independent teams who provide a "fresh set of eyes" and continue to improve the content to ensure that each item on the exam and the overall exam meet quality criteria. They also verify that the documentation is complete in accordance with the NCEES exam development requirements. Each exam must also have a pre-test where a group of committee members take the exam in as if they were actual test takers. After working all the items on the exam, they answer a series of questions about the items designed to predict item performance, then perform a distracter evaluation. Comments from the pre-testers are incorporated into the exam. In the final meeting before administration, a quality team conducts one last review of the exam verifying that all preceding checks (and associated documentation) are complete, all comments have been addressed, specification distribution is correct, and that author diversity limits and other NCEES criteria are met.

When the exam is administered and scored, data is compiled that can be used to evaluate the quality of the exam as a whole, and individual items. The data is analyzed and used to help the committee to learn what makes items perform effectively, and improve our performance prediction skills. Authors are provided a 'report card' so they know how items they wrote performed and have the opportunity to see how those items evolved in the review process. At each meeting, the committee will review items that had great statistics and try to find out what made them perform well. Similarly, we will review items that didn't perform well and identify traits that are undesirable in future items.

Item Statistics

Item quality is ultimately measured by the statistics generated from the candidate responses after an exam administration. Before discussing the statistical analysis of the examinee responses, one must first consider what makes an item a "good" item. The definition of a good item is one that is capable of distinguishing between "minimally competent" and not "minimally competent" candidates, does not discriminate against any particular group of test takers, and performs consistently in multiple exam administrations. To achieve those goals, it must be relevant, not confusing, and measure competency.

Most importantly, the content of the item must be relevant. Relevance is defined as the appropriateness of the topic and the difficulty of the item. The topic of the item must test knowledge or skills that are directly related to the exam specifications. For example: a question on high voltage power distribution may be relevant for an electrical engineer, but will not provide information on engineering competency of a *chemical* engineering candidate – no matter how well written it is. If the item does not relate to the exam specifications, it does not belong on the exam. Second, the difficulty level and complexity of the item must be such that a "minimally competent" engineer with 4 years experience can solve it in about 6 minutes. When writing items, authors often think about the interesting and challenging problems that are encountered in practice; however, if an engineer with 20 years experience requires several hours or days to solve the problem, what are the chances that a junior engineer will get the answer in 6 minutes? These type of problems, as interesting as they may be, do not make good exam items.

Another important principle to consider is that items should not be confusing to people who know the answer. When an item is confusing, a competent engineer who understands the concept being tested may not select the key, thus the question does not help to distinguish between competent and not competent. The so called "trick" question does not evaluate a candidate's competence and is considered a "bad" item to be weeded out in the exam development process.

It is equally undesirable to use items that give the answer away to people who don't know the answer. If the item includes clues to the key, then an incompetent engineer can choose the right answer without understanding the subject matter; thus the item does not measure competency. Clues can take many different forms:

- If some of the answer choices can be ruled out as impossible (like negative absolute pressures), the chance of guessing the right answer goes up.
- Grammar can give clues when one of the answer choices "stands out", i.e. is more detailed or does / does not match the grammatical structure of the item statement

For example, an item that asks about the 19th president of the United States, where three of the four answer choices are well known pop stars, does not require the candidate to know anything about US history to pick the correct answer.

Following an exam administration, a psychometric analysis is performed (psycho – metric = measuring the mind) on examinee responses. The analysis is solely based on information provided on the marked answer sheets without regard to the actual content of each item, but quite a bit of information about an item can be observed just from examinee responses. The psychometric analysis will evaluate a number of things. Of course, the simplest analysis is to determine how many candidates selected the key. Statistically, there is a 25% chance of guessing the correct answer because there are four choices. An item where less than 25% of the candidates selected the key may be flawed because it may have directed candidates towards preferentially selecting a distracter over the key. The next question is to ask *who* got the question right. Ideally, the candidates who got the question right were the same people who performed well on the exam overall. The analysis looks at the distribution of answers between the distracters and who selected them. This analysis also considers if all distracters were selected equally or if there was one distracter that was particularly attractive; maybe it even got selected more frequently than the key.

The objective of this analysis is to confirm the validity of each item as a tool capable of measuring engineering competence. An item that follows the basic three criteria (relevance, no confusion and no clues) can be expected to distinguish between competence and incompetence; however, sometimes clues, confusion or complexity can creep into an item unintentionally and be missed during the review before the exam is given. Statistics are then used to identify potential flaws. One such flaw is a mis-key. Mis-key is when the committee identified for example option A as the correct right answer, but option B is actually correct. Another flaw is an ambiguous item where more than one of the options could be correct. Despite the committee's efforts, such flawed items occasionally make it through the review process and are placed on an exam. The rigor of review process prior to exam administration minimizes the likelihood of a flawed item; however, it cannot totally eliminate them. The statistics after the exam will reflect item imperfections and provide an opportunity to compensate for item flaws in exam scoring so that candidates are not negatively impacted by flawed items.

Figure 5 is an example of the statistical analysis report for an item prepared by the psychometric consultant. One such sheet is prepared for each of the 80 items on an exam. The top left corner shows the "number of cases", i.e. the number of examinees who took the exam in this administration. In this example, there were 2910 candidates. So, this example is not from any Chemical PE exam where typically about 300 candidates sit per exam. The number below the case number (highlighted in yellow) is the Mean (or the P+ value). This number shows what percentage of the candidates selected the key; in this case, 65%. P+ alone indicates that the question is of appropriate difficulty for the average test taker; not too easy and not too hard. Statistically, the higher the P+, the easier the item.

Figure 5: Item Statistics

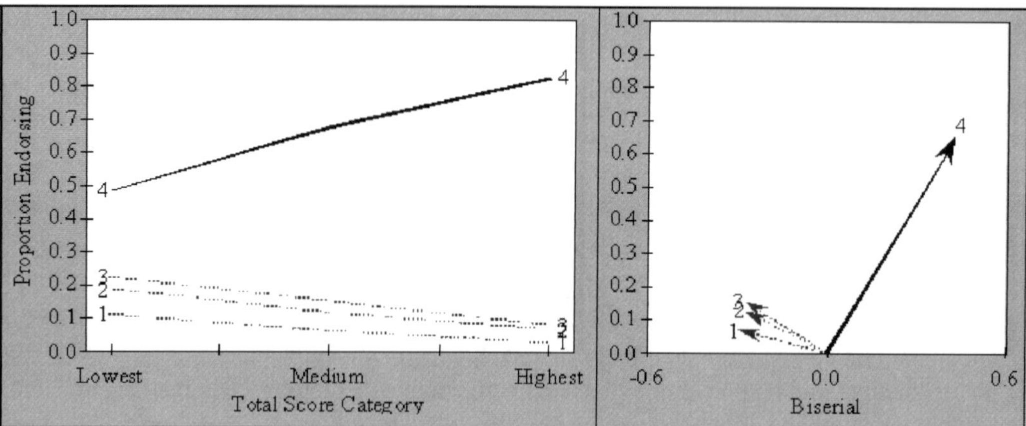

Number of Cases	2910
Mean (P+)	0.65
Delta	11.44
Pearson	0.33
Biserial	0.43
Mean of Passers	0.78
Mean of Failers	0.53
Mean Near Cut	0.68
Phi	0.26
DIF Grade	

Alt.	Lowest	Low	Medium	High	Highest	%	N	Mean Scr	Biserial
1	0.11		0.06		0.03	6.7%	195	43.0	-0.29
2	0.19		0.12		0.07	12.5%	365	44.1	-0.27
3	0.22		0.15		0.08	15.4%	449	44.3	-0.27
4*	0.48		0.67		0.83	65.2%	1896	50.1	0.43
OM	0.00		0.00		0.00	0.2%	5	42.4	-0.20
Mn Scr	38.68		48.50		58.27				
Grp N	1025		1022		863				

The statistics not only evaluate how many candidates selected the key, but how selecting the key on this item correlates to overall exam scores. Test takers are divided into three groups of roughly equal size based on overall score on the exam; the low scoring third, the medium scoring third and the high scoring third. For each of these groups, the report in the table under the left graph shows how many selected which answer choice. For example on this question, "4" (or "D") was the key as indicated by the asterisk. Of the low scoring third, only 48% got the right answer, while 83% of the high performers selected the key. Thus, someone who performs well on the exam overall (which is an indication of competence) is significantly more likely to select the key than someone who scored low (may indicate lack of minimal competence). This spread shows that the question is capable of testing competency.

Figure 5 also illustrates that the incorrect answers were relatively evenly distributed between the distracters, indicating that none of the distracters appears to be overly attractive or easily ruled out. Another way of looking at the same data is presented on the right of the page. The table shows average overall scores for the people who selected each of the answer choices. In this case, the people who selected the key had an average overall score of 50, while the people who selected the distracters had overall scores in the range of 43 to 44. Five candidates with an average score of 42 did not answer the item. This clearly demonstrates that the people who selected the key were more likely to do well on the exam overall, and indicates that the item can effectively measure competency.

The data is evaluated in yet another way by the cumulative statistics in the lower left: Mean of Passers indicates that of those candidates who passed the exam, 78% selected the key. Mean of Failers shows that only 53% of the people who failed the exam selected the key. The larger the spread between these two numbers, the better the item is at measuring competency.

A handy way of summarizing the "who got the question right" analysis is the R-Biserial number. Basically, this number is related to the slope of the lines in the graph on the left. For a good item, this slope for the key should be positive and the larger the better because it expresses the difference in performance of the low and the high scorers. Theoretical range for the R-Biserial number is from -1 to +1, positive values indicate that the candidates who selected the key also scored higher on the overall exam. A negative R-Biserial number indicates that higher scoring candidates selected one of the distracters, which could be an indication of a mis-key or ambiguous item. The R-Biserial number is highlighted in yellow on the left and shown in the picture on the right. The R-Biserial number is important because it evaluates how well an item meets the three basic criteria. We previously observed that the sample item is of appropriate difficulty based on the percentage of correct answers (P+ value). The R-Biserial number shows that we did not confuse the competent candidates (a high percentage of high performers got it right) and apparently it did not give any clues to the less competent candidates (. It also appears to be relevant with regard to the topic since the high performers (those with better understanding of chemical engineering) did better on this than the low performers.

To further illustrate the concept of the R-Biserial number, consider an example of an item that might produce undesirable statistics. The item is asked on the final exam for a graduating class of chemical engineers: "Which bar in town has the most beers on tap?" One would expect a certain percentage of students to select the correct answer, but are those candidates the most knowledgeable students who spend a lot of time in the lab, or the "party crowd"? This question would likely have a negative R-Biserial number, indicating the overall least knowledgeable students are more likely to select the key, which means this item is not an effective measure of a graduating chemical engineer competency, nor is it relevant to chemical engineering knowledge and skills.

Determining Competency

The individual item statistics are the building blocks for determining the overall exam performance. To make the move from the individual item statistics to determining competency of the candidate, a passing score, also called cut score, needs to be established. Figure 6 on the next page illustrates the philosophy behind relating the score distribution to the cut point, or passing score. Total score is simply the number of items answered correctly. For the Chemical Engineering PE Exam, the total score has a theoretical range of 0 to 80. Most candidates have at least some knowledge and get lucky guessing on some questions, so scores on the very low end are uncommon.

If it were known upfront which candidates are competent and which ones are not, the scores of the two groups could be plotted as shown in Figure 6. The at least minimally competent engineers have higher scores, the not minimally competent engineers have lower scores. However, there will be some overlap between the two groups. The better individual items differentiate between competency and incompetency, the smaller the overlap. However, because there is a range of competency and because a person's performance can vary, there will always be some overlap and thus the cut point decision is not clear cut.

Figure 6: Score Distribution

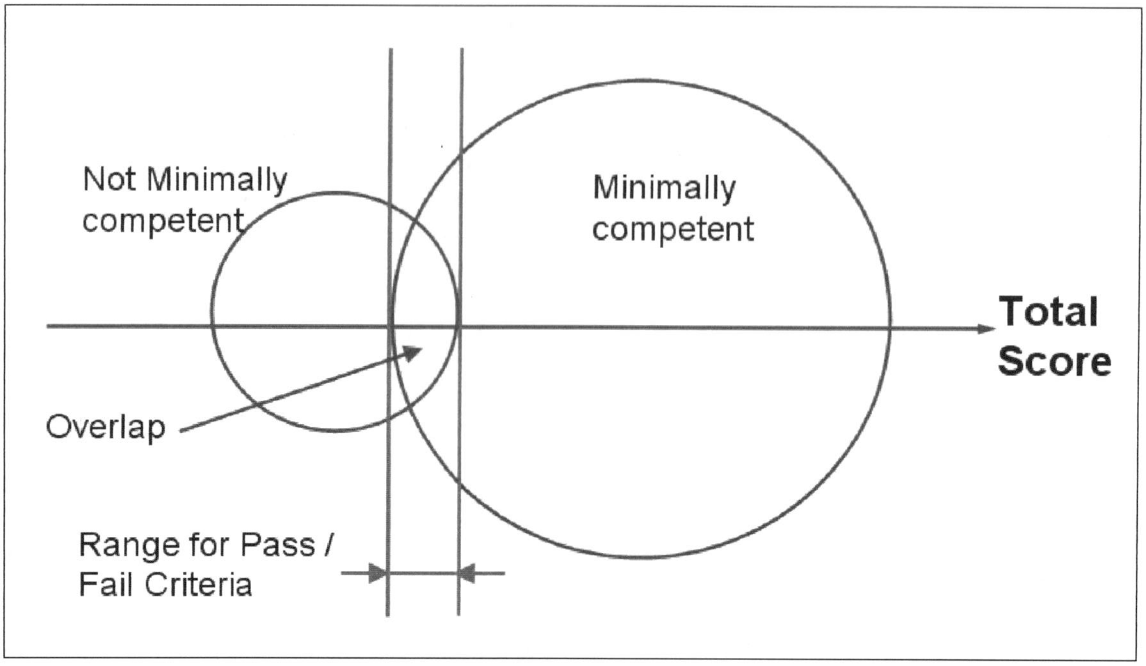

The overlap represents the range where the pass/fail decision should be made. All the at least minimally competent candidates on the right side should pass, all the not minimally competent candidates on the left should fail. With any decision point within that overlap range, there is a potential for competent engineers to fail and/or for less than minimally competent engineers to pass. Failing a competent engineer means that they will need to re-take the exam and includes the risk that they may give up on licensure. Passing a less than minimally competent engineer may allow an under-qualified person into public practice. Since the second type of error is more consequential, there is a tendency towards raising the bar when determining the pass/fail criteria.

Without knowing upfront which candidates are "minimally competent' and which ones are not (after all, it is the purpose of this exam to establish this), statistical measure are used to determine the pass/fail criteria, i.e. establish the "cut score". The cut score is the number of correct answers that a candidate needs to have in order to be considered "minimally competent". One way of determining the cut score is through panel of licensed PE's. This process is fairly involved, since it requires about 15-20 PE's who can volunteer 2-3 days of their time and travel to attend to cut score session. The panel must have good diversity (age, gender, area of practice, geographic location, demographics, etc…) and the cut score study must be completed before the results of the exam can be reported back to the state boards and ultimately to the candidates who are eagerly waiting for the scores. For these reasons, a cut score panel is usually only done when there was a change in exam specifications where the exam results cannot be directly compared to recent prior exams. The first exam where the format or specification is changed is called the "anchor exam" and cut point is established by a formal study.

For a cut score study the diverse panel of volunteers (all licensed PE's) meet and spend about a day to work the exam in a timed environment just like the actual administration. They can use any references, cannot consult with one another and they don't have access to the exam solution. The next day, they receive their score, and review and discuss each item as a group. They look at the statistics similar to those described above, but lacking the pass/fail metrics, review the solution, the distracters and discuss the topic and difficulty of each item. On the basis of that discussion, the panel members make a judgment call on how many "minimally competent" engineers out of 100 would answer this question correctly. A psychometrician guides the session and collects these estimates from each panel member. This data is then used in statistical analysis, generating a proposed passing score. The proposed passing score can then be adjusted (for example, to err on the "safe side"). This then becomes the basis for the pass/fail decision for the anchor exam.

Such a cut score study is not practical for every exam at every administration. So as long as there is no change in the exam specification, it is possible to "equate" the current exam to a previous exam. The equating process, heavily supported by a psychometrician, compares examinee performance on the current exam with examinee performance on previous exams. From this comparison it can be determined if the current exam is harder or easier than the previous exams, and cut point adjusted accordingly. If the data indicates that the exam is easier, the cut score is adjusted upward and a candidate will need to answer more questions correctly in order to pass. Conversely, candidates are not penalized if they take a harder exam because the cut point is revised downward and fewer questions must be answered correctly to pass. Equating is NOT the same as grading on a curve. When grading on a curve, the passing percentage would be fixed and the cut score adjusted to achieve that percentage. With equating, the competency of a candidate is measured on personal performance, and not dependent on the individual candidate's position within the test taking population, nor on the difficulty of the exam.

The psychometricians calculate other measures from the data. Coefficient Alpha measures the internal consistency or statistical homogeneity of the data, thus basically expresses the reliability of the exam. A measure called "Subkoviak" provides an estimate of the consistency of the pass/fail decision or an estimate of the probability that an individual would receive the same pass/fail decision on multiple administrations of the exam. All these measures are regularly reviewed to ensure that the exams can consistently measure competency.

After an exam is administered and all the answer sheets are scored, the statistical item analysis is performed. Exam Committees need to review potentially flawed items in case the key needs to be adjusted and the exam rescored. Then the cut point is determined, either through a cut score study with a panel or through equating. Only then can the pass / fail decision be made. State boards are notified with the results, perform state specific pass/fail assessments and then notify candidates of the results. NCEES recommends that only pass/fail should be reported to the candidates, though some states provide the actual score (or a "scaled percentage" based on a fixed passing percentage) to candidates. Candidates who fail get a diagnostics report so that they know how they performed in the various subject areas. NCEES recommends to only report the pass/fail decision for two reasons: The cut score can vary from exam to exam, so a candidate may pass one year and another candidate taking the exam another year may fail with the same score. The other reason is that the exam is designed to make a pass/fail decision on minimum competency. In that regard, the PE exam is different from an exam like the SAT, which is designed to measure "aptitude" over a wide range of abilities. The PE exam is not designed to measure ability accurately at the far ends of the scale. The PE exam does not distinguish between the "outstandingly competent" and the merely "extraordinarily competent engineer". Similarly, it cannot distinguish between the "totally incompetent" and the "absolutely clueless". Thus, a candidate who received a score of 75 out of 80 is not necessarily a better engineer than a candidate with a score of 72 out of 80 – both are "minimally competent" and pass.

The skeptical reader asks; "Does this really work?" Can an exam such as this really measure competency as the statistical "number magic" appears to indicate? The answer is found in the historical exam pass rates. NCEES does not report an overall pass rate for the exam, but breaks it up into a pass rate for first time takers and for repeat takers. Table 2 includes pass rates for exams given under the current Chemical PE Specification. The most recent pass rates for other exams can be found on the NCEES website.

One specific item to note about the data in Table 2 is that there is a gap between the pass rate for first time takers and for repeat takers. The pass rate for first time takers is typically rather high; in the 70 – 80% range for the Chemical PE exam. This high pass rate is because the examinee pool is already a pre-selected group. In order to be admitted to the exam, a candidate had to pass the FE exam, had to prepare an application with documentation of engineering experience, and had to be approved by the state board, a process that weeds out some of the less qualified potential candidates. Also, especially in chemical engineering where there are many jobs that come under the industrial exemption and do not require licensure, the people who decide to pursue licensure are already a non-random subset of all engineering graduates. This contributes to the relatively high pass rates for first time takers. The portion of the candidate population that passes on a repeat attempt includes in candidates that are minimally competent but may have been inadequately prepare in their first exam sitting.

Table 2: Chemical PE Exam Pass Rates

Chemical PE Exam Pass Rates	April 2011	October 2011	April 2012	October 2012
1st Time Pass Rate	78%	75%	74%	71%
Repeat Pass Rate	31%	29%	31%	20%

How to Prepare for the Exam

All the fascinating information presented in this paper aside, chances are the reader really wants to know how to prepare for the exam. Preparation philosophy will differ for a young candidate near the 4 year experience mark from that for a veteran engineer. One theory is that any minimally competent engineer can pass the examination without any preparation if he or she had unlimited time - time to look up references, read up on concepts and topics outside one's familiar area of practice, find the right equations and conversion factors, etc. But there is a time limit, and 6 minutes per question is not a lot of time.

Exam preparation should focus on building speed and efficiency for solving items. Through study and preparation, a candidate should become familiar with reference content, placing tabs to help find frequently used tables or compile a personal reference of frequently used information. Common equations and frequently used conversion factors should be memorized. Candidates should reactivate the habit of using math: solving equations, working with variables and using a scientific calculator instead of a computer. Speaking of calculators, it is strongly recommended to prepare for the exam using one of the approved models. Precious time can be wasted searching for the logarithm button on an unfamiliar calculator. Approved calculator models can be found on the NCEES website.

Exam readiness is achieved through lots and lots of practice. Preparation classes are useful for refreshing material basics, but are no substitute for rehearsal. A candidate should prepare by solving as many sample problems as one can find. By solving many, many problems, one will automatically remember the frequently used equations and conversion factors, enhance math skills and remember where to find information. The process also develops test day confidence because the items encountered on the exam have been encountered in similar form during exam preparation. Candidates should conduct one or more mock exam day(s) to practice pacing and practice maintaining focus for 8 hours. NCEES publishes a sample exam for each specification that has been developed by the exam committees in the same manner as a "real" exam. One author of this paper frequently recommends to potential exam candidates that they practice with readily available materials, complete a timed trial with the NCEES exam, then continue practicing with more materials.

Finally each candidate should prepare for the test environment and stage to be at peak performance on test day. Exams are administered in finite locations within a state and often require candidates to travel. Plan a travel itinerary that will allow arrival at the test site early, rested and unstressed. Have a satisfying but not heavy breakfast and make a plan for a relaxing lunch. Consider the test room conditions, dress comfortably and in layers. Be aware of the materials that are allowed in the test hall, and prepare accordingly. Candidates will be turned away without a valid government photo ID. Cell phones and other forbidden materials may be confiscated or admission refused. The exam is worked in a printed booklet, which contains sufficient space for working the questions and pencils are provided. Candidates need to bring an approved calculator, unapproved calculators will be rejected. The exam is open book, so candidates also need to bring any references that they want to use during the exam. References must be bound (no loose paper), but can be marked with flags. Review the test day policies on the NCEES website and plan accordingly (http://ncees.org/exams/exam-day-policies/).

Closing Comments

Professional Engineer Licensure is a personal decision and requires one to undertake a dedicated and challenging journey. Candidate should embark on this journey knowing that the exam development process is designed to ensure each exam measures for minimal competency in a consistent manner, and items are statistically validated. Finally, the factors influencing exam score processing timelines are understood. The authors of this paper encourage engineers to pursue licensure. We speak for ourselves and our fellow PEs when we say that nobody ever regrets the resources and effort spent obtaining a PE, and we appreciate the industry approval and instant expert acceptance granted by those two little letters following our names. P.E. Professional Engineer.

(78d) The Current Hot Button Issues in Domestic Professional Licensing

Amos Holt, Southwest Research Institute, San Antonio, TX

Amos Holt, a past president of ASME and current VP for Environmental, Safety and Quality Systems will review the issues currently receiving the greatest attention by the PE community which includes all of the various state and territorial licensing boards, the National Council for Examiners of Engineers and Surveyors as well as the numerous professional society stakeholders and educational organizations and societies. The discussuion will touch on the Model Law and its currect sections on the Masters or Equivalent (MOE) requirements as well as Industrial Exemptions. The various positions of key players covered including the position of a consortium of professional societies called "Licensing that Works" or LTW will be summarized. This group includes both ASME and AIChE and represents about 240,000 engineers thtroughout the world. Amos will tell you what you will likely be hearing more about in the next few years and how it might affect both current PE licensees and prospective PE licensees.

The California Divide – And Why You Need To Engage Your Legislators

Diane Spencer, PE, Lawrence Livermore National Laboratory, CA
and Emmett Miller, PE, consultant and AIChE Fellow, CA

Abstract

California is crippled by a great divide that has nothing to do with tectonic plates. California's professional engineering statute gives civil engineers free reign, allows electrical and mechanical engineers to work in their disciplines so long as that does not conflict with civil engineering, and acknowledges by title only that other engineering disciplines exist.

This division presents problems for engineers whether they have a professional engineering license or not. For example, Spill Prevention Control and Countermeasure (SPCC) plans are mandated by federal regulations. Such plans are prepared by chemical engineers all over the country, but NOT in California. SPCC plans are part of the exclusive domain of civil engineers.

A fire protection engineer was once told he could not design a fire suppression system because that was the exclusive domain of mechanical engineers. A chemical engineer was once told he could not submit a Risk Management Plan (RMP) (required by Cal EPA) without a civil engineer stamp because even the hazard analysis on a facility that uses chemicals is the civil engineer's domain. These are just a few examples where engineers are not allowed to work on projects that similarly trained engineers are expected to do in the rest of the country and the rest of the world.

So how do engineers in California navigate this regulatory nightmare? And how do we influence the legislators, engineering board, and others to nudge the statute in the right direction?

California's answer included forming a coalition of professional engineering societies that work together to present a united voice. The California Legislative Council of Professional Engineers (CLCPE) not only represents the interests of engineering professions to the legislature, but serves as a pro bono resource for the legislature on technical issues as well.

This paper will discuss: (1) our lessons learned from working with legislators; (2) California's professional engineering statute and why it is written as it is; and (3) why one needs to be engaged in the legislative process. Professional engineers are entrusted with protecting the health, safety, and welfare of the public. Getting engaged has the benefit of letting legislators understand the impact of their decisions and allows engineers to serve the communities where they live, and provide a safer, better community for everyone.

California's Problem

While most other states utilize some form of a generic registration model for engineers, California's model is discipline-specific, with civil engineering held on a pedestal. The definition of "civil engineering" in the Professional Engineers Act (California Business and Professions Code Sections 6700-6799.) is so broad that it encompasses nearly the entire spectrum of engineering. This overly broad definition, coupled with a mandate that only civil engineers can do civil engineering, means there is little room for those in other engineering disciplines to legally work in their own professions. The law allows overlap from civil engineering into all other disciplines, but other disciplines cannot overlap into civil engineering. So, not only are chemical engineers impacted, the other engineering disciplines are impacted as well.

Section 6731 states the following, with parts underlined for emphasis:

> "Civil engineering embraces the following studies or activities in connection with fixed works for irrigation, drainage, waterpower, water supply, flood control, inland waterways, harbors, municipal improvements, railroads, highways, tunnels, airports and airways, purification of water, sewerage, refuse disposal, foundations, grading, framed and homogeneous structures, buildings, or bridges:
>
> (a) The economics of, the use and design of, materials of construction and the determination of their physical qualities.
>
> (b) The supervision of the construction of engineering structures.
>
> (c) The investigation of the laws, phenomena and forces of nature.
>
> (d) Appraisals or valuations.
>
> (e) The preparation or submission of designs, plans and specifications and engineering reports.
>
> (f) Coordination of the work of professional, technical, or special consultants.
>
> (g) Creation, preparation, or modification of electronic or computerized data in the performance of the activities described in subdivisions (a) through (f)."

For many years the conventional wisdom has been that the term "fixed works," as used in the statute, (see Section 6731 above) somehow limits the scope of civil engineering, leaving room for other engineers to practice their craft. In order to clarify where that boundary lies, Senator Mimi Walters asked the California Legislative Counsel Bureau for a legal opinion (2011). The findings were shocking. "Fixed works" includes not only the structures, but also all the attachments to the structure. So, if a mechanical engineer designs a pumping system or a chemical engineer designs a fractionating column, and that hardware and the hardware is secured to a foundation, those engineers have crossed into the exclusive domain of the civil engineer. Other engineers are similarly impacted.

When a reasonable person reviews this list of studies or activities, it becomes apparent that the list precludes even scientists from legally studying the phenomena and forces of nature. And in most jobs, engineers are expected to produce engineering reports or to coordinate the work of other professionals.

How Did California Get Into This PE Mess?

The perverse nature of the current Professional Engineers Act in California is a result of good intentions gone awry.

In March 1928, a dam northeast of Castaic, CA, broke and flooded the valley below, wiping out the town of Santa Paula and flooding other towns downstream. Hundreds of people died and millions of dollars' worth of property was damaged.

The Professional Engineers Act was then created to ensure that only qualified people (civil engineer PEs) could work on dams and other fixed works. The legislators made the law as inclusive as possible to make sure there were no loopholes. Later, the Legislature expanded the Act to include Chemical, Electrical, Mechanical, Petroleum, Industrial and Metallurgical Engineering as Title Act Disciplines. With the passage of time, the Board for Professional Engineers, Land Surveyors, and Geologists (BPELSG, aka "Engineering Board") was empowered by the Legislature to create Title Disciplines on its own authority, which it did, eventually increasing the count to 15 title protected disciplines (Title Act disciplines), These are defined by Board Rules (16CCR 400-476) and have minimal stature in law. Along the line, Mechanical and Electrical Engineering were converted by statute to practice protection (Practice Act discipline)[1].

This patch job to add mechanical engineering and electrical engineering to the Professional Engineers Act has been ineffective (per Legislative Counsel's review). Inclusion of Title Act disciplines constitutes an even lower tier where those engineers are not able to take responsible charge of work within their field of expertise. In summary, the field is far from level.

The problem with the Professional Engineers Act has been known for many years. The difference now is the Legislative Counsel's legal opinion that was issued in July 2011 (www.CLCPE.org). According to that opinion, companies that rely on engineering services are on notice that the only engineering discipline allowed to conduct engineering work (not just supervise) in California is civil engineering. If a project requires a chemical engineer, a dual-licensed engineer will be needed to meet the competency requirements of both a chemical engineer and the legal requirements of a civil engineer. A dual license is needed for the other engineering disciplines as well.

---

[1] Four of the title disciplines (Corrosion, Manufacturing, Quality and Safety) were dropped ~1999 for lack of an NCEES exam.

The Industrial Exemption

Common to most states, California has an Industrial Exemption that allows companies engaged in manufacturing to use unlicensed engineers in their businesses— except for civil engineering work. The justification for the Industrial Exemption has been that such employers are capable of and competent in evaluating their own engineering talent, that they do not need interference from the state in doing so, and that failure that will impact the public is subject to strict liability.

There are exemptions for people who perform some specific and narrow civil engineering services. Architects performing architecture and building contractors following building codes are examples. But no general exemption exists for engineers who are not civil engineers if they perform "civil engineering." The Industrial Exemption even states that it does not apply to civil engineering.

With the recent legal opinion, however, all engineering practice in connection with facilities that are geographically fixed is now regulated as civil engineering, and the Industrial Exemption is therefore meaningless.

Engaging the Legislature

In the late 1940's, engineering societies in California organized the California Legislative Council for Professional Engineers (CLCPE) in order to work together as a resource to assist the California legislature, support laws that benefit the public and profession, and help keep society members informed about legislation that might affect them. Application of the Professional Engineers Act was not a problem to engineers in the state so long as the Engineering Board and regulators maintained a liberal interpretation. Likewise, CLCPE did not intensify its effort to change the law until the literal interpretation ultimately reinforced by the Legislative Counsel ruling came into focus.

CLCPE's efforts to change the law began in the 1980s. Our members helped draft a bill and found a senator to carry it. In California, the legislator who carries the bill is the "author" and organizations that support the bill, such as CLCPE, are "sponsors" regardless of who actually drafts the bill. Drafting a bill and finding "authors" is a process CLCPE repeated about a dozen times.

Let's reframe the process we used in California in terms that may help others take advantage of our lessons learned.

1) Coalition. We had a consortium of engineering societies working together as a coalition. This provided more credibility in that we were able to speak with one voice instead of individual organizations promoting separate proposals. A coalition also allows a sponsoring organization such as CLCPE to represent a larger portion of a legislator's constituency. AIChE local section leaders met with a senator to promote the one bill that chemical engineers sponsored without going through CLCPE. The senator listened intently about the impact of current law, asked a few clarifying questions, and then asked about how many chemical engineers were in the state. She was not satisfied with our answer, stood up and thanked us for our time.

Creating a coalition of several engineering societies enabled us to afford a lobbyist. AIChE and local sections are a 501(c)(3) nonprofit organization, which restricts how much money can be spent on lobbying. CLCPE however, was organized as a 501(c)(6) nonprofit, which allows for lobbying, but since the societies and CLCPE are all volunteer organizations, we do not have an intense political presence, and that restricts campaigning for or donating to candidates or political parties. Unfortunately, the opposition we face in California has major campaign funding and they openly support candidates that vote as the opposition wants them to vote.

Another advantage of working within a coalition of engineering societies goes beyond promoting legislation. The conversations we had when we met allowed us to gain insight into the problems each of the different disciplines were facing and we heard personal testimony about those various issues. This perspective has been invaluable.

2) Hiring a lobbyist. Over its lifetime, CLCPE has successively employed three lobbyists who helped with efforts to change the Professional Engineers Act, in addition to supporting the overall mission. A good lobbyist helps focus the discussion and facilitate consensus on language for a given bill. They also know how to navigate through the legislature, explain the bill or other proposals to legislators, legislative staff, and to others, and to find legislators who will author a bill. Typically engineers have full-time jobs, and working a bill is very time-consuming. The first of CLCPE's lobbyists was an engineer who later became Executive Officer of the Engineering Board. The other two were professional lobbyists with legal backgrounds.

3) Language for the bill. Earlier this paper discussed drafting language for a bill and then entering that bill into the legislative process (i.e., finding an author). "Entering" is the key word here. In the late 1990s, our bill attracted attention from the Professional and Technical Consultants Association (PATCA). In our quest to expand the law to allow engineers to practice their own profession, we needed to be careful not to adversely impact scientists (e.g., chemists) and other technical professionals. After several meetings with PATCA, we were able to work out compatible language. Going forward on subsequent bills, we deliberately engaged with the chemists (PATCA) early in the process.

4) Bill's "author". Over the decades we have been pursuing change, our bills have been authored by different senators and assemblymen. In several cases, the legislator had been an engineer in his/her former career. One such senator, who had been an aeronautical engineer, had commented that it took multiple disciplines to design an aircraft. A legislator with a technical or science background readily grasped the issue we were dealing with. One of the first committees to hear a bill dealing with engineering (i.e., the bills we were promoting) was either the Business and

Professions Committee (B&PC) in the Senate or the B&PC in the Assembly. When seeking an author we often approached members of these committees as well. The intent was to get early support and move the bill through the legislative process.

5) Finding support. In addition to sponsors, it helps to find other organizations that will support a bill through (1) letters, (2) testimonies, and (3) lobbying efforts. We collected binders of letters from individuals, municipalities, regulators, and corporations. These included Chevron, the Chemical Industry Council of California, and Genentech. The letters provided a lot of anecdotes on how this perverse law affected people. The volume of the letters attracted the attention of legislators.

Once a bill is scheduled for a vote, testimonies at the hearing can be very compelling. The city of San Diego, California Fire Prevention Officers Association, a division of the California Fire Chief Association, along with various individual department fire chiefs and fire marshals, individuals impacted by current law, and others have testified at legislative hearings over the years. The Dean of Engineering at University of California—Davis testified on behalf of the University of California at the hearing in January 2012, stating that the state is educating engineers who cannot work in their own professions. The dean was shocked to find out that the legislators still would not vote to support changing the current law.[2]

The third level of support is when companies/ organizations involve their lobbyists. However, we found it very difficult to convince companies that had lobbyists in Sacramento to get engaged at this higher level.

6) Other curve balls. We experienced numerous unexpected challenges. One such curve ball happened in 1997, when chemical engineers tried independently (without CLCPE) to change the law. The Engineering Board required that we provide a Sunrise Package[3]. Essentially converting engineers from Title Act (defined in by BPELSG Board Rules) to Practice Act (defined in the Professional Engineers Act) is viewed as creating a new engineering discipline. We were given one week to prepare this. Fortunately, CLCPE had prepared something similar 10 to 15 years earlier, which we were able to mine and then create the required Sunrise Package. The Engineering Board still did not support our bill. Note: Keeping a good archive of material produced in supporting legislation could have unexpected value in the future. The key is being able to find the needed archived information when it is needed. Another advantage is that many of the same people remained engaged in CLCPE for long periods.

---

[2] Incidentally, the western region of ASCE sent their lobbyist to testify in opposition at the same hearing where the UCD dean testified.

[3] Sunrise is a process under which an occupation or profession wishing to receive state certification or licensure must propose the components of the legislation, along with cost and benefit estimates of the proposed regulation

What We Proposed and Why

Our approach to changing legislation evolved over the years. As mentioned earlier, chemical engineers made an attempt in 1997 to raise chemical engineering to a Practice Act (defined by statute). Our opposition protested that our attempt was part of a turf battle among the engineering disciplines. Legislators do not want to be drawn into controversies (e.g., infighting). Our next attempt involved CLCPE, with the intent to define all the Title Act disciplines as Practice Act disciplines. Because we had a coalition involving more than 15 engineering societies, we were able to combat this accusation. Still we did not succeed.

We also shifted our approach and proposed legislation that was closer in alignment with generic engineering. Specifically, we proposed that engineers should be allowed to do work based on their education, experience, or testing. In talking to legislators and to our lobbyist, we learned that we needed to approach any change in engineering statute with baby steps. Adopting generic licensure language (similar to other states or similar to Model Law proposed by the National Society of Professional Engineers (NSPE) was too much at one time. Because of the heavy influence by civil engineering companies in the state, we drafted bills that would continue all the advantages afforded to civil engineering. Unfortunately, our opposition would not negotiate.

In 2000, CLCPE had caught the attention of the legislature sufficient enough that they commissioned a study of the two-tier approach (Title Act/Practice Act) used in California. The study considered natural overlap between the professions, whether there was enough distinction between the engineering disciplines to support this unique regulatory structure, and whether the three Practice Act disciplines (civil, mechanical, and electrical engineering) posed more of a risk to the public (relative to the other engineering disciplines) and thus justified the existing two-tier structure. (See **www.CLCPE.org** for study by the Institute for Social Research.)

The study strongly supported changing the existing regulatory structure. The task of making this change fell to the Engineering Board. The bill it sponsored was opposed by American Consulting Engineers Council (ACEC) - California (a contractors' association) and Professional Engineers in California Government (PECG, a public employee union). Both were well-funded with strong ties to legislators — and the bill failed. [PECG is allowed to donate directly to candidates and its 13,000 members individually campaign for legislators. That has a big influence on those legislators who might support our position.]

The study marked a shift in how we approach changing legislation; CLCPE did not sponsor bills after the study. Instead, CLCPE and our individual engineering societies supported a succession of bills that were sponsored by the Engineering Board or by companies and organizations that use engineering services.

The disadvantage of this approach is that we had less influence on making amendments to the language in the bills. So while sponsoring our own bills had not been successful, this other approach allowed us to continue to pursue fixing the Professional Engineers Act.

Another challenge we had was finding companies and organizations that would commit their lobbyists. Many of these companies were very willing to prepare letters of support, but were not ready to involve their lobbyists.

End Results

This paper provides numerous lessons learned from decades of experience in trying to change California law to allow all engineers to practice in their own professions without finding a civil (typically) or other Practice Act engineer to take responsible charge of the work. Despite years of effort, we have not succeeded in changing the Professional Engineers Act. We are debating our next steps. It is our hope that the lessons we learned will be helpful to others who find the need to engage their state legislators.

(82a) Simulation Through the Eyes of a Practitioner Turned Academic

Kenneth R. Cox, Chemical and Biomolecular Engineering Department, MS-362, Rice University, Houston, TX

Teaching process design in the age of computation poses unique challenges. Typically, students entering a design course expect to find a process simulation course where the title clearly says "Chemical Engineering Design". These are really very different entities. Why is there such confusion that a single tool so dominates the student expectations?

On the other hand, proficiency with process simulation is critical to successful design. It almost sounds paradoxical, but the best simulations are generally produced by those who have learned the need to think outside the simulator. This talk explores what we mean by this and provides insight of the difference in approaches of experienced engineers and those who are newly introduced to design. With this, this paper will explore how even the student can take advantage of some of the tricks of the trade that can help guide their simulations. Some common errors by students learning simulation will also be highlighted.

In the end, an understanding of the role of simulation design and the ability to "think outside the simulation" provides an important education perspective that can help students to a quick start for simulations contributing to a successful design.

(82b) Expectations and Responsibilities When Purchasing Process Simulation Software: Working Together During the Software Sales Cycle

Benjamin Wincure, ProSim, Inc., Philadelphia, PA

Many young chemical engineers are excited to apply process simulation software, on the job, to improve and better understand their plants and processes. But your company does not have the software you require. So, as a young engineer, you search online for the right software, which might be different from what you used at university. But process simulation web sites don't provide enough engineering detail and rarely show prices. You need more information to convince yourself and your supervisor (...and their supervisor!) that you have found an appropriate process simulation tool. This talk will discuss effective ways to work together with any process simulation software vendor to get the essential information you require, and to explain what they might be expecting from you. Benjamin Wincure is a PhD chemical engineer with over 20 years of experience developing, testing, using, and now, selling process simulation software. This talk will give you his perspective on some of the best practices that he has found for meeting the customer's needs and the vendor's needs. Welcome to an insider's look at the sales cycle for process simulation software!

(82c) Process Simulator Basics...That No One Teaches

David Hill, CHEMCAD Technical Support, Chemstations, inc., Houston, TX

Most chemical engineers encounter process simulators in their senior design class, although some Universities use them as early as Thermo II. When you start a job where you need to use a process simulator you discover that correctly using the tool is more complicated than just specifying numbers and pressing the calculate button. Unfortunately you don't realize you need to learn more...until you start making mistakes.

This presentation discusses common mistakes, misunderstandings, and oversights with use of simulators. Some best practices are also presented. The presentation draws upon more than a decade of experience providing technical support to users of process simulators.

(84a) Chemical Engineering Careers At Independent Research Institutes

Chee Kai Tan, Div. of Chemistry and Chemical Engineering, Southwest Research Institute, San Antonio, TX

Independent research institute is established to serve industries, government, and the public in areas that need assistant in problems identification, improvement, and implementation. Chemical Engineers are always faced with challenging science and technology questions related to chemical reactions, process control and development, and scaling up laboratory bench scale to pilot plant studies. In order for chemical reactions to work properly for producing desired products, the professionals will understand the basic theories and involve in controlling the chemicals in proper quantity ratio as feedstock. In addition, other associated factors such as catalyst selection and catalyst development, controlling the processes reaction temperature and pressure, and ensuring the processes are probably designed, evaluated and safe to operate.

Chemical engineers at independent research institutes mostly serve the manufacturing industries varied from petroleum refining, natural gas processing, chemical intermediates, alternative fuels, renewable fuels, and pharmaceutical industries. The Chemical Engineer professionals will offer services in process development and implementation, troubleshooting, pilot plant operation, and small-scale laboratory tests. The Institutes must have extensive, real world expertise in chemistry, chemical engineering, as well as other specialties such as corrosion engineering and safety engineering.

To support these areas of research and development, the institutes must have extensive laboratory facilities in hydrocarbon, organic, and inorganic chemistry for analysis, synthesis and development. In addition, the professionals must be equipped with knowledge in lab tests, pilot plant and demonstration scale operations, chemical and refining process modeling, process flow diagram, safety review, and debottlenecking, etc. Communication skill and team work are also necessary for a successful Chemical Engineer career.

Managing a Process Development Career in Industy

William Hollar
Chief Scientist - MBB
SABIC
1 Lexan Lane
Mount Vernon, IN 47620

Abstract

A career in process development is an interesting and rewarding way to use your chemical engineering training. Proper management of one's process development career can make the difference between a lifelong journey of technical successes and growth or unhappy career experiences. While there are no recipes or ideal paths to a successful process development career, there are factors which can enhance your probability of success.

Important Aspects in Career Management

1. What Do You Want to Do in Your Career?

Managing your career properly requires that you understand what you want to accomplish in your career and which aspects of process development that you truly enjoy. Setting the direction and focus of your career requires understanding your answers to several questions.

What aspects of process development do you prefer?

Process development activities can cover an enormous range of scope from the bench scale all the way to the manufacturing plant scale. Understanding where your interests lie (from both a career and technical standpoint) provides insight into the aspects of career development for focus and development. For example if your interests lie in process modeling and simulation, this choice defines both skills you will need to build/develop and also impacts where in the process development cycle you would contribute. Alternatively if you enjoy working in a pilot plant or plant environment, a different training and development focus would be more appropriate. In case 1, you may focus on training in process simulators, CFD codes and economic evaluation software. Case 2 would lead to greater focus in the details of unit operations, process measurement and control, scale-up and environmental/safety considerations.

Based on your preferences you should look for assignments which build and expand your chosen skillset through experience. This may involve taking on specific assignments or incorporating desired skills into existing assignments. As an example, some process development programs may not require modeling, but the assignment results may be improved by you personally including modeling efforts as part of the scope. Keep in mind this choice may require extra effort on your part.

Do I want to manage or remain technical?

This aspect of career management is not limited to a process development career. It is an important decision for managing the steps in achieving your desired goals. Many chemical companies have both technical and management ladders for career development. While not mutually exclusive, increasing managerial responsibilities will lead to less focus on technical aspects of the job and more on the people management, program planning and communication skills. Ultimately your choice will determine how much focus you want to maintain on the technical aspects of process development.

Some of the other papers in this session go into more detail about this decision and the training and development needs to be successful.

One unique situation can arise when you reach advanced levels on the technical ladder. Many engineers in these roles will lead teams of people on projects that do not report to them. This type of project requires strong leadership skills and an ability to influence people to achieve your goals. Success in this type of role can be both challenging and very satisfying. In this case, learning interaction skills to lead and influence can help insure your success in such a role.

Technical Breadth versus Depth

For the purposes of this discussion, technical breadth refers to competencies/experience in a wide range of chemical engineering topics such as unit operations, heat and mass transfer, separation processes, etc. Depth refers to very deep knowledge in a more limited topic, such as distillation, two phase reactor design or similar more narrowly focused fields.

My experience has been that some combination of both breadth and depth can enhance a process development career. Many process development projects require knowledge of multiple unit operations, process and cost modeling, and an ability to propose alternatives and narrow down to optimum choices. This requires breadth in the field of chemical engineering. In many cases depth can be developed by focus on process technologies. In my case I've done multiple projects on processes involving solids unit operations like drying, so I have developed depth in this area that I leverage and translate where possible.

2. Tools to Build your career

Project management

Any career in process development requires strong project management skills. Process development programs are typically large and complex projects with tight time lines, cross-functional interactions and resource constraints. Effective project management will improve the chances of success. Project management training is widely available and should be considered early in your career.

Problem solving skills and systems

Engineers are by definition problem solvers. A strong problem solving approach and toolset can improve the quality and speed of resolving process development problems.

Many chemical companies use problem solving systems such as Lean Six Sigma to help provide a common language and set of tools for solving business and technical problems. Most of these tools relate back to common best practices for solving a problem efficiently – define the problem, identify solutions, test solutions and validate improvements. The name or type of the approach is less critical than it being efficient in properly defining the problem to be solved and quickly identifying and demonstrating viable solutions that are sustainable in the long term.

Mentors

A technical mentor early in your career can be a huge benefit in launching a process development career. Some companies may assign a mentor at the start of your career. Typically the mentor has significant technical experience within the company and can support you in two ways:

1. Acting as a consultant or sounding board for your early technical assignments and helping you avoid time-consuming pitfalls
2. Assisting in accomplishing goals within the company's operating systems (i.e. other functions such as Sourcing, Legal, EHS)

Getting the right mentor is important. Look for one who enjoys the mentoring process and does not treat it as another task on an already full schedule.

Keep in mind that while the goals of having a mentor change as you get more experience, there continues to be value in having either mentors or other colleagues who you can use to discuss technical or career questions. This remains true throughout your career.

Continuing Growth

Conferences

Attending conferences is an excellent way to develop new contacts, stay current in your field and in general get a fresh viewpoint from the internal work processes and viewpoints of your company. You can also take the opportunity at these meetings to attend sessions in unfamiliar areas. Many times new concepts through analogies or pieces of the unrelated work can bring innovative ideas to your assignments. I typically target 1 – 2 sessions outside my field of interest/expertise at each conference I attend.

Participation in a local or national professional society can often bring the benefit of widening your professional network and exposing you to new areas of expertise.

Technical Journals

E-Journals and e-books allow you to stay current on the latest developments in process development topics very easily. Early in my career I would devote 2 – 4 hours each Friday to scanning a wide range of technical journals. Staying current has become even easier as a corporate or University library is no longer required to access the latest books and journals. The ability to set automatic alerts in some technical data base systems makes the effort even easier by letting keyword queries do the work for you.

Conclusions

Proper management of a career in process development can be aided by two key elements:

1. Choosing your desired path for your career
2. Taking advantage of opportunities to continuously learn and grow throughout your career

The suggestions presented in this paper can help you formulate your strategy for succeeding in the path you choose for your career.

Innovation as a Pathway for Industry Career Enhancement

Joseph B. Powell, Ph.D.
Shell Chief Scientist – Chemical Engineering, Houston, TX

Abstract

Most organizations actively solicit new ideas, and most career professionals have knowledge, competencies and interests that extend beyond their current job assignment. Actively engaging in "innovation" is a means to extend one's role and visibility in the organization, and to extend one's network. It also provides a mechanism which drives lifelong learning and expansion of knowledge base, by providing a justification and outlet for professional society involvement (AIChE) and conference attendance, as well as reading of technical literature.

At Shell, innovation has been formalized into a Gamechanger™ process, which provides support for innovation and investigation of technology and opportunities beyond one's current job assignment. Informal "Hunters" networks also exist in a number of locations, created to facilitate matching of ideas with identified challenges and new opportunities. These networks provide initial screening and enhancement for inputs to the formal idea review process. Network collaboration provides the energy and support often needed to enable the pursuit of new ideas and opportunity, despite barriers and inertia that can act to suppress organizational growth.

This seminar will present an overview of corporate innovation methodologies and the use of networks and online communities to encourage development and pursuit of new ideas. It will also examine how innovation can be used to grow one's personal network, and broaden one's skill base and value to the organization. Finally, it will conclude with a review of grand challenges for chemical engineering in future energy and chemicals, which provides the impetus for new idea generation and resultant opportunities.

Career Enhancement Challenges and Opportunities

An academic career path may reward in-depth knowledge of a specific discipline area. While this may be true also of industry, quite often the successful industry career requires broadening to consider a variety of experiences or job assignments. Management and especially senior management training will typically entail a set of job assignments designed to broaden experience in research and technology, operations, and business functions.

If you are identified early on as a fast track candidate for senior management, broadening may happen with a great deal of help from the organization, in a structured way. In other situations, career broadening may occur by happenstance, via termination of projects or assignments. AIChE CheNected is currently seeking examples of career transitions and impacts on career [1].

One of the reasons a broadened career background is considered valuable both in industry and in academy, beyond its use in managing a technology space or project, is the fact that much innovation occurs at boundaries or overlaps between disciplines. In last decade universities have realized this in creation of multidisciplinary centers of excellence, which resemble those deployed in industry work teams and departments.

Broadening can be self-directed, to a degree. Involvement in AIChE through volunteer opportunities, or attending national conferences or local section meetings to gain exposure to new developments in areas related or adjacent to one's main discipline or job assignment, is one good way to self direct one's career broadening. Large companies have both internal and external speaker seminar series, which provide opportunities to both learn, and network. Setting time aside to read the relevant publications in ones broader discipline is a surprisingly simple, yet often difficult to accomplish method for career and expertise broadening.

Most companies have an explicit interest in innovation, either directly in new technology, or in more efficient systems to execute existing work processes. "Quality" [2] or "Six Sigma" [3] training involves assembly of work team subgroups to identify streamlined and more efficient ways of working. Volunteering for one of these groups or initiatives provides an excellent way to network, and learn about other functions within the company.

Disruptive innovation often requires a less structured approach, and may find resistance from existing organizational structures. In Shell, the Gamechanger™[4] program was created to drive innovation beyond the interest of existing business units. An on-line idea submission site is maintained to encourage idea development. Ideas are reviewed, and those showing sufficient promise are subjected to a panel review to provide feedback, and suggestions for next steps. Many universities are forming incubation centers to encourage cross discipline innovation, and provide business and legal advice to burgeoning idea generators and entrepreneurs. In the absence of formal innovation processes or groups, there is the self-empowered option of a business lunch with friends and colleagues, to exchange ideas and interests. Several prominent ideas are said to have been first sketched on a lunch table napkin. At Shell, informal Hunters networks exist for idea sharing, networking, as well as brainstorming specific problems or targets, typically via a lunchtime setting. These events provide not ideas for the Gamechanger program or business sponsors, but also provide networking and broadening for the participants themselves. Online communities and bulletin boards or blogs are also used to create global virtual collaboration on innovation themes.

A central theme for innovation is the connection with business value. "Innovation" means developing and progressing a new ideas into a commercial reality, often requiring a set of innovations. Participation in innovation activities can thus help instill business acumen, which can be applied either in development of the idea or concepts, or to one's current assignment. A second theme is collaboration and broader knowledge of the organization with which one is associated, which quite often can be translated into career advancement via awareness of additional opportunities, or opportunities to practice ones expertise in solving problems beyond ones current assignment.

Current innovation opportunities in chemical engineering are extensive, given new advantaged sources of feedstocks for energy and chemicals resulting from shale gas and tight or unconventional oil, development of renewable energy solutions to address energy security, climate change and the environment, advances in sensors and computing leading to new opportunities in big data management and optimization, advances in materials lead by approved ability to manipulate and model matter at the nanoscale.

Summary

The concept of the single inventor working meticulously in the lab or with theory to discover the "next big thing" may be partially true in some instances. Innovation is more likely to occur via cross fertilization and collaboration, however, and a broader group will almost inevitably be required to complete the additional steps and deliver the complementary innovations needed to achieve commercial reality. Much innovation occurs at the interface between disciplines, and may entail both technical and social factors. Participation in cross-functional innovation is a great way to succeed at innovation, while also broadening your own knowledge base, acumen, organizational visibility, and value.

References

1. Loraine Kasprzak, *ChemE's and Career Transitions, AIChE CheNected,* *http://chenected.aiche.org*
2. *Deming, W. Edwards (1986). Out of the Crisis. MIT Press*
3. *M. Barney, T. McCarty, The New Six Sigma: A Leader's Guide to Achieving Rapid Business Improvement and Sustainable Results, Prentice Hall (2002).*
4. *Shell Gamechanger™: http://www.shell.com/global/future-energy/innovation/game-changer.html*

(84d) Bridging Between Engineering Communities: A Hitchhiker's Guide

Kenneth R. Cox, Chemical and Biomolecular Engineering Department, MS-362, Rice University, Houston, TX

Research and application opportunities are continually lost by ineffective communication among engineering communities. Many of the great innovations are cross-over technologies where great ideas from one discipline finds new application with another. Communications between academia and industry are often hindered by a different sense of mission and very different reward systems. Yet, the ability to bring new technology into practice is greatly slowed by this well-documented industry-academia divide. Lack of interactions also limit the effectiveness of engineering societies which far too often function as weakly-joined collection of smaller societies.

Learning to span these divides is an important career skill every student should strive to develop. Professor Cox will share his first-hand experience working with different bodies of engineers and offer his perspective about the challenges, opportunities, trends, and strategies for bridging between the various communities.

(84e) Industry or Academia? Ph.D. or MBA? - Factors to Consider in Your Decision Making

Santiago Faucher, Technologies, Hatch, Mississauga, ON, Canada

We define our careers through our passions, interests, actions, efforts, work, and drive; but most importantly we define it by the decisions we make. For a young engineer who is early on in their career, the first fork in the road forcing a decision is often related to the pursuit or not of further education, the second fork relates to the type of academic education (MBA or Ph.D. or other) and the third (for those who have chosen a Ph.D. path) addresses the pursuit of an academic or industrial career. In this talk, the author will review the factors and benefits associated with the above potential choices and use his own career as an example to provide the audience with better insight into career decisions and outcomes.

(110a) Acid Gas Absorption Fundamentals

Clint P. Aichele[1], James R. Whiteley[1], Yash Tamhankar[1] and M.R. Resetarits[2], (1)School of Chemical Engineering, Oklahoma State University, Stillwater, OK, (2)Fractionation Research, Inc., Stillwater, OK

Absorption is a standard unit operation that has been studied by chemical engineering students for decades. One of its primary uses is for the removal of acid gases, like CO_2 and H_2S, from natural gas streams. Historically, aqueous amine and hot carbonate solutions have been used to perform these absorptions. Many different amines, solution strengths, activators and passivators have been tested and employed industrially. Approximately half of the industrial towers are trayed; half, packed. Foaming adversely affects the hydraulic performance in these systems.

Acid gas absorption is receiving appreciable new attention as concerns rise regarding CO_2 in the atmosphere. In addition, the United States has recently experienced significant growth opportunities in shale gas production which require improvements in gas treating technologies. This presentation reviews acid gas absorption and regeneration fundamentals. Several PFD's will be shown to illustrate important concepts. New absorbent solutions will be discussed, including ionic liquids. Flue gas applications will be discussed, including the difficulties associated with column pressure drops and very large column diameters. Rate-based modeling approaches will also be reviewed.

(110b) Sorption and Diffusion of Ethanol and Water in Fluoride Mediated Silicalite-1

Ke Zhang[1], Michelle Dose[1], Ryan P. Lively[2], William J. Koros[3] and Ron R Chance[4], (1)School of Chemical and Biomolecular Engineering, Georgia Institute of Technology, Atlanta, GA, (2)Algenol Biofuels, Bonita Springs, FL, (3)School of Chemical & Biomolecular Engineering, Georgia Institute of Technology, Atlanta, GA, (4)Chemical and Biomolecular Engineering, Georgia Tech, Atlanta, GA

Sorption and diffusion of ethanol and water have been studied in silicalite-1 crystals synthesized via a fluoride mediated route. The near-perfect hydrophobic silicalite (F^-) crystals have very few internal silanol defects and show unusually attractive ethanol and water transport behavior. Full isotherms (0.05 – 0.95 activity) over the range of 25- 55 ^oC are presented. At a water activity level of 0.95 (35 ^oC), the total water uptake by silicalite-1 (F^-) was found to be 0.263 mmol/g, a factor of 8 lower water uptake than in silicalite-1 (OH^-), thereby giving an ideal sorption selectivity of 35 for 5 mol% ethanol solutions. Uptake rates for water and ethanol have been measured in silicalite-1 (F^-) over a wide range of activities between 25-45 ^oC. The transient uptake curves are consistent with intra-crystalline diffusion control and yield transport diffusivities (D_t) of order 10^{-9}-10^{-10} cm^2s^{-1} for ethanol and water, with the values for ethanol being about three times larger than those for water at comparable low activity conditions. Thermodynamically corrected diffusivities (D_o) for ethanol, which reflect the intrinsic mobility, are almost an order of magnitude smaller than the values for water, as is expected from the difference in molecular size. Such results yield ethanol/water permselectivities of 7-27 for 1-5 mol% ethanol solutions and estimated ideal separation factors of 130-310 over the same concentration range. For binary ethanol-water mixture tests with competitive sorption, lower actual separation factors are expected, since the sorbed ethanol molecules may act as water sorption site "seeds" to decrease ethanol/water sorption selectivity. However it is clear that silicalite (F^-) is promising for ethanol extraction from water due to its significantly depressed water adsorption compared to silicalite (OH^-).

(110c) Distillation and Absorption Equipment

Mike Resetarits, Anand N. Vennavelli and A. Y. Ogundeji, Fractionation Research, Inc., Stillwater, OK

Distillation is almost as old as boiling, and yet, it remains the primary technique whereby chemicals are separated. Absorption is the primary technique whereby acid gases like CO2 are removed from gas streams; carbon capture is in the news. Distillation and absorption columns still dominate the skylines of refineries, chemical plants, and gas plants. The columns almost always contain trays or packings. Fractionation Research, Inc. (FRI) is a global consortium of companies that fund tray and packing research in Stillwater, OK. New generations of these devices have allowed column owners to revamp existing columns to achieve capacity increases, separation improvements, and energy consumption decreases. Augmented crossflow, counter-flow, and co-current flow trays have exhibited capacity increases. New generations of random packings and the associated distributors haven exhibited separation improvements. Structured packings and new high capacity versions have revolutionized vacuum towers where low-pressure drops are critical. This tutorial session reviews the newest trays and packings and FRI's testing of them

(110d) Radioisotope Methods for Obtaining Process Information

Lowell Pless, Tracerco, Pasadena, TX

In all chemical plants and refineries there generally is not enough instrumentation in the right places in our processes. During the design phase, decisions were made between knowing more about the process, which means buying more instruments, and saving capital funds. When a problem appears with the operation of a vessel, there are always gaps in the knowledge of what is happening due to a lack of instruments in the places where knowledge is not routinely required. Process engineers and operations personnel need to know what is happening so they can take corrective action. When knowledge is missing, they often take actions that make matters worse.

For distillation and separation processes Chemical Engineers have historically relied upon plant process measurements such as flow rates, temperatures, and pressures and model or simulation results for data with which to solve troubleshooting or re-design problems. An array of on-line diagnostic services are available that can provide additional data offering real-time information on how pieces of process equipment are actually operating.

The most common application is G***amma Scanning***. This test is primarily applied, but not necessarily limited to, distillation or separation columns. Gamma scans provide a density profile of the internal process of operating distillation columns and other process vessels. The density profile can be used to diagnose the hydraulic operating conditions of mass transfer devices such as damage to internals, flooding, degree of entrainment or weeping, liquid levels on trays and distributors, liquid distribution through packed beds, etc. The presenter will show case studies where scanning revealed vital process information that helped solve an operating problem, or helped make a revamp/re-design successful.

Additional derivatives from gamma scanning of detecting flow distribution patterns will be demonstrated. The ***CAT-Scan*** technique is a specialized horizontal gamma scan used to generate a topographic profile of the internal cross-sectional density of process equipment. This profile is useful for the detailed study of liquid flow distribution through packed columns as well as having applications beyond distillation columns. A hybrid application involving tracers and scanning on fixed bed catalytic reactors will also be discussed.

Another technique involving radioactive materials is the ***Tracer Test***, in which a radioactive tracer is introduced into the process for investigating the process flows. External radiation detectors are placed at strategically chosen locations outside vessels and piping to monitor the tracer flow through the process. Either liquid, gas or a solid tracer can be utilized depending on process conditions and the fluid to be studied. This tracer technique can identify liquid or gas maldistribution in fixed or fluidized bed reactors. The residence time through a tank or reactor can be measured and compared with the calculated residence time. Observed differences can relate directly to the degree of mixing or channeling inside the reactor. These tracer techniques have also been used in identifying leakage of heat exchangers and valves.

(110e) Liquid Extraction Processes & Equipment

Mike Resetarits, A. Y. Ogundeji and T.J. Cai, Fractionation Research, Inc., Stillwater, OK

What is the best way to remove C_6, C_7 and C_8 compounds from a feed stream containing C_6, C_7 and C_8compounds? How do you take a lubricating oil and remove certain compounds and improve its viscosity index (so that its viscosity does not change excessively with the oil's temperature)? What is propane deasphalting? The answers to all of these questions in "liquid extraction processes."

What technology was supported and funded by the actor Kevin Costner to help clean up the large 2010 Gulf of Mexico oil spill? What are Karr reciprocating plate columns, Oldshue-Rushton contactors and Podbielniak centrifuges? What the heck is a Graesser Raining Bucket? The answer to all of these questions is "liquid extraction equipment."

Imagine heavy and light fluids flowing vertically past each other with only a very small density difference. Imagine rotating equipment that works better without rotating. Liquid extraction is different – very different. Do you know of any experts? The authors are looking for one.

AUTHOR INDEX